Drinking Water: Distribution, Treatment and Management

Drinking Water: Distribution, Treatment and Management

Edited by Herbert Lotus

CLANRYE
INTERNATIONAL
www.clanryeinternational.com

Clanrye International,
750 Third Avenue, 9th Floor,
New York, NY 10017, USA

ISBN: 978-1-63240-658-3

Cataloging-in-Publication Data

Drinking water : distribution, treatment and management / edited by Herbert Lotus.
p. cm.
Includes bibliographical references and index.
ISBN 978-1-63240-658-3
1. Drinking water. 2. Drinking water--Standards. 3. Drinking water treatment units.
4. Water quality management. 5. Water--Distribution. 6. Water-supply engineering. I. Lotus, Herbert.
TD353 .D75 2018
628.1--dc23

For information on all Clanrye International publications
visit our website at www.clanryeinternational.com

CLANRYE INTERNATIONAL

Printed in China.

Contents

Preface

The main aim of this book is to educate learners and enhance their research focus by presenting diverse topics covering this vast field. This is an advanced book which compiles significant studies by distinguished experts in the area of analysis. This book addresses successive solutions to the challenges arising in the area of application, along with it; the book provides scope for future developments.

Water that is safe to drink or to use in the preparation of food is known as drinking water. It is essential for the existence of life. Polluted water can result in infections such as cholera and typhoid. The common types of water treatment methods are filtration systems, water softeners, distillation systems and disinfection. This book is appropriate for students seeking detailed information in this area as well as for experts. It elucidates the varied aspects related to the distribution, treatment and management of drinking water. It includes contributions of experts and scientists which will provide innovative insights into this field.

It was a great honour to edit this book, though there were challenges, as it involved a lot of communication and networking between me and the editorial team. However, the end result was this all-inclusive book covering diverse themes in the field.

Finally, it is important to acknowledge the efforts of the contributors for their excellent chapters, through which a wide variety of issues have been addressed. I would also like to thank my colleagues for their valuable feedback during the making of this book.

Editor

1

Metals releases and disinfection byproduct formation in domestic wells following shock chlorination

M. Walker and J. Newman

University of Nevada, Department of Natural Resources and Environmental Science, MS 370/FA 132, 1664 N. Virginia Street, Reno, NV 89557, USA

Abstract. Shock chlorination is used for rapid disinfection to control pathogens and nuisance bacteria in domestic wells. A typical shock chlorination procedure involves adding sodium hypochlorite in liquid bleach solutions to achieve concentrations of free chlorine of up to 200 mg L^{-1} in the standing water of a well. The change in pH and oxidation potential may bring trace metals from aquifer materials into solution and chlorine may react with dissolved organic carbon to form disinfection byproducts. We carried out experiments with four wells to observe and determine the persistence of increased concentrations of metals and disinfection byproducts. Water samples from shock chlorinated wells were analyzed for Pb, Cu, As, radionuclides and disinfection byproducts (haloacetic acids and trihalomethanes), immediately prior to treatment, after sufficient treatment time with chlorine had elapsed, and at intervals determined by the number of casing volumes purged, for up to four times the well casing volume.

Elevated concentrations of lead and copper dissipated in proportion to free chlorine (measured semi-quantitatively) during the purging process. Trihalomethanes and haloacetic acids were formed in wells during disinfection. In one of two wells tested, disinfection byproducts dissipated in proportion to free chlorine during purging. However, one well retained disinfection byproducts and free chlorine after 4 WV had been purged. Although metals returned to background concentrations in this well, disinfection byproducts remained elevated, though below the MCL. This may have been due to well construction characteristics and interactions with aquifer materials. Simple chlorine test strips may be a useful method for indicating when purging is adequate to remove metals and disinfection by-products mobilized and formed by shock chlorination.

1 Introduction

Shock chlorination is an in-situ method for disinfecting water wells contaminated with pathogens and nuisance bacteria. Much guidance is available for treatment of domestic wells (e.g. Schnieders, 2005; Driscoll, 1986) and http://www.unce.unr.edu/publications/files/nr/2006/FS0668.pdf, last accessed April 2010). The guidelines recommend CT values (free chlorine concentrations × resting time of the solution) that are very high relative to those used for public water supply treatment, which is appropriate given that treatments occur sporadically, usually in response to perceived problems with water or the health of those who consume water from a well. The procedure typically involves adding sodium hypochlorite solution directly to a well followed by mixing and a resting period of 12–24 h. The chlorinated water must be purged prior to resuming use. Recommendations for the amount of purging vary significantly, with Schnieders (2005) recommending 10–20 well volumes, and University of Ohio Cooperative Extension and others recommending purging until water no longer smells of chlorine (http://ohioline.osu.edu/aex-fact/0318.html, http://srwqis.tamu.edu/media/2553/shockwells.pdf, last accessed April 2010). Cooperative Extension is part of the land-grant system of universities throughout the United States. In most states, Cooperative Extension maintains offices in each county seat in a state. Specialists and educators based in each county provide advice and information on a wide range of topics, including management and treatment of private domestic water supplies.

Table 1. Well characteristics and water physical and chemical characteristics immediately prior to shock chlorination; DTW is depth to water from the land surface, WD is total depth of the well, Static vol. refers to the standing volume of water in a well after water level recovery following the end of pumping. The measurements reported are also referred to as IP (Initial Purge) in Table 4. Initial Purge refers to purging prior to treatment, to replace water stored in the casing with water from the surrounding saturated formations. Well diameter, case material, depth to water from the land surface and well depth were recorded at each site. The screened interval length was obtained from well logs available on the Nevada Division of Water Resources (http://water.nv.gov/, last accessed April 2010).

US fish and wildlife service site designation (year finished)	Well dia cm (in)	Case material	DTW from land surface m (ft)	WD from land surface m (ft)	Screened interval length m (ft)	Static vol. l (gal)
182 (1975)	15 (6)	PVC	3.4 (11.1)	9.4 (31.0)	4.6–7.9 (15.0–26.0)	110.6 (29.2)
167N (1996)	15 (6)	Steel	2.4 (8.0)	9.0 (29.6)	8.2–9.1 (27.0–30.0)	120.0 (31.7)
51 (1994)	15 (6)	Steel	1.7 (5.5)	20.4 (67)	18.9–20.4(62.0–67.0)	171.2 (45.2)
142 (1993)	15 (6)	Steel	3.6 (11.8)	8.1 (26.7)	6.4–7.9 (21.0–26.0)	84.8 (22.4)
US fish and wildlife service site designation	**pH**	**Temperature °C**	**Conductivity mS cm^{-1}**	**Oxidation-reduction potential mV**	**Bleach added l (gal)**	**Treatment duration (h)**
182	8.50	14.9	0.740	not taken	1.0 (0.3)	20 h
167N	7.14	16.9	0.345	114	1.1 (0.3)	15 h
51	7.27	15.57	0.454	270	1.5 (0.4)	24 h
142	7.67	16.86	2.093	−123	0.7 (0.2)	17 h

The odor threshold for chlorine gas in air is approximately 0.3 mg L^{-1} (Amoore and Hautala, 1983). The relationship between the amount of chlorine degassing from a solution and the amount that can be detected by smell varies, depending upon the individual sensitivity and the degassing rate, which is in turn related to mixing dynamics within a water body, contact time, temperature of water, and changes in temperature and barometric pressure. Because of these several factors, guidelines about purging based on scent are subjective and the consistency and efficacy of application may vary considerably between people.

Aqueous chlorine (as hypochlorite ion and hypochlorous acid) cleaves carbon-carbon bonds in organic molecules to form two classes disinfection byproducts (DBPs), haloacetic acids (HAAs) and trihalomethanes (THMs) (Westerhoff et al., 2004). HAAs and THMs include carcinogenic organic compounds and have Maximum Contaminant Levels (MCLs) of 0.060 mg L^{-1} for HAA5 (the sum of concentrations of monochloroacetic acid, dichloroacetic acid, trichloroacetic acid, monobromoacetic acid and dibromoacetic acid) and 0.080 mg L^{-1} for total THM, as specified in the Stage 1 Disinfectants/Disinfection Byproducts Rule (40 CFR, Parts 9, 141 and 142) (http://water.epa.gov/drink/contaminants/index.cfm#Byproducts, last accessed August 2010). Trihalomethanes include chloroform, bromodichloromethane, dibromochloromethane, and bromoform.

Shock chlorination may also increase the concentrations of lead and other trace elements following treatment (Seiler, 2006) and change arsenic concentrations (Gotkowitz et al., 2008). Both lead and arsenic have toxicological effects, with an Action Level and Maximum Contaminant Level of <15 µg L^{-1} and 10 µg L^{-1}, respectively.

This paper describes the changes in concentration of Pb, Cu, As, U, gross-α and gross-β radiation, HAA5, THM, and free chlorine (semi-quantitatively measured) from shock chlorination of four domestic water wells. The study also sought to demonstrate that simple test strips for semi-quantitative measurement of concentration of free chlorine (used for pool and spa maintenance) indicates when purging has returned concentrations of metals and disinfection byproducts to pre-treatment background levels. Test strips may be a better indicator than the scent of water that post-treatment purging is complete, because they provide a semi-quantitative indicator of total chlorine concentrations.

2 Methods and materials

2.1 Study site

Four wells (Table 1) were selected in the Lahontan Valley, in Nevada (Fig. 1). Each well was used for domestic supply prior to being retired for water right acquisition by the US Fish and Wildlife Service within the Stillwater National Wildlife Refuge. Three were cased with steel (ASTM A135 SCH40 ERW low-carbon steel, based on inspections at the sites) and one was cased with polyvinyl chloride casing (specifications unknown). These wells were chosen for two reasons. First, they conformed to standards and practices commonly used for domestic water well

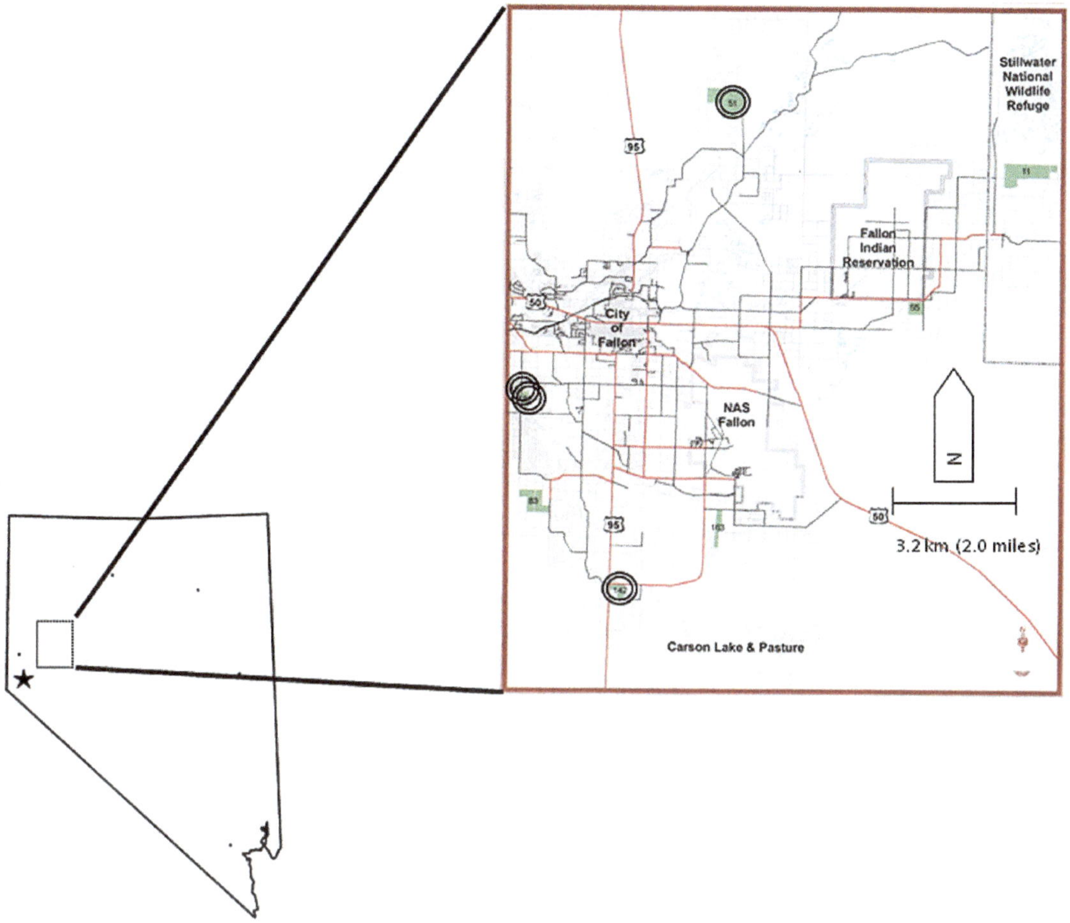

Figure 1. Location of study area, with locations of test wells indicated in circles.

construction and had well logs available through the Nevada State Engineer's office. Second, because they were no longer in service we did not run the risk of exposing people to metals or disinfection by-products released or created by shock chlorination. Well logs for each indicated that the exteriors of screened intervals on the casing were packed with gravel for wells 182 (PVC-cased), 167N (steel-cased) and 142 (steel-cased). The construction log for well 51 (steel-cased) contained incomplete information about how the screened interval was finished. The wells pumped water from a stratum of Quaternary valley-fill sands in the inland terminus of the Carson River from depths of less than 15 m (50 ft) from the land surface. Infiltration from irrigation and the Carson River has been identified as the main source of recharge (Glancy, 1986). Although the groundwater system in the Lahontan Valley is nominally comprised of three geochemically separated systems (shallow (<15.2 m (50 ft) from the land surface); intermediate (15.2–<152.4 m (50–<152.4 ft); deep (≥152.4 m (≥500 ft)), aquifer material composition, yields and chemistry vary highly throughout the region (Glancy, 1986). Water in the aquifer has high but spatially variable concentrations of arsenic (as much as 2100 mg L^{-1} (Walker et al., 2005)) and uranium (as much as 290 mg L^{-1} (http://www.atsdr.cdc.gov/HAC/PHA/fallonleukemia2/fln_p1.html, last accessed April 2010) from contact with sediments and from long-term evapoconcentration (Welch and Lico, 1998). Depth to water (DTW) in the four wells ranged from 1.7 to 5.5 m (5.5 to 18.0 ft) with total well depth ranging from 8.0 to 9.5 m (26.7 to 31.0 ft) (Table 1).

2.2 Well treatment and purging

A ½ horsepower portable jet pump fitted with a 7.6 m (25 ft) long, 1.9 cm (0.75 in) interior diameter suction line was used for each trial. The pump and suction line were rinsed with distilled, deionized water between uses and allowed to air dry. Each well was chlorinated and purged as a separate experiment, to avoid cross-contamination of wells. The pump had a fiberglass-reinforced thermoplastic housing and impeller, with Buna-N seals and ceramic bearings. The outflow line was fitted with a GPI electronic inline flow meter, a flow control valve, and a tee that divided flow between a discharge hose and a flow-through chamber for

Table 2. Sampling intervals used, with associated chemical constituents measured.

Sample type	Initial Purge (IP)	Post-Chlorination (PC)	Post-Chlorination Purging (PCP, for ½, 1, 2, and 3 well volumes)	Final (4F: 4 well volumes purged)
gross α, gross β, Uranium	√	√		√
Metals (Cu, Pb, As)	√	√	√	√
HAA5, THM (wells 142 and 182)	√	√		√
Total Organic Carbon (wells 142 and 182)	√			
Chlorine Test Strips		√	√	√

a YSI model 556MPS (Yellow Springs Instruments, Yellow Springs OH), for real-time measurement of temperature (T (°C)), pH, oxidation-reduction potential (ORP (mV)), and conductivity (C (mS cm^{-1})). The multi-probe was calibrated immediately prior to each stage of field trials, using pH 4.00, 7.00 and 10.00 standards (Fisher Scientific Buffer-Pac, Cat# SB105), a conductivity standard (1000 mS cm^{-1} at 25 °C (Yellow Springs Instrument Co. Cat #3167) and an oxidation-reduction potential reference solution (Equipco Inc., part #3682500).

Prior to conducting trials, each well was purged of stagnant water at a rate of 9.5 to 18.9 liters per minute (2.5 to 5.0 gpm) until temperature, pH, ORP, and conductivity readings stabilized (<5% variation in continuous readings), then purged an additional four times the well's standing volume of water. The standing well volume (WV) was estimated as $WV = H \times A$, with H as the measured height of the water column in the well after the level stabilized following pumping and A as the cross sectional area of the interior of the well casing. Using this as a basis for purging following treatment differs from the dynamic well volume, which represents the steady state water level achieved during pumping. This differs from the static water level due to drawdown during pumping. The water levels used as a reference in this paper are standing well volumes (WV), which do not reflect the influence of pumping. Pre-chlorination (designated as IP) water samples for all wells were collected for As, Cu, Pb, U, gross-α and gross-β radiation, total organic carbon, and for a steel-cased well and a PVC-cased well, water samples were collected for HAA5 and THM analyses. Table 1 presents IP conditions in each well.

2.3 Shock chlorination

Following initial purging, each well was chlorinated to an estimated 200 mg L^{-1} as total Cl using household bleach (unscented, regular strength Clorox® labeled as containing 6% sodium hypochlorite (NaOCl), which is 64.6% Cl by mass) by adding 8.9 ml (0.3 fl oz) bleach per liter (0.26 gallons) of well volume (Table 1). Pump discharge was circulated back into the well casing for 15 min to disperse bleach into solution. Wells were re-capped for a resting period of at least twelve hours.

Table 3. Methods used for sample analysis. "EPA" refers to a standard analytic method published by the US Environmental Protection Agency (available at http://www.epa.gov/sam/index.htm, last accessed April 2010), used by the Nevada State Health Laboratory. "SM" refers to standard method 5310 C (Persulfate-ultraviolet or Heated Persulfate Oxidation Method), published in Standard Methods for the Examination of Examination of Water and Wastewater (Clesceri et al., 1998). Total Organic Carbon samples were analyzed using a Shimadzu Spectrophotometer at the University of Nevada.

Analyte	Method	Analytic detection limits (mg L^{-1})
Copper	EPA 200.7	0.020, 0.050
Arsenic, Lead, Uranium	EPA 200.8	0.003, 0.002, 0.002
gross α, gross β	EPA 900.0	3 pCi L^{-1}
THM	EPA 524.2	0.005
HAA5	SM 6251B	0.001
TOC	SM 5310C	1.0*

* The analytic detection limit is based on linear regression analysis of the calibration curve, conducted with three replicates each of 4 standards, ranging from 0.274–20.270 mg L^{-1}.

2.4 Sampling

Immediately prior to and during purging temperature, pH, oxidation-reduction potential, and conductivity were measured. Immediately prior to purging post-chlorination (PC) samples from all wells were tested for As, Cu, Pb, U, and gross-α and gross-β radiation. Samples from two wells (142 and 182) were tested for disinfection byproducts (HAA5 and THM). All metals and disinfection by-products samples were collected using a PTFE dip bailer.

Post chlorination purging samples were collected for As, Cu, and Pb at intervals defined by the volume of water purged from each well, including ½, 1, 2, 3, and 4× WV (designated as 1/2, 1, 2, 3, 4). After 4 WV, samples were collected for As, Cu, Pb, U, gross-α and gross-β radiation, HAA5 and THM (Table 2).

Table 4. Physical and chemical characteristics of water in test wells observed during post-chlorination purging (T: temperature, C: conductivity, ORP: oxidation-reduction potential, ND: not measured). Final column displays the percentage difference between values at IP (prior to chlorination) and post-chlorination, with 4 well volumes purged.

Well volumes purged→	0	1/2	1	2	3	4	(IP-4)/IP%
182 (discharge rate = 14.4 liters per minute (3.8 gpm))							
T (°C)	16.7	15.7	15.7	15.7	15.6	15.7	5%
C (mS cm^{-1})	0.68	0.49	0.56	0.56	0.54	0.55	−26%
pH	7.97	7.57	7.70	7.52	7.47	7.58	−11%
ORP (mV)	772	817	810	827	837	828	no IP
167N (discharge rate = 18.9 liters per minute (5.0 gpm))							
T (°C)	ND	17.1	17.1	17.1	17.1	17.1	1%
C (mS cm^{-1})	2.12	1.23	0.53	0.45	0.42	0.44	27%
pH	8.78	8.38	7.82	7.56	7.27	7.51	5%
ORP (mV)	732	787	793	783	791	777	581%
51 (discharge rate = 14.6 liters per minute (3.9 gpm))							
T (°C)	16.4	14.8	15.0	15.1	15.3	15.2	−3%
C (mS cm^{-1})	1.95	1.07	1.08	0.79	0.81	0.77	70%
pH	8.65	8.8	8.5	8.27	8.41	8.42	16%
ORP (mV)	867	957	962	956	831	586	117%
142 (discharge rate = 2.5 liters per minute (0.7 gpm))							
T (°C)	22.4	16.7	16.7	16.6	16.6	16.6	−2%
C (mS cm^{-1})	2.26	1.64	2.07	2.04	1.97	1.94	−7%
pH	8.56	8.48	8.13	7.71	7.74	7.73	1%
ORP (mV)	704	719	719	−109	−124	−124	1%

All samples were unfiltered. Samples were collected, immediately placed in a portable cooler with blue ice to avoid exposure to sunlight and changes in temperature and submitted within 24 h to the Nevada State Health Laboratory (University of Nevada School of Medicine, Reno, NV – a certified drinking water analysis laboratory) for analysis (Table 3). Samples for arsenic, copper, lead and uranium were collected in 500 ml high density polyethylene bottles with 5.0 ml of 15% nitric acid as a preservative in containers provided by the laboratory. After sample collection, the final concentration of nitric acid was 0.15% nitric acid.

2.5 Chlorine test strips

Free chlorine was measured semi-quantitatively with test strips for swimming pool and spa maintenance (Arch Chemicals, Inc. – HTH line). The test strips indicated free chlorine concentrations in ranges rather than absolute numbers, similar to pH indicator strips. In order to determine the range, a user dips the strip in a solution and compares colors appearing in segregated rectangles with a key. The ranges reported include undetectable (indicated as 0 on the test strip), >0–1 mg L^{-1}, >1–2 mg L^{-1}, >2–3 mg L^{-1}, >3–5 mg L^{-1}, >5–10 mg L^{-1} and >10 mg L^{-1}. We tested the accuracy of the strips using dilutions of sodium hypochlorite solution and determined that they were adequate for distinguishing between the classes noted above (data not shown).

3 Results and discussion

3.1 Conductivity, pH, oxidation-reduction potential and temperature

Post-chlorination conductivity measurements were elevated from initial values, and returned to near initial values following purging 4 WV (Table 4). pH rose in wells 167N and 51 and decreased slightly in wells 182 and 42. Increases in pH conform to observations that concentrations of chlorine of 200 mg L^{-1} can be expected to increase pH by up to two units (Schnieders, 2005). Oxidation-reduction potential increased above pre-chlorination levels in all wells, as would be expected with the addition of an oxidizer. Water in well 142 returned to background levels for temperature, pH, conductivity, and oxidation-reduction potential after four WV had been purged, though measurements from the other wells indicated that 4 WV of purging was not sufficient to return to background conditions (Table 4). The oxidation-reduction potential in wells 51 and 167 remained above background

Table 5. Trace metal concentrations from pre-treatment (PT) through purging, expressed in number of well volumes pumped. The Maximum Contaminant Level for arsenic is 0.010 mg L^{-1}. Action Levels for lead and copper are 1.3 and 0.015 mg L^{-1}, respectively. Samples that contained analytes in concentrations less than the reporting limit (RL) are reported as "<RL." The reporting limit for lead varied according to results of internal laboratory quality control assessments and chemical quality of water samples.

	Well volumes purged						
Metals (mg L^{-1})	PT	0	½	1	2	3	4
182							
Arsenic	0.020	0.016	0.016	0.020	0.021	0.021	0.021
Copper	< 0.02	0.04	< 0.02	< 0.02	< 0.02	< 0.02	< 0.02
Lead	< 0.001	0.011	0.001	< 0.001	< 0.001	< 0.001	< 0.001
Cl	< 1	> 10	> 10	> 10	> 10	> 10	>3–5
167N							
Arsenic	0.019	0.024	0.018	0.019	0.019	0.019	0.019
Copper	< 0.02	0.03	0.04	< 0.02	< 0.02	< 0.02	< 0.02
Lead	< 0.001	0.013	0.008	< 0.002	< 0.001	< 0.001	< 0.001
Cl	< 1	> 10	> 10	> 10	>3–5	>2–3	>1–2
142							
Arsenic	0.580	0.440	0.460	0.750	0.620	0.590	0.600
Copper	< 0.02	< 0.02	0.02	< 0.02	< 0.02	< 0.02	0.02
Lead	< 0.010	0.010	< 0.010	< 0.010	< 0.010	< 0.010	0.010
Cl	< 1	> 10	> 10	< 1	< 1	< 1	< 1
51							
Arsenic	0.410	0.390	0.380	0.410	0.460	0.460	0.460
Copper	< 0.02	0.08	0.08	0.06	< 0.02	< 0.02	< 0.02
Lead	0.003	0.022	0.016	0.012	0.003	0.002	0.003
Cl	< 1	> 10	> 10	> 10	>1–2	< 1	< 1

levels, though temperature and pH were <16% of starting values. This suggests that the oxidizing effects of chlorination led to short-term changes in the immediate vicinity of the well screen or that were stagnant zones within the well, which would have required more purging to eliminate.

3.2 Chlorine concentration

Decreases of free chlorine concentration were hypothesized to be an indicator of purging of mobilized trace metals and disinfection byproducts. The concentration of free chlorine decreased in wells 51, 167N and 142 to >1–2 ppm free chlorine after 4 WV were pumped (Table 5). Site 182 required that 5 WV be purged before free chlorine decreased to >3–5 ppm (data not shown).

3.3 Mobilization of trace metals

Concentrations of lead and copper in well water increased following shock chlorination, which corresponds with results reported by Seiler (2006). Lead concentrations increased up to thirteen-fold and copper concentrations increased up to four-fold following treatment. Concentrations of both decreased to initial levels within 2 WV of purging. The return to background levels corresponded with the decline in free chlorine to >3–5 mg L^{-1} (Table 5).

All wells contained arsenic in concentrations that exceeded the MCL (0.010 mg L^{-1}) prior to treatment. The decline in arsenic concentration in the chlorine solution prior to purging was similar to results reported by Gotkowitz et al. (2008) and would be expected with the observed changes pH and oxidation-reduction potential. However, the decline was followed by an increase in arsenic above background levels during pumping. After 4 WV, water from one well returned to initial arsenic concentrations, while the others remained 3–12% higher than initial concentrations.

3.4 Uranium and radionuclides

Samples were analyzed for uranium and gross-α and gross-β radionuclides (Table 6). Uranium concentrations increased at sites 51 and 142 but remained the same or decreased in wells 167N and 187. Concentrations returned to approximately the same as starting levels in all wells after purging 4

Table 6. Uranium, gross-α and gross-β results for three stages of treatment: IP (prior to chlorination), PC (post chlorination but prior to purging) and 4 WV (after purging 4 well volumes). The Maximum Contaminant Levels for uranium, gross-α and gross-β are 30 µg L^{-1}, 15 pCi L^{-1} and 4 mrem yr^{-1}, respectively, (see http://www.epa.gov/safewater/radionuclides/pdfs/qrg_radionuclides.pdf, last accessed April 2010).

182	IP	PC	4 WV
Uranium, µg L^{-1}	32	22	35
gross-α, pCi L^{-1}	18	21	24
gross-β, pCi L^{-1}	18	14	20
167N			
Uranium, µg L^{-1}	13	13	13
gross-α, pCi L^{-1}	8	15	12
gross-β, pCi L^{-1}	19	17	19
142			
Uranium, µg L^{-1}	560	620	580
gross-α, pCi L^{-1}	120	92	127
gross-β, pCi L^{-1}	173	168	158
51			
Uranium, µg L^{-1}	4	29	9
gross-α, pCi L^{-1}	13	43	16
gross-β, pCi L^{-1}	6	8	5

WV. Gross-α concentrations changed from the IP to PC sampling steps, increasing in wells 51 and 167N, decreasing in well 142 and remaining approximately the same in well 182. Gross-β concentrations remained approximately the same in all wells at each sampling stage, with concentrations appearing to decrease slightly between the IP and 4 WV samplings. The maximum change in gross-β concentrations was approximately −9% in well 142.

3.5 Total organic carbon and disinfection byproducts

Total organic carbon (TOC) concentrations in wells 182 and 142, collected prior to treatment, are shown in Table 7. Water from wells obtained after treatment, but prior to purging contained concentrations of THM up to ten times the MCL (Table 7). Following purging of 4 WV, concentrations in well 142 decreased to below the detection limit, indicating that the increase in concentration was temporary and could be remediated by purging 4 WV after treatment. Well 182 retained disinfection byproducts and free chlorine, with the concentration of free chlorine (indicated by test strips) >10 mg L^{-1} after purging 4 WV. Testing for free chlorine at the fifth well volume purged indicated a concentration of >3–5 ppm.

Table 7. Total Organic Carbon (TOC) concentrations following initial purging of test wells (IP), and HAA5 and THM concentrations following IP, post chlorination (PC) and after 4 well volumes (WV) had been purged.

Site	TOC (IP) mg L^{-1}	HAA5 (IP,PC,4 WV) µg L^{-1}	THM (IP,PC,4 WV) µg L^{-1}
182	3.9	<1.0, 101.0, 128.0	<0.5, 43.7, 70.1
142	15.1	<1.0, 394.0, <1.0	<0.5, 747.0, <0.5

4 Conclusions

This study confirmed temporary mobilization and changes in concentration of trace metals from shock chlorination treatment, as previously reported by Seiler (2006) and Gotkowitz et al. (2008). It also demonstrated that disinfection byproducts can be formed and can persist beyond 3 WV of purging. The concentration of disinfection byproducts exceeded drinking water MCLs in the two wells tested. In one well, concentrations decreased to below the detection limits for HAA5 and THM with purging. In the other, concentrations of disinfection byproducts remained elevated, with final concentrations of THM and HAA5 at highest levels after purging 4 WV. Metals, though mobilized by shock chlorination, decreased to near pre-treatment background levels after 4 WV were purged in all wells.

The concentrations of DBP precursors and peak concentrations of HAA5 and THM differed in the two wells sampled. Although concentrations of THM and HAA5 were lower in well 182, samples collected after 4 WV indicated that DBPs persisted. This may have been due to differences in well construction and interaction between chlorine solutions and aquifer materials. With respect to well construction, the depths of wells and depths to water from the land surface for wells 182 and 142 were similar, but lengths of screened intervals and static volumes differed (Table 1). The screened interval in well 182 spanned 3.3 m (11 ft), compared with the screened interval of 1.5 m (5 ft) in well 142. Well 182 also had approximately 1.3 m of casing below the screened interval, which may have served as a reservoir for residual chemicals because volume exchange in this portion of the casing would be difficult to accomplish with a surface pump. The stagnant portion of well 182 may also have contained sediments, which would have further prolonged release of HAA5 and THM into pumped water. In contrast, screens in the other wells were mounted at the bottom of the casing. Although this does not assure that complete mixing took place throughout post-treatment purging, it is more likely that purging led to complete exchanges of water within the well casing than in well 182. It is also possible that disinfection byproducts formed within the saturated formations immediately adjacent to the well screen.

Although well logs indicated that the screened intervals were set in sands, well yields appeared to be very different based on the discharge rates used for purging and drawdowns observed. In order to maintain a steady discharge rate from well 142, the pumping rate was <20% of the purging rate used in well 182, even though lift distances from the water table to the surface were slightly less for well 142 than well 182. This suggests that the saturated formations in the screened intervals pumped for well 142 were less permeable than those in well 182, which in turn suggests that aquifer materials may have been composed of fine-textured soils including silts and clays. In the absence of significant advective mixing, formation of disinfection byproducts beyond the immediate well volume would be limited to chemical dispersion. In well 182, however, it is likely that a larger volume of aquifer material was exposed to chlorine. This hypothesis is, in part, supported by the persistent indication of oxidizing conditions in water withdrawn after purging 4 WV. A prior study in the region (Fram et al., 2005) demonstrated that free chlorine released significant amounts of carbon from aquifer materials. They concluded that carbon was likely released from clay minerals in proportion to the amount of available chlorine in solution. This suggests that aquifer material and interactions with high concentrations of available chlorine were a potential source of precursors to HAA5 and THM formation.

The mixing procedure used to disperse chlorine throughout the entire well volume was unlikely to have created a gradient that would advect chlorine into aquifer materials, but chemical diffusion through the well screen may have taken place during the resting phase. Given the differences in screened intervals, this may have led to accumulation of DBPS's in aquifer materials in contact with the screened interval of well 182, which could have required more thorough purging to remove than those in contact with the shorter screened interval in well 142. Although purging appeared to be adequate based on number of well volumes removed and stability of temperature and pH, test strips indicated the presence of free chlorine. Consequently, the chemical reactions that led to metals releases and DBP formation may have been due to chemical diffusion during the resting time following introduction of bleach and likely due to reactions with aquifer materials and gravel packing in the immediate vicinity of the screened interval of the well casing.

Publications about shock chlorination recommend purging varying numbers of standing well volumes post treatment, before returning a well to service. Guidance also is based on detecting the scent of chlorine in water. Without metering equipment most domestic well owners have no accurate means of determining when pumping is sufficient to remove a specific number of well volumes. Also, determining the sufficiency of purging by scent is subjective and may not be consistent, especially with respect to avoiding exposure to metals and disinfection byproducts. The use of chlorine test strips shows promise as a simple and accurate means of determining when purging is complete, though this should be verified with further experimental work. This technique has the potential to be a reliable guideline for public health officials and informal educators (including county-based educators of Cooperative Extension), who provide advice about proper procedures for shock chlorination, especially when well owners cannot measure discharge rates and discharge volumes.

Acknowledgements. This material is based upon work supported by the National Institute of Food and Agriculture, US Department of Agriculture, under Agreement No. 2009-51130-06048. Any opinions, findings, conclusions, or recommendations expressed in this publication are those of the author(s) and do not necessarily reflect the view of the US Department of Agriculture.

References

Amoore, J. E. and Hautala, E.: Odor as an ald to chemical safety: Odor thresholds compared with threshold limit values and volatilities for 214 industrial chemicals in air and water dilution, J. Appl. Toxicol., 3(6), 272–290, 1983.

Clesceri, L., Greenberg, A., and Eaton, A. (Eds): Standard Methods for Examination of Water and Wastewater. American Public Health Association, Washington, D.C., 1325 pp., 1999.

Driscoll, F. G.: Groundwater and Wells, 2nd ed., Johnson Division, St. Paul, MN, 1108 pp., 1986.

Fram, M. S., Maurer, D. K., and Lico, M. S.: Potential for formation of disinfection by-products from storage of chlorinated surface water in the basalt aquifer near Fallon, Nevada, US Geological Survey, Carson City, NV, 23 pp., 2005.

Glancy, P.: Geohydrology of the Basalt and Unconsolidated Sedimentary Aquifers in the Fallon Area, Churchill County, Nevada, Institution, Alexandria, Va., US Geological Survey Water Supply Paper 2263, 60 pp., 1986.

Gotkowitz, M., Ellickson, K., Clary, A., Bowman, G., Standridge, J., and Sonzogni, W.: Effect of well disinfection on arsenic in ground water, Ground Water Monit. R., 28(2), 60–67, 2008.

Schnieders, J. H.: Use of Chlorine to Disinfect Water Wells, Water Well Journal, 7, 52–53, 2005.

Seiler, R. L.: Mobilization of lead and other trace elements following shock chlorination of wells, Sci. Total Environ., 367(2–3), 757–768, 2006.

Walker, M. J., Benson, M., and Shaw, W. D.: Significance of private water supply wells as a route of exposure to aqueous arsenic, J. Water Health, 3(3), 305–312, 2005.

Welch, A. H. and Lico, M. S.: Factors controlling As and U in shallow ground water, southern Carson Desert, Nevada, Appl. Geochem., 13(4), 521–539, 1998.

Westerhoff, P., Chao, P., and Mash, H.: Reactivity of natural organic matter with aqueous chlorine and bromine, Water Res., 38(6), 1502–1513, 2004.

2

Predicting the residual aluminum level in water treatment process

J. Tomperi[1], **M. Pelo**[2], **and K. Leiviskä**[1]

[1]University of Oulu, Control Engineering Laboratory, P.O. Box 4300, 90014 University of Oulu, Finland
[2]Finnsugar Ltd, Sokeritehtaantie 20, 02460 Kantvik, Finland

Correspondence to: J. Tomperi (jani.tomperi@oulu.fi)

Abstract. In water treatment processes, aluminum salts are widely used as coagulation chemical. High dose of aluminum has been proved to be at least a minor health risk and some evidence points out that aluminum could increase the risk of Alzheimer's disease. Thus it is important to minimize the amount of residual aluminum in drinking water and water used at food industry. In this study, the data of a water treatment plant (WTP) was analyzed and the residual aluminum in drinking water was predicted using Multiple Linear Regression (MLR) and Artificial Neural Network (ANN) models. The purpose was to find out which variables affect the amount of residual aluminum and create simple and reliable prediction models which can be used in an early warning system (EWS). Accuracy of ANN and MLR models were compared. The new nonlinear scaling method based on generalized norms and skewness was used to scale all measurement variables to range $[-2\ldots+2]$ before data-analysis and modeling. The effect of data pre-processing was studied by comparing prediction results to ones achieved in an earlier study. Results showed that it is possible to predict the baseline level of residual aluminum in drinking water with a simple model. Variables that affected the most the amount of residual aluminum were among others: raw water temperature, raw water $KMnO_4$ and PAC/$KMnO_4$ (Poly-Aluminum Chloride/Potassium permanganate)-ratio. The accuracies of MLR and ANN models were found to be almost the same. Study also showed that data pre-processing affects to the final prediction result.

1 Introduction

In water treatment processes surface waters are most commonly treated with chemical coagulation. Aluminum salts are widely used as a coagulant to reduce the organic matter, color and turbidity of raw water. Using aluminum salts in a water treatment process may lead to an increased concentration of aluminum in drinking water if aluminum is overdosed or the water treatment process is dysfunctional. The residual aluminum increases the water turbidity, may have some health effects on consumers and aluminum hydroxide may deposit on the walls of the pipes decreasing flow capacity (Driscoll and Letterman, 1995; WHO, 2008). Reported minor symptoms of the high level of residual aluminum in drinking water are nausea, vomiting, diarrhea, mouth and skin ulcers, rashes and arthritic pain (WHO, 2003). Symptoms are generally mild and short-lived. More serious health effects of aluminum in drinking water have been studied widely and the results are conflicting. A Canadian study of health and aging claims that residual aluminum in drinking water does not increase the risk of Alzheimer's disease (Leakey, 2004). However, several researches that showed relationships between aluminum in drinking water and Alzheimer's disease have been found in George et al. (2010), Mclachlan et al. (1996), WHO (2008).

The total intake level of aluminum from drinking water varies according to the aluminum level in raw water and whether aluminum coagulants are used in a water treatment process. The concentration of aluminum in natural waters can vary significantly depending on various physicochemical and mineralogical factors. The aluminum intake from food and water is unavoidable but only 5 % of the total intake is from drinking water. The major part ($5\,mg\,day^{-1}$) of the total intake comes from food and its additives (WHO, 2003, 2008).

Juntunen et al. (2010) found that the most important parameters that affect the amount of residual aluminum in drinking water were the raw water temperature, Al/$KMnO_4$-ratio, turbidity and silicate concentration. The treated water pH, the pH of coagulation and the temperature and turbidity of water were found to have an effect on the amount of residual aluminum in Driscoll and Letterman (1995). WHO (2008) reported that the residual aluminum concentration in drinking water can be minimized by optimizing pH, avoiding excessive dosing of aluminum, good mixing of coagulants, optimum paddle speed in the flocculation process and efficient floc filtration.

Artificial Neural Networks (ANN) has been reported to have many benefits against traditional data modeling methods. Data-driven ANN can capture relationships using the desired input output mapping and physical processes do not have to be known explicitly like when using mechanistic models. In a drinking water treatment process modifications can occur frequently and very often micro-scale interactions are poorly understood. This makes it impossible to develop a useful mechanistic model. Using an ANN model gives the ability to quickly modify process models using full-scale operation data without necessity to understand all micro-scale interactions (Baxter et al., 2001; Maier et al., 2004).

The performance of Multiple Linear Regression (MLR) and Artificial Neural Network models has been compared in several studies (Juntunen et al., 2010; Maier et al., 2004; Bowden et al., 2006; Ibarra-Berastegi et al., 2007; Areerachakul and Sanguansintukul, 2009; Kulkarni and Chellan, 2010) and ANN models have been found to outperform the MLR models in training, testing and validation of different prediction cases. However, the difference was not always significant. The supremacy of ANN models indicates nonlinear relationships in used datasets. In the water treatment and drinking water production, ANN models have been successfully used to among others modeling and predicting chlorine residual in a water distribution system (Bowden et al., 2006), drinking water quality (Baxter et al., 2001), contaminant removal (Shetty et al., 2003), fouling and backwash efficiency in ultrafiltration (Delgrange-Vincent et al., 2000), optimal aluminum doses (Maier et al., 2004) and residual aluminum (Juntunen et al., 2010) in the water treatment process. Baxter et al. (2001) created also a prediction model to provide plant operators with an early warning system (EWS) for raw water quality changes and to improve treatment efficiency.

In this paper the data of a water treatment plant is analyzed and prediction models are created using traditional linear and nonlinear methods. The purpose was to find out which variables affect the amount of residual aluminum in drinking water and to study if it is possible to predict reliably the residual aluminum level using only a few important measurement variables. Information of the reliable simple prediction model could be used for a EWS at the plant, to activate inhibitory actions by process operators or on-line control systems to avoid increased values of residual aluminum. The performance of MLR and ANN models is compared with each other and to earlier prediction results presented in Juntunen et al. (2010). Same modeling methods and the same data source were used to study the effect of data pre-processing (scaling, interpolation, averaging) to modeling accuracy.

2 Material and methods

2.1 The water treatment plant

The data was collected from the water treatment plant (WTP) of Finnsugar Ltd. in Kirkkonummi, Finland. This chemical treatment plant uses surface water from a lake nearby (Humaljärvi), an artificial lake (Pikkala) or the mixture of these two sources as raw water. Before adding the coagulation chemical the pH of raw water is adjusted to the optimal value with calcium hydroxide. Aluminum based coagulation chemical PAX-14 (Kemira Kemwater) is used in the coagulation process. The coagulation chemical dose is controlled as a function of raw water $KMnO_4$ (potassium permanganate) value. After the filtration, water pH is again adjusted to the optimal level for distribution. UV-radiation and sodium hypochlorite are used for disinfection. The process stages of the WTP is shown in Fig. 1.

During the period of data collection, the long-term mean value of residual aluminum in drinking water produced at Finnsugar Ltd. WTP was less than half of the maximum target value of the quality recommendation ($0.2\,mg\,l^{-1}$) defined in the Health Protection Act of the Finland's Ministry of Social Affairs and Health (FINLEX, 2000). Thus, a high amount of residual aluminum in drinking water is not a serious concern in this water treatment plant. Even so, Finnsugar Ltd. has a great interest to find out which variables have an effect on the amount of residual aluminum and use this information to keep it at minimum level.

2.2 Dataset

The quality of the developed model depends highly on the quality of the source dataset. The dataset used in analyses and modeling the water treatment process must be fully representative of the full spectrum of all possible conditions. The temperature of surface water, for instance, varies depending on the season of the year. Therefore the source dataset must encompass at least one full year of measured data (Baxter et al., 2001).

The dataset used in analysis and modeling in this study covered a period of 16 months. The dataset included online process measurement variables, and laboratory measurements of raw water and drinking water. Only the measurements which could be used for the prediction of residual aluminum in drinking water were used, thus the measurements

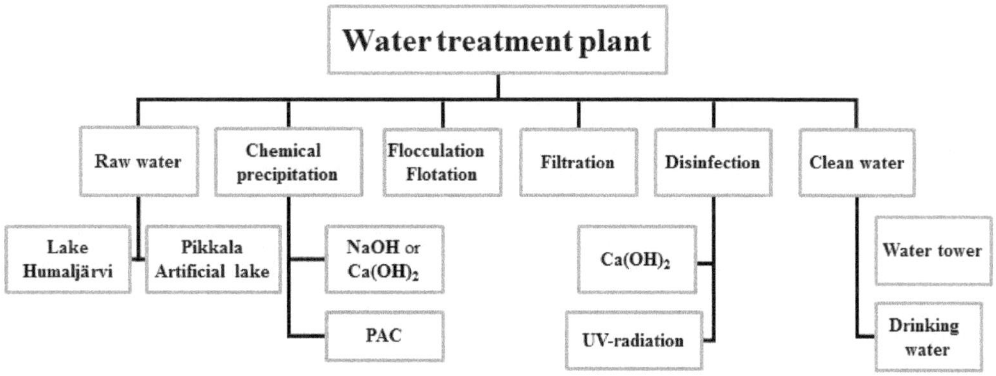

Figure 1. Processing stages of the Finnsugar Ltd. water treatment plant.

of drinking water were not used in modeling the residual aluminum. Measured laboratory variables were pH, potassium permanganate, turbidity, hardness, color, conductivity, smell, chlorine, bacteria and aluminum.

The laboratory measurements of raw and drinking waters were done at least once in every working day. If some measurement results showed anomalous values, new samples were collected and analyzed. For the data analysis, all on-line measurements, which were originally stored at 5 min intervals, were averaged to one hour data. Laboratory measurement values were combined to the corresponding hour of on-line measurement data. Few evident outliers (values that were not realistic) were manually filtered out and missing data values were added by linear interpolation using inpaint_nans-Matlab function created by D'Errico (2004).

2.3 Nonlinear scaling

A dataset which contains several different measurement variables has to be scaled before data analysis to facilitate analysis and avoid incorrect conclusions. The inspection of the raw dataset may not reveal all the noteworthy changes or states. After scaling different variables can easily be used in calculations and the real values of measurements are not revealed. In this work, the dataset was scaled between $[-2\ldots+2]$ using the new nonlinear scaling method based on generalized norms and skewness.

Nonlinear mapping function has been developed in Juuso (2010) and Juuso (2011) to extract the meanings of variables from measurement signals. These functions are called membership definitions. Membership definitions map the real values of variables to the range of $[-2\ldots+2]$. Thus, a normal scaling to range $[-1\ldots+1]$ is combined with handling of warnings and alarms. A trapezoidal membership function which is based on the support and core areas defined by fuzzy set theory is used to define the concept of the feasible range. The support area is defined by the minimum, $\min(x_j)$, and maximum, $\max(x_j)$, of the values of the variable. The value range x_j is divided into two parts by the central tendency value c_j. The core area $[(c_l)_j,(c_h)_j]$ is limited by the central tendency values of the lower and upper part. The mapping function contains one monotonously increasing function for the values between −2 to 0 and one monotonously increasing function between values 0 to +2. Membership functions consist of two second order polynomials: one for the negative values and one for the positive values presented in Eq. (1).

$$\begin{aligned} f_j^- &= a_j^- X_j^2 + b_j^- X_j + c_j, X_j \in [-2,0), \\ f_j^+ &= a_j^+ X_j^2 + b_j^+ X_j + c_j, X_j \in [0,2]. \end{aligned} \tag{1}$$

Because the scaling idea is based on the membership functions of fuzzy set systems these values are called linguistic values. The coefficients of the polynomials are defined by the corner points

$$\left\{\min(x_j),-2),((c_l)_j,-1),c_j,0),((c_h)_j,1),\max(x_j),2\right\}. \tag{2}$$

The detailed description of this new nonlinear scaling method is presented in Juuso (2010, 2011).

2.4 Variable selection

Variable selection is one of the most important steps in the model development process. The greater number of variables does not necessary mean better prediction results. Some input variables may be correlated, noisy or have no significant relationship with the output variable and thus will not be informative. Selecting non-essential inputs only increases computational complexity, makes the training process more difficult and prediction results worse (Bowden et al., 2006).

In this work, variable selection was done using the stepwise regression, a forward selection method, which adds the best variable to, or deletes the worst variable from a variable subset at each round. Adding and deleting is based on variables statistical significance (p value) in regression. The forward selection method starts with an initial model and continues until either no further model changes occur over one complete round or a preset number of variable selections and deletions occur. Manually selected variables were also used in modeling to test if it is possible to create a good and simple model using only a few important variables. Another goal was to compare the results with ones presented in Juntunen et al. (2010) and see the effect of data pre-processing on modeling accuracy.

2.5 Multiple Linear Regression

Multiple Linear Regression can be used to describe a quantitative relationship between several independent variables and a dependent variable as a linear system, to predict future scores on the dependent variable or to test specific hypotheses based on a scientific theory or prior research. A linear equation is fitted to observe independent variables. MLR equation is a weighted linear combination of the independent variables and can be written as presented in Eq. (3) (Areerachakul and Sanguansintukul, 2009; Matlab, 2011; Tranmer and Elliot, 2008; Audone and Giunta, 2008).

$$Y = b_0 + b_1 X_1 + b_2 X_2 + ... + b_n X_n + e \tag{3}$$

where b_o is a constant value, $b_1 \dots b_n$ are regression coefficients, $X_1 \dots X_n$ independent variables and e is the error.

The major limitations of MLR are that it may not be useful with nonlinear features and that one can only ascertain relationships, but not be sure about underlying causal mechanism.

2.6 Artificial Neural Network

An Artificial Neural Network typically consists of at least three layers: an input layer, one or more hidden layers and an output layer. External inputs of the network are received by neurons in the input layer. Inputs are multiplied by interconnection weights and sent forward to the hidden layer where they are summed and processed by a nonlinear transfer function. Each value from the input layer is sent to every neuron in the hidden layer. If the network has more than one hidden layer, data is multiplied by interconnection weights, summed and processed by a transfer function in every layer before it is sent to the output layer. The output of the network is given by the neurons on the output layer (Maier et al., 2004; Dayhoff, 1990).

The multilayer perceptron (MLP) is the most common neural network model. MLP is a feedforward ANN which utilizes a supervised learning technique called back-propagation for training a network. Neural networks are trained by examples using historical data. The three-layer back-propagation network is one of the most used architectures in process modeling. Back-error propagation, or back-propagation, is widely and successfully used in Neural Network paradigms because it is easy to understand. The aim of the training process is to minimize the output error by adjusting the interconnection weights which are set at random values at the beginning of the training. The error is defined to be the difference between the predicted output and measured output. The calculated error is back-propagated to the neural network through each layer and the weights are adjusted to decrease the error. The training process is continued until the error is minimized and the network has learned the data (Baxter et al., 2001; Maier et al., 2004; Delgrange-Vincent et al., 2000; Dayhoff, 1990; Pal and Mitra, 1992; Beale et al., 2010).

The number of hidden layers, the number of neurons, the learning rate and initial weights, for instance, can influence the network training and prediction accuracy (Maier et al., 2004; Bowden et al., 2006). The optimum number of hidden layers and nodes are often found by trial and error. It has been proven that one hidden layer can give sufficient degree of freedom but using more than one hidden layer provides greater flexibility and enables the approximation of complex functions with fewer connection weights (Maier et al., 2004; Delgrange et al., 1998).

In this work the Neural Network consisted of measured variables as inputs, the predicted value of residual aluminum as output and one hidden layer (5 neurons). Resilient back-propagation was used as the training function and the mean squared error (MSE) as the performance function. Hyperbolic tangent sigmoid was used as the transfer function for the hidden layer, and the linear transfer function for the output layer. The configuration was selected similar as used in the study of Juntunen et al. (2010) so that the comparison of prediction results would be easier and more reliable.

Hyperbolic tangent sigmoid is defined as Eqs. (4) and (5).

$$f(s) = \frac{1 - e^{-2s}}{1 + e^{-2s}}, \tag{4}$$

where

$$e = \sum_{i=1}^{n} w_i x_i + b, \tag{5}$$

where w_i are the weights, x_i are the inputs of neurons, b is a bias and n is the number of variables.

Performance of ANN and MLR models can be evaluated for example using Root Mean Square Error (RMSE), Mean Absolute Error (MAE) and coefficient of determination (R^2). MAE can be used to determine whether model predictions are suitable for process control. R^2 value can be used to compare the relative performance of the models (Baxter et al., 1999).

The coefficient of determination value R^2 is defined as in Eq. (6), RMSE is defined as in Eq. (7) and MAE is defined as in Eq. (8).

$$R^2 = 1 - \frac{\sum (y_{\text{meas}} - y_{\text{pred}})^2}{\sum (y_{\text{meas}} - \frac{\sum y_{\text{meas}}}{k})^2} \tag{6}$$

where y_{meas} is a measured value, y_{pred} is a predicted value and k is the number of values.

$$\text{RMSE} = \sqrt{\frac{1}{k} \sum ((y_{\text{meas}} - y_{\text{pred}})^2)} \tag{7}$$

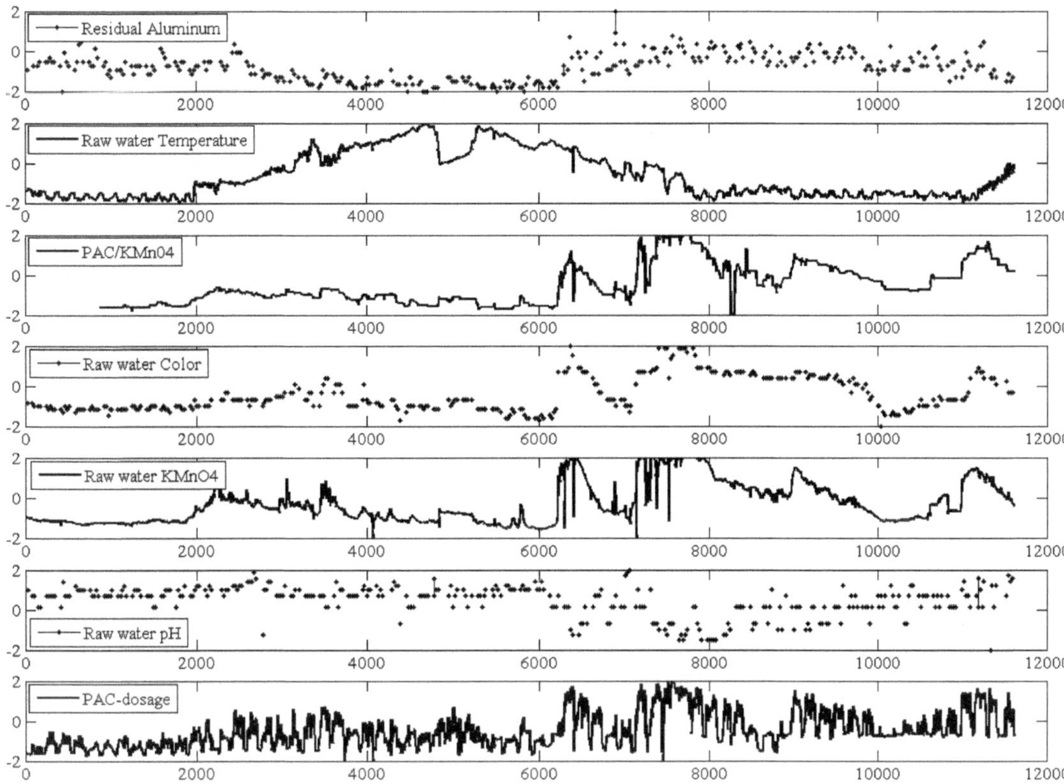

Figure 2. Trend lines of residual aluminum and important measurement variables. From top to down: residual aluminum, raw water temperature, $PAC/KMnO_4$, raw water color, raw water $KMnO_4$, raw water pH and PAC-dosage.

where y_{meas} is a measured value, y_{pred} is a predicted value and k is the number of values.

$$\mathrm{MAE} = \frac{1}{k}\sum \left| y_{preds} - y_{meas} \right| \qquad (8)$$

where y_{meas} is a measured value, y_{pred} is a predicted value and k is the number of values.

3 Results and discussion

3.1 General study

The combined on-line and laboratory measurement dataset was studied to find out significant correlations between measured variables and residual aluminum. The highest correlated variables to residual aluminum are shown in Table 1 sections (A), (B), (C) and (D). It can be seen that results of correlation determination varied depending on the order of the nonlinear scaling and data interpolation. Section (A) shows the correlation coefficients of the original, unprocessed, dataset. Only the laboratory measurements of raw water had high correlation to residual aluminum and all on-line measurements had very low correlation coefficients. In section (B) correlation coefficients are presented when the dataset was first scaled and then interpolated. Good correlations of some on-line measurements were now revealed. The new nonlinear scaling method clearly improved the ability to identify interactions of measurement variables and changes in the specific trend line compared with original data trends. The number of variables with good correlation and values of correlation coefficients decreased if the original data was at first interpolated and then scaled, which can be seen in section (D). This indicates that there are several Not a Number, in other words missing values, in the measurement data of that variable. Section (C) shows the correlation coefficients of residual aluminum from the dataset which consists only of the on-line and laboratory measurement values at the exact time of drinking water sampling. Interpolation did not affect the results of the correlation calculation. Selecting the pre-processing method of the dataset is important because the quality of the data affects the accuracy of prediction models.

It can be seen in Table 1 that certain variables, like raw water temperature, $PAC/KMnO_4$ (Poly-Aluminum Chloride/Potassium permanganate)-ratio, raw water color, raw water $KMnO_4$ and pH, always have high correlation with residual aluminum. The temperature, pH and $PAC/KMnO_4$-ratio were found to be affecting variables to residual aluminum also in earlier studies of Driscoll and Letterman (1995) and Juntunen et al. (2010).

Trend lines of residual aluminum and the highest correlated variables are shown in Fig. 2. Seasonal changes in raw water temperature are clearly seen at the second subfigure.

Table 1. Correlation coefficients of the residual aluminum.

(A) Original Dataset Correlation coefficient	Variable	(B) Scaled and interpolated dataset Correlation coefficient	Variable
0.82	Raw water color (lab)	−0.61	Raw water temperature (on-line)
0.81	Raw water pH (lab)	0.51	Raw water color (lab)
0.79	Raw water $KMnO_4$ (lab)	0.50	PAC/$KMnO_4$ (on-line)
0.78	Raw water smell (lab)	0.50	Raw water $KMnO_4$ (on-line)
0.76	Raw water smell, heated (lab)	−0.37	Raw water pH (lab)
0.37	Raw water conductivity (lab)		
(C) Scaled, sampling moment dataset Correlation coefficient	**Variable**	**(D) Interpolated and scaled dataset Correlation coefficient**	**Variable**
−0.65	Raw water temperature (on-line)	−0.68	Raw water temperature (on-line)
0.60	PAC/$KMnO_4$ (on-line)	0.38	PAC/$KMnO_4$ (on-line)
0.50	Raw water color (lab)	0.37	Raw water color (lab)
0.49	Raw water $KMnO_4$ (on-line)	0.37	Raw water $KMnO_4$ (on-line)
−0.43	Raw water pH (lab)	0.26	Level of Humaljärvi
0.38	PAC-dose (on-line)		

Table 2. Prediction models and goodness values of the models.

Model	Variables	RMSE	MAE	R^2
Model 1	– Raw water temperature – Raw water $KMnO_4$ (lab) – Level of surface water – Raw water coliform bacteria – WTP returning flow – Sodium Hypochlorite – Pressure of filter 1	ANN: 0.490 MLR: 0.477	ANN: 0.388 MLR: 0.377	ANN: 0.244 MLR: 0.228
Model 2	– Raw water temperature – PAC/$KMnO_4$ – Raw water turbidity – Raw water $KMnO_4$	ANN: 0.700 MLR: 0.561	ANN: 0.532 MLR: 0.441	ANN: 0.001 MLR: 0.078
Model 3	– Raw water temperature – Raw water $KMnO_4$ – Raw water pH – PAC-dosage	ANN: 0.570 MLR: 0.537	ANN: 0.435 MLR: 0.427	ANN: 0.088 MLR: 0.086
Model 4	– Raw water temperature – Raw water $KMnO_4$ – PAC-dosage	ANN: 0.544 MLR: 0.547	ANN: 0.429 MLR: 0.430	ANN: 0.132 MLR: 0.071

The amount of residual aluminum is high when raw water is cold even if the raw water $KMnO_4$ and poly-aluminum chloride (PAC) dosage are at the low level. When the raw water temperature is higher and raw water $KMnO_4$ rises the amount of residual aluminum is relatively low. This shows the fact that the effectiveness of the water treatment process is better when raw water is warmer which is very common in most of the WTP. It can be seen that autumn rains and snow melting in the spring affects the quality of raw water and the efficiency of the water treatment process. The raw water color and $KMnO_4$ are at a higher level in the autumn, late winter and spring seasons due to heavy raining and snow melting.

3.2 Prediction models

Prediction models were trained using 3/4 of the data and tested using the final 1/4 of the data. Scaled and interpolated dataset was used in modeling and the same variables were used in both MLR and ANN models. Created models and calculated goodness of fit values (RMSE, MAE and R^2) are listed in Table 2. Model 1 was created using variables

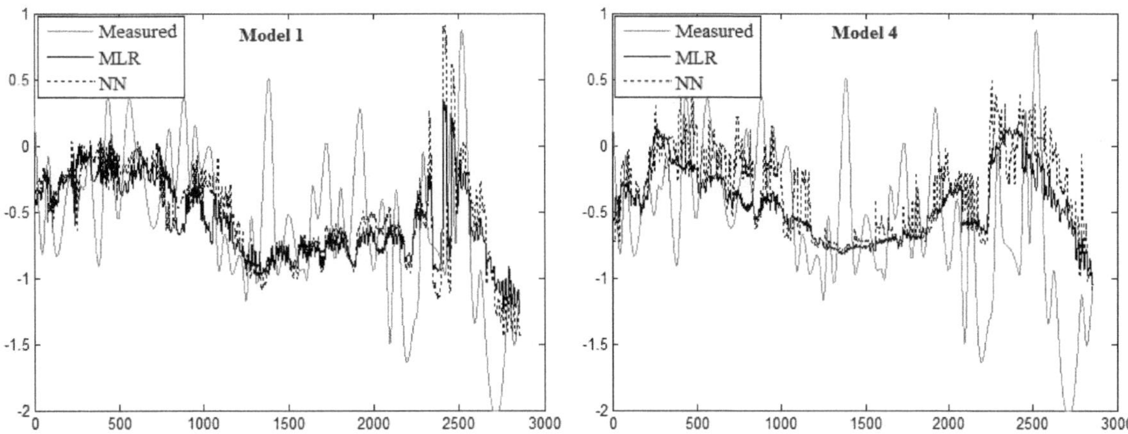

Figure 3. The testing periods of Model 1 (left) and the testing periods of Model 4 (right).

selected by a forward variable selection method. Model 2 was created using the best four variables of variable selection presented in Juntunen et al. (2010) where the best MLP model was achieved with four variables. Manually selected variables were used in Model 3 and Model 4.

During the modeling session it was noticed that there are several different combinations of variables that could be used in modeling the residual aluminum with fairly good accuracy. It was also noticed that the peak values of residual aluminum could be predicted and the accuracy of the model improved only if some laboratory measurements of drinking water were used in the model. This, however, does not give any extra value for real-life application (EWS). The best prediction result was achieved with Model 1 and in Model 2 both ANN and MLR had the lowest accuracy of the models presented in this paper. The difference between ANN and MLR is in Model 2 notably bigger than in Model 1. As it can be seen from Table 2, the error values of the ANN and MLR models were nearly the same in Models 1, 3 and 4, but ANN models seem to be slightly better in every variable set except in Model 2. The accuracy of MLR was better in Model 2 than the accuracy of ANN model. This is an opposite result than presented in Juntunen et al. (2010), where MLP showed a better performance than the MLR method. Modeling methods were quite similar in both studies. Discrepancy between the results could be explained by the significant differences in data pre-processing. In Juntunen et al. (2010) the process data was averaged to 1 day data and combined with daily laboratory data, the measuring period was 275 days, and the dataset was not scaled to range $[-2 \ldots +2]$ using the new nonlinear scaling method.

The results of Model 1 and Model 4 testing periods are shown in Fig. 3 for both ANN and MLR methods. The baseline of the residual aluminum prediction is fairly good with both models. Modeled residual aluminum follows the changes of measured residual aluminum but the peak values could not be predicted. As the calculated results in Table 2 showed, the difference between ANN and MLR models is minor. Model 4 was created using only three variables and the error values were not higher than in Model 1. This is an encouraging result for creating the EWS or on-line control. Almost the same accuracy can be attained with fewer variables.

4 Conclusions

The purpose of this work was to analyze the data of the water treatment plant, find out which variables affect the amount of residual aluminum in drinking water, create as simple and reliable prediction models for residual aluminum as possible using ANN and MLR methods and to compare the accuracy of models with each other and to earlier presented results. Same modeling methods and the same data source were used to study the effect of data pre-processing to modeling accuracy. Clear correlations to residual aluminum were found after the dataset was scaled using the new nonlinear scaling method based on generalized norms and skewness. Variables that had the highest correlation to the amount of residual aluminum were among others: the raw water temperature, raw water $KMnO_4$ and PAC/$KMnO_4$-ratio.

Decent prediction models were created using only a few important variables. The baseline of residual aluminum in drinking water can be predicted with fairly good accuracy with both MLR and ANN models. MLR and ANN methods gave almost the same results. Comparison to earlier results of modeling the residual aluminum at the same water treatment plant was done. In the earlier study the overall modeling accuracy was slightly better. However, in this study the MLR method was found to be better than the ANN method when using the same variables as in the earlier study. Discrepancy between the results can be explained by different kind of data pre-processing. The accuracy of created prediction models could be improved by using different variable selection or using different modeling methods. Yet, the results were promising and an early warning system could be created based on these models to give additional information to the process personnel of the water treatment plant.

Acknowledgements. This work was done as a part of the project "Development of a comprehensive water quality management system (POLARIS)". Contribution of relevant persons in Finnsugar Ltd. is greatly acknowledged. The funding of the TEKES (the Finnish Funding Agency for Technology and Innovation) is also acknowledged.

References

Areerachakul, S. and Sanguansintukul, S.: A Comparison between the Multiple Linear Regression Model and Neural Networks for Biochemical Oxygen Demand Estimations, SNLP '09, Eighth International Symposium on Natural Language Processing, 11–14, 2009.

Audone, B. and Giunta, G.: Multiple Linear Regression to Detect Shielding Effectiveness Degradations. International Symposium on Electromagnetic Compatibility – EMC EUROPE 2008, 8–12 September 2008, 1–6, 2008.

Baxter, C. W., Stanley, S. J., and Zhang, Q.: Development of a full-scale artificial neural network model for the removal of natural organic matter by enhanced coagulation, J. Water Supply Res. T., 48, 129–136, 1999.

Baxter, C. W., Zhang, Q., Stanley, S. J., Shariff, R., Tupas, R.-R. T., and Stark, H. L.: Drinking water quality and treatment: the use of artificial neural networks, Can. J. Civil Eng., 28 (Suppl. 1), 26–35, 2001.

Beale, M. H., Hagan, M. T., and Demuth, H. B.: Neural Network Toolbox TM 7 User's Guide, Matlab MathWorks Inc., September 2010.

Bowden, G. J., Nixon, J. B., Dandy, G. C., Maier, H. R., and Holmes, M.: Forecasting chlorine residuals in a water distribution system using a general regression neural network, Math. Comput. Model., 44, 469–484, 2006.

Dayhoff, J. E.: Neural Network Architectures: An introduction, Van Nostrand Reinhold, New York, ISBN: 0-442-20744-1, 259 pp., 1990.

Delgrange, N., Cabassud, C., Cabassud, M., Durand-Bourlier, L., and Laine, J. M.: Neural networks for prediction of ultrafiltration transmembrane pressure – application to drinking water production, J. Membrane Sci., 150, 111–123, 1998.

Delgrange-Vincent, N., Cabassud, C., Cabassud, M., Durand-Bourlier, L., and Laine, J. M.: Neural networks for long term prediction of fouling and backwash efficiency in ultrafiltration for drinking water production, Desalination, 131, 353–362, 2000.

D'Errico, J.: "inpaint_nans-function", Matlab Central, File Exchance, http://www.mathworks.com/matlabcentral/fileexchange/4551 (last access: 26 October 2011), February 2004.

Driscoll, C. and Letterman, R.: Factors regulating residual aluminium concentrations in treated water, Environmetrics, 6, 287–309, 1995.

FINLEX: Decree of the Ministry of Social Affairs and Health: Relating to the quality and monitoring of water intended for human consumption, No. 461/2000, Issued in Helsinki on 19 May 2000, 16 pp., http://www.finlex.fi/en/laki/kaannokset/2000/en20000461.pdf, last access: 26 October 2011, 2000.

George, S., Pandit, P., and Gupta, A. B.: Residual aluminium in water defluoridated using activated alumina adsorption – Modeling and simulation studies, Water Res., 44, 3055–3064, 2010.

Ibarra-Berastegi, G., Elias, A., Arias, R., and Barona, A.: Artificial Neural Networks vs Linear Regression in a Fluid Mechanics and Chemical Modelling Problem: Elimination of Hydrogen Sulphide in a Lab-Scale Biofilter, 2007 IEEE/ACS International Conference on Computer Systems and Applications, 584–587, 2007.

Juntunen, P., Liukkonen, M., Pelo, M., Lehtola, M., and Hiltunen, Y.: Modelling of residual aluminum in water treatment process, in: Proceedings of the 7th EUROSIM Congress on Modelling and Simulation, 6–10 September 2010, Prague, Czech Republic, Vol. 2: Full Papers, 5 pp., 2010.

Juuso, E. K.: Data-based development of dynamic models for biological wastewater treatment in pulp and paper industry, Preprints of the 51st Conference on Simulation and Modelling, 14–15 October 2010, Oulu, Finland, 1–9, 2010.

Juuso, E. K.: Intelligent Trend Indices in Detecting Changes of Operating Conditions, in: Proceedings of UKSim 13th International Conference on Modelling and Simulation – UKSim 2011, Cambridge, UK, 30 March–1 April 2011, 162–167, 2011.

Kulkarni, P. and Chellam, S.: Disinfection by-product formation following chlorination of drinking water: Artificial neural network models and changes in speciation with treatment, Sci. Total Environ., 408, 4202–4210, 2010.

Leakey, H.: Aluminium residual management in drinking water using selection criteria of aluminium based primary coagulants, M.Sc thesis, Royal Roads University Victoria, ISBN: 0-494-05113-2, 2004.

Maier, H. R., Morgan, N., and Chow, C. W. K.: Use of artificial neural networks for predicting optimal alum doses and treated water quality parameters, Environ. Modell. Softw., 19, 485–494, 2004.

Matlab Statistics Toolbox™ 7 User's Guide, Matlab MathWorks Inc., April 2011.

McLachlan, D. R. C., Bergeron, C., Smith, J. E., Boomer, D., and Rifat, S. L.: Risk for neuropathologically confirmed Alzheimer's disease and residual aluminum in municipal drinking water employing weighted residential histories, Neurology, 46, 401–405, 1996.

Pal, S. K. and Mitra, S.: Multilayer perceptron, fuzzy sets and classification, IEEE Transactions on neural networks, 3, 683–697, 1992.

Shetty, G. R., Malki, H., and Chellam, S.: Predicting contaminant removal during municipal drinking water nanofiltration using artificial neural networks, J. Membrane Sci., 212, 99–112, 2003.

Tranmer, M. and Elliot, M.: Multiple Linear Regression, Catie Marsh Centre for Census and Survey Research, Teaching papers, 47 pp., 2008.

WHO – World Health Organization: Aluminium in Drinking-water: Background document for development of WHO Guidelines for Drinking-water Quality, 2003.

WHO – World Health Organization: Guidelines for drinking-water quality [electronic resource]: incorporating 1st and 2nd addenda, Vol. 1, Recommendations, 3rd Edn., 2008.

3

Leakages and pressure relation: an experimental research

F. De Paola and M. Giugni

Department of Hydraulic, Geotechnical and Environmental Engineering, University of Naples "Federico II", Italy

Correspondence to: F. De Paola (depaola@unina.it)

Abstract. Leaks in water systems are presently a frequent and increasing event that involves cost increase and poor service, not compliant to quality standards and modern management criteria. The most recent data available in Italy, resumed into the report issued by Control Committee for Water Resources Use (CONVIRI), shows leakages with an average value of 37 %. It is therefore important, for maintenance perspective, to investigate occurrence and evolution of water leaks and the analytical link between leaks Q_p and network pressure P, for a reliable calibration of water networks quali-quantitative simulation models. The present work reports the first results of an experimental campaign started at Laboratory of Hydraulics of Department of Hydraulics, Geotechnical and Environmental Engineering of University of Naples Federico II in order to analyze the features of $Q_p(P)$ relation, which are compared with other results issued in literature.

1 Introduction

Leaks in water systems are presently a frequent and increasing event causing cost increase and poor service levels not compliant to quality standards and modern management criteria. The most recent data available in Italy are resumed into the report issued by Control Committee for Water Resources Use (CONVIRI, 2010) which shows leakages between 21 % and 61 %, with an average value around 37 %. The recovery of part of these lost water volumes would allow to acquire "new" resources and also economical benefits due to adduction, treatment and distribution cost savings. Losses reduction could obviously be achieved by well planned maintenance campaigns, both ordinary and extraordinary: they are, however, expensive actions from a financial perspective and are often not rewarded by the economical value of the recovered resource.

A different approach is to control network operation pressure by installing PRVs (Pressure Reducing Valves) or realizing District Meter Areas (DMAs). For this purpose it is important to analyze the forming and evolution mechanisms of water leaks and the analytical link between the leaks Q_p and the network pressure P. An experimental research on these topics has been started at the Hydraulics Laboratory of Department of Hydraulics, Geotechnical and Environmental Engineering of University of Naples Federico II (De Paola et al., 2010; De Paola and Giugni, 2010), in order to investigate the parameters affecting the leakage/pressure relationship for different pipe materials. The present work gives a global summary of all the experimental tests conducted so far. In particular, results of the dynamic tests performed on steel and ductile iron pipes are presented, and compared to the other ones present in literature (Sendil and Al Dhowalia, 1992; Lambert, 2000; Thornton and Lambert, 2005; Greyvenstein and Van Zyl, 2006; Milano, 2006; Coetzer et al., 2006; Ferrante et al., 2010; Massari et al., 2012).

2 Experimental set-up

Experimental tests were carried out in the Hydraulics Laboratory high pressure circulation plant. The water discharge, detected by an electromagnetic flow meter, is led to a system of two parallel air vessels which allows to keep a constant pressure up to 10 bars, and from this to a cylindrical steel adductor (800 mm diameter) connected to the test plumbing

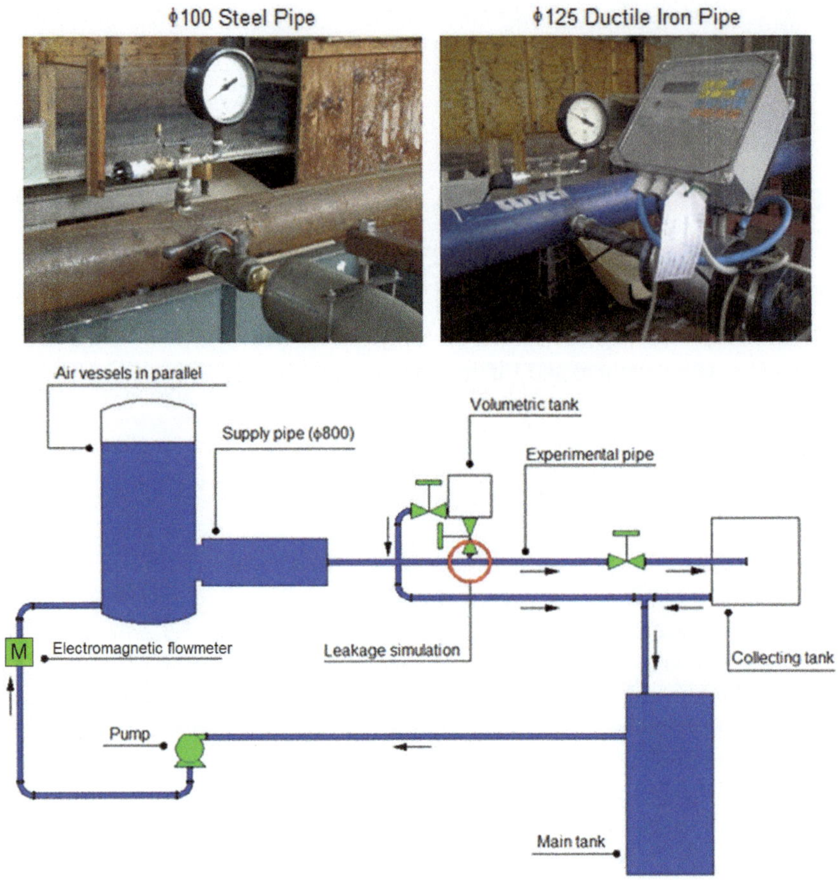

Figure 1. Experimental installation layout.

(Fig. 1). This latter in the first tests was a steel pipe 100 mm nominal diameter (DN 100) and thickness $s = 4$ mm, 6.70 m long and flat; in the second tests was a last generation ductile iron pipe (BLUTOP – PAM Saint Gobain) nominal diameter (DN 125) and thickness $s = 3$ mm. The choice of the pipe materials at this early stage of the research is due to the the large use of steel and ductile iron pipes in Italian water distribution networks.

At the downstream end of the experimental duct was fitted a rubber wedge sluice valve (Fig. 1), connected to a PVC ⌀110 pipe which collects discharge to a recirculation tank. At 3.35 m downstream of the experimental pipe was installed a junction with an interception spherical valve at the end of which there is a nozzle used to simulate the occurrence of a leak (detail of Fig. 1).

This layout is assumed to reliably reproduce private connections in water distribution networks, where leakages due to metal pipe damages are more likely to occur. Furthermore the flow that is established in the pipe branch is surely a function of the geometry and the roughness of the main pipe. The boundary conditions must be taken into account and they are established only by the pressurized pipeline.

During the tests the water discharge flowing out of the nozzle is collected into a tank suitably calibrated, measuring filling time or, as alternative (for the ductile iron pipe), is measured by a small size electromagnetic flow meter (DN 25) (Fig. 1). Pressure measure is run by a WIKA transducer with a 0–10 bar range and sample frequency equals to 10 Hz.

In the tests the following parameters were varied:

1. feeding pressure, between 2 and 7 bars;
2. water discharge between 3 and 45 $l\,s^{-1}$;
3. nozzle geometry: the presence of leaks with different shape and size was simulated by using suitable nozzles: round, rectangular, square, irregular (Fig. 2);

aiming to achieve, by tests analysis, the water discharge and feeding pressure relationship.

3 Experimental results

Experimental tests were run with both *static* and *dynamic* approach, in order to simulate various operating conditions of a water system. The static tests simulate the night operating conditions (suitable to verify more precisely leaks occurrence), while the dynamic tests simulate day running.

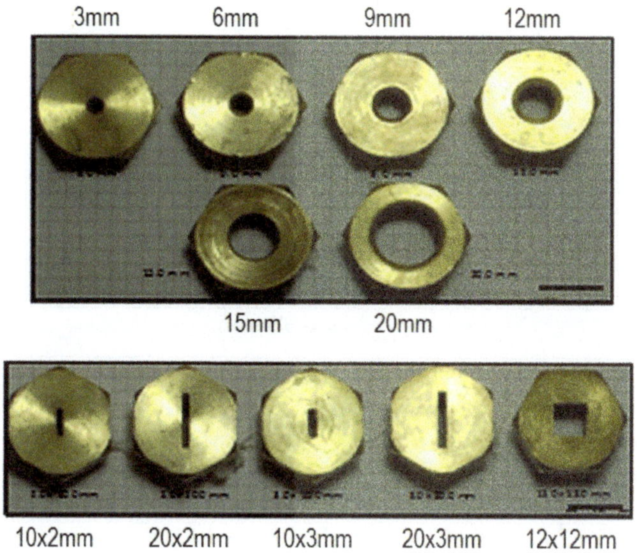

Figure 2. Test nozzles with different shape and size.

During each test, pressure was detected at least for 60 s, in order to evaluate pressure variation features due to the leak.

3.1 Steel pipe – static tests

Generally a power law function is used to interpret the relation between water leaks (Q_P) and pressure (P) as follows:

$$Q_P = a \cdot P^b \tag{1}$$

In static tests conducted on steel pipe the driving pressure P was varied in the range 2–7 bar, while the discharge Q between 1 and 30 lps. The tests results allowed to calculate for the considered nozzle shapes, with a very high correlation rate (R^2 always above 0.99), the values of coefficient a and exponent b in Eq. (1), as reported in Fig. 3. The coefficient a in both cases increases, as obvious, according to nozzle size. The exponent b, instead, is always little far from the theoretical value 0.50.

3.1.1 Steel pipe – dynamic tests

Dynamic tests allow to analyze, as already mentioned, the daily demand variation, by generating different discharge values (varying from about 3 to 45 lps) by the downstream valve. The tests duration was always about several minutes.

In these tests, leaks increase, as expected, as pressure and nozzle size increase. The exponent b tends to a constant value, close but not equal to the theoretical value 0.50, also for dynamic test conditions (Fig. 4).

It can be also highlighted that coefficient a has an almost linear trend as surface ratio A_n/A_p (A_n = nozzle area; A_p = pipe cross sectional area) varies for nozzles having either square or rectangular section, while for round section it shows a linear trend according to d^2/D^2 ratio and a parabolic trend according to d/D ratio (d = nozzle diameter; D = pipe diameter) (Fig. 4) as also shown for the static tests.

3.2 Ductile iron pipe

In the first phase of experimental tests (steel pipe) the discharge was detected by a volumetric tank of small volume (see Fig. 1) and the duration of dynamic tests was therefore limited to few minutes. An evolution in the laboratory equipment was brought by introducing an electromagnetic flow meter on the pipe branch which allowed a continuous record of leakage discharge. The electromagnetic flowmeter, provided by Automation Progetti s.r.l., measures discharge in the range 0.047–5.55 lps with a precision of ± 0.5 % of the upper limit.

Experimental tests for ductile iron pipes were run in the already described way but with longer duration (almost 1 h), in order to analyze leaks variations as discharge and pressure vary in the feeding pipe.

Figure 5 shows the close correlation between outflow discharge and pipe pressure for a rectangular (10 × 2 mm) and a round (6 mm diameter) nozzle. In Fig. 6 is also shown the time progress of flow experimental values compared to the points given by Eq. (1) after calibrating a and b coefficients using a Genetic Algorithm (GANetXL software, CWS, 2010). The GA was run varying at each iteration the a and b values in order to minimize the sum of the square differences between measured discharge at time t and that obtained by Eq. (1) at the same time.

The model resulted very effective in the calculation of losses, and calibration leads to exponent b values slightly lower than the theoretical value 0.50.

The results of the dynamic tests performed on ductile iron pipe (Fig. 7) confirmed the findings of the tests carried out on steel pipe. For the b exponent, tests indicate a substantial invariance, with values almost always slightly below 0.50.

4 Conclusive remarks

Starting from the first experiments run in UK (Goodwin, 1980), the relation between water leaks and pressure ratio was analyzed through the monomial Eq. (1). Some research works afterwards developed in Saudi Arabia (Sendil and Al Dhowalia, 1992), UK (Burnell and Race, 2000) and Iran (Ardakanian and Ghazali, 2003) by the analysis of field data on different kind of pipes and different discharge range, confirmed such equation: available data are summarized in Table 1, showing that exponent b resulted systematically higher than 0.50.

The experiments run so far in Naples for metal pipes, with pressure varying within a wide range not investigated in other experimental works, have shown as follows:

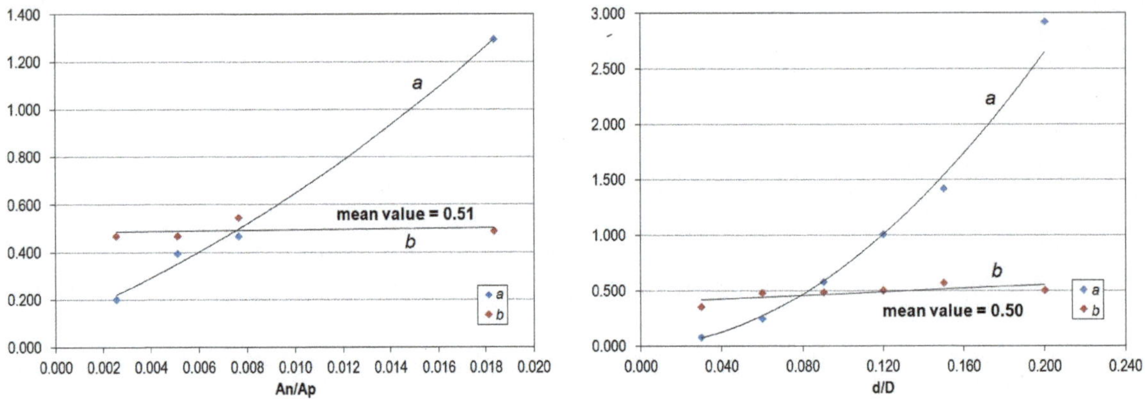

(3a) (3b)

Figure 3. Discharge coefficients according to: **(a)** surface ratio A_n/A_p (rectangular nozzles); **(b)** diameter ratio d/D (round nozzles). Static tests on steel pipe.

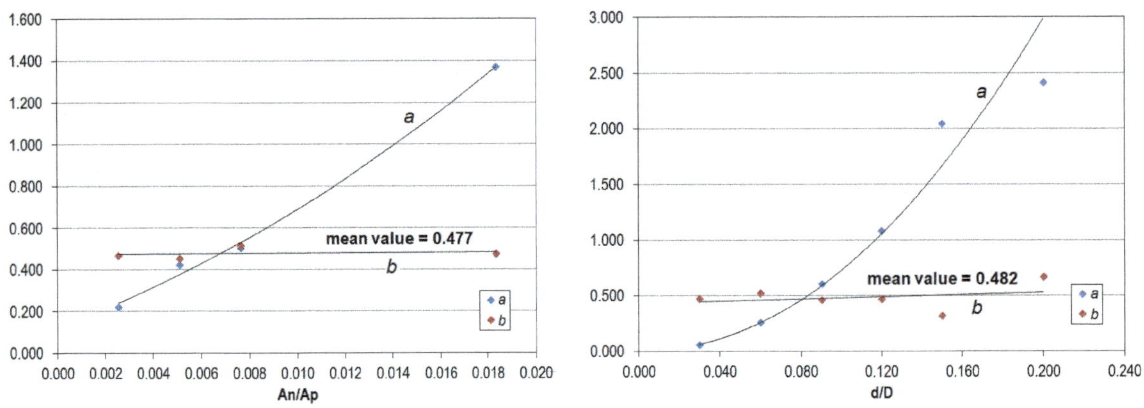

(4a) (4b)

Figure 4. Discharge coefficients according to: **(a)** A_n/A_p ratio (rectangular nozzles); **(b)** d/D ratio (round nozzles). Dynamic tests on steel pipe.

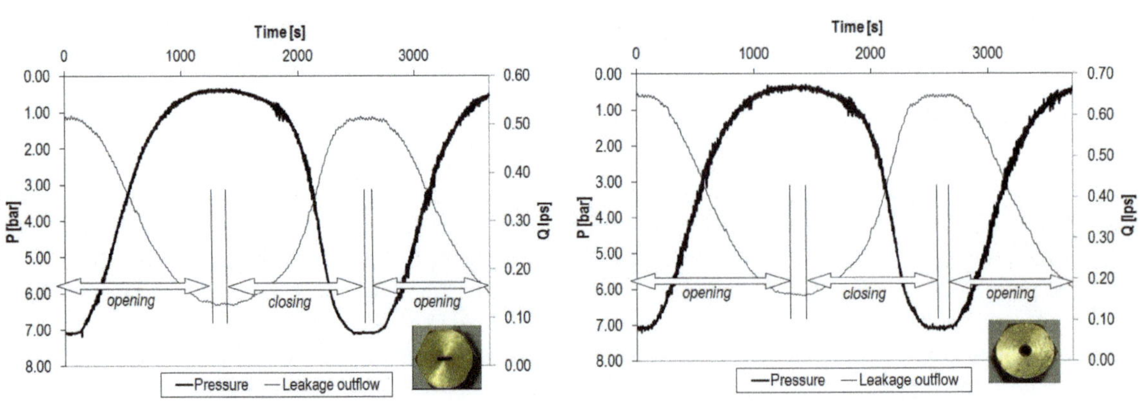

(5a) (5b)

Figure 5. Trend of detected pressures and outflow discharge for a rectangular **(a)** and a round **(b)** nozzle.

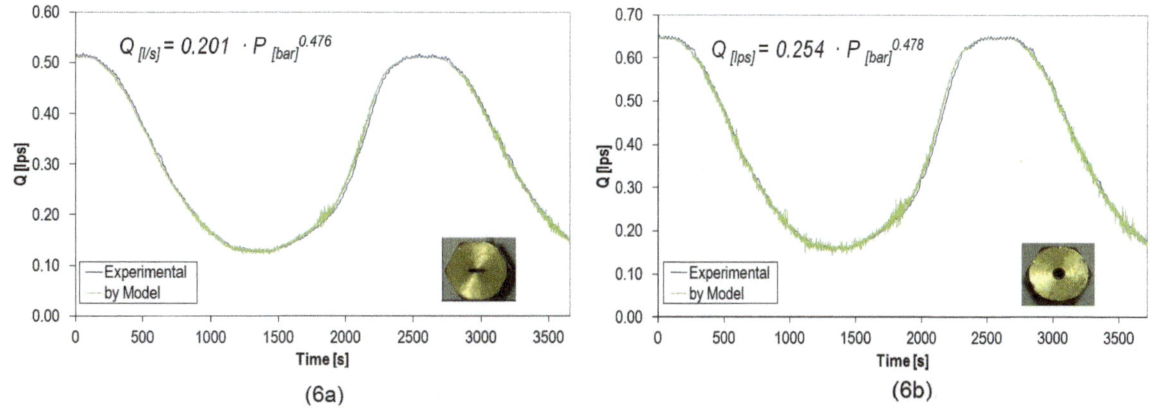

Figure 6. Comparison between time series for measured outflows and model results after calibration (rectangular nozzles, **a** – round nozzle, **b**).

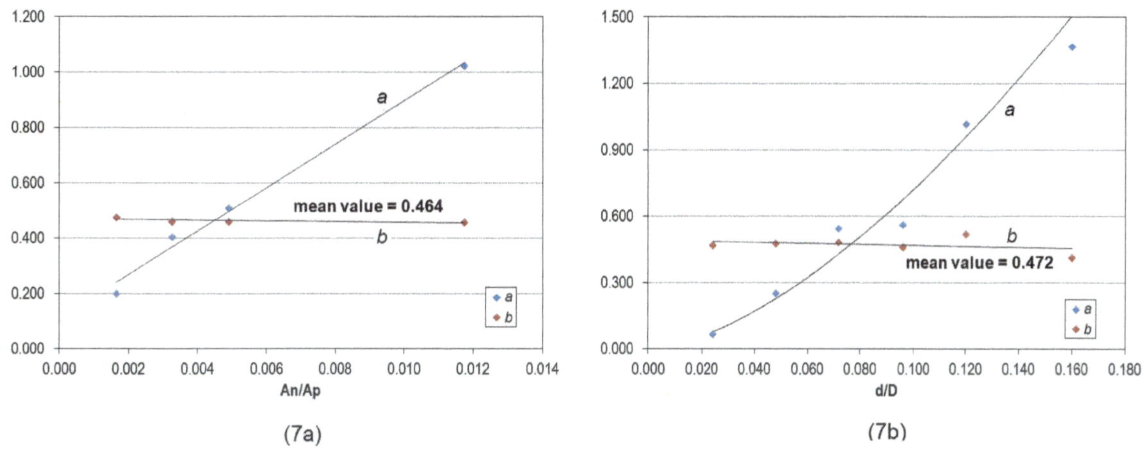

Figure 7. Discharge coefficients according to: **(a)** A_n/A_p ratio (rectangular nozzles); **(b)** d/D ratio (round nozzles). Dynamic tests on ductile iron pipe.

- the relation (1) allows a reliable forecast of leaks depending on pressures, if (*a*) and (*b*) coefficients are suitably defined;

- *a* values show an increasing trend with the size of the outflow orifice. Since the combination of the pressure and the pipe stiffness for metal pipes produces negligible strains around the hole there is a linear dependence from A_n/A_p ratio, also for circular nozzles for which in Figs. 3, 4, 7 the influence of the only characteristic dimension (diameter *d*) was highlighted.

- *b* exponent resulted little far from the theoretical value 0.50 for both round and rectangular nozzles;

- *b* values resulted slightly dependent from test characteristics (static or long-term dynamic).

Experimental data collected in Naples so far, therefore, confirm the results achieved by other tests for steel pipes (Fig. 7) with round nozzles (Coetzer et al., 2006), with reference to a range of pressures slightly smaller.

The *b* values significantly higher than 0.50 reported in literature are related instead to steel pipes affected by corrosion and asbestos cement or plastic pipes with longitudinal cracks. Probably they are due to a reduced consistence of material around the holes (due to corrosion) (Greyvenstein and Van Zyl, 2006) or to the remarkable flexibility of plastic pipes (Farley and Trow, 2003; Walski et al., 2006; Ferrante et al., 2010; Bovolin and Picciotti, 2008).

For pipes characterized by high elasticity, then it would be appropriate to refer to an equation that takes into account the pressure/area link of the hole (such as FAVAD equation, Cassa et al., 2010). The analysis is, however, further complicated by the viscoelastic characteristics of materials such as HDPE.

Nevertheless, in the authors opinion these topics require more investigations to discern if they depend on the physics of the problem or maybe there is a distortion induced by the

Table 1. Exponent *b* values as reported in literature.

Source	*b*	Installation type
UK (Goodwin, 1980)	1.18 (average)	Operating districts with little night metered consumptions
SaudiArabia (Sendil and Al Dhowalia, 1992)	0.54–1.61	Ten operating districts in Riyadh
ThamesWater (Burnell and Race, 2000)	1.00	Supply-pipe leakage patterns in over 2000 accurately-logged domestic sites
Iran (Ardakanian and Ghazali, 2003)	1.10–1.18	Old cast iron with high leakage in Sanadaj (Kurdistan)

Figure 8. Corroded steel pipes (Greyvenstein and Van Zyl, 2006).

analysis of experimental data. For example, for steel pipes with holes caused by corrosion, the coefficients *b* identified experimentally (Greyvenstein and van Zyl, 2006) for the pipes 4, 5 and 6 shown in Fig. 8 are, respectively, 0.67, 1.96 and 2.3; for an undamaged pipe, with the same mechanical properties, with an artificial hole diameter of 12 mm, the identified coefficient is equal to 0.518.

Further experiments are still in progress in the Hydraulics Laboratory of Naples on pipes of different materials (HDPE, GRP), in order to acquire new elements about the influence of the characteristics of the pipe and the features of the hole simulating the leak on the outflow mechanisms and, therefore, on the leakage law.

Acknowledgements. The Authors would like to thank the Editor Guangtao Fu, Kobus Van Zyl and the other anonymous reviewer for their suggestions, which improved the quality of this paper. They also thank Enzo Galdiero, Alfonso Stefanelli, Francesco Pace, Oriana Milito, Emanuele Mellucci for their contribution in performing the experimental tests.

References

Ardakanian, R. and Ghazali, A. A.: Pressure-leakage relation in urban water distribution systems, Proceedings of the ASCE International Conference on Pipeline Engineering and Construction, Baltimore, 2003.

Bovolin, V. and Picciotti, M. V.: Analisi dei meccanismi relativi alle perdite idriche nelle reti in pressione, Proceedings of the 31th Conference of Hydraulics and Hydraulic Engineering, Perugia, 1–10, 2008 (in Italian).

Burnell, D. and Race, J.: Water distribution systems analysis: patterns in supply-pipe leakage, Proceedings of the Joint Conference on Water Resources Engineering and Water Resources Planning & Management, Minneapolis, 2000.

Cassa, A. M., Van Zyl, J. E., and Laubscher, R. F.: A numerical investigation into the effect of pressure on holes and cracks in water supply pipes, Urban Water J., 7, 109–120, 2010.

Centre for Water System (CWS): GANetXL User Manual, University of Exeter, 2010.

Coetzer, A. J., Van Zyl, J. E., and Clayton, C. R. I.: Experimental study of the hydraulics of small circular holes in water pipes, Proceedings of the 8th Annual Symposium on Water Distribution Systems Analysis Symposium, Cincinnati, Ohio, 27–30 August 2006.

Control Committee for Water Resources Use (CONVIRI): Rapporto sullo stato dei servizi idrici – Anno 2009, Rome, 2010 (in Italian).

De Paola, F. and Giugni, M.: Analisi sperimentale della relazione tra perdite e pressioni, Proceedings of the 32th Conference of Hydraulics and Hydraulic Engineering, 2010 (in Italian).

De Paola, F., Giugni, M., and Ranucci, A.: Analisi sperimentale della relazione tra perdite idriche e pressioni, Proceedings of the 4th Workshop "La ricerca delle perdite e la gestione delle reti di acquedotto", Aversa, 18–19 settembre 2009 (in Italian), 2010.

Farley, M. and Trow, S.: Losses in Water Distribution Networks, IWA Publishing, London, 2003.

Ferrante, M., Massari, C., Brunone, B., and Meniconi, S.: Analisi sperimentale della legge di perdita in condotte in acciaio e polietilene, Proceedings of the 32th Conference of Hydraulics and Hydraulic Engineering, 2010 (in Italian).

Goodwin, S. J.: The Results of the Experimental Programme on Leakage and Leakage Control, Technical Report TR 154, Water Research Centre, UK, 1980.

Greyvenstein, B. and Van Zyl, J. E.: An Experimental Investigation into the Pressure Leakage Relationship of some Failed Water Pipes, Water Institute of Southern Africa Biennial Conference, Durban, South Africa, 22–24 May 2006.

Lambert, A.: What do we know about Pressure-Leakage Relationship in Distribution System?, Proceeding of the AWWA Conference on System Approach to Leakage Control and Water Distribution Systems Management, 2000.

Massari, C., Ferrante, M., Brunone, B., and Meniconi, S.: Is the head-discharge relationship in polyethylene pipes a bijective function?, J. Hydraul. Res., 50, 409–417, 2012.

Milano, V.: Dipendenza delle perdite di una tubazione dalla pressione di esercizio, L'Acqua, 4, 47–54, 2006 (in Italian).

Sendil, U. and Al Dhowalia, K. H.: Relationship between pressure and leakage in a water distribution network, Proceedings of the AWWA Conference, 1992.

Thornton, J. and Lambert, A.: Progress in practical prediction of pressure/leakage, pressure/burst frequency and pressure/consumption relationships, Proceedings of IWA Special Conference "Leakage 2005", Halifax, Canada, September, 2005.

Walski, T., Betzts, W., Posluszny, E. T., Weir, M., and Whitmann, B. E.: Modeling leakage reduction through pressure control, J. AWWA, 98, 147–155, 2006.

4

Natural organic matter removal by ion exchange at different positions in the drinking water treatment lane

A. Grefte[1], M. Dignum[2], E. R. Cornelissen[3], and L. C. Rietveld[1]

[1]Delft University of Technology, Faculty of Civil Engineering, P.O. Box 5048, 2600 GA Delft, The Netherlands

[2]Waternet, P.O. Box 94370, 1090 GJ Amsterdam, The Netherlands

[3]KWR Watercycle Research Institute, P.O. Box 1072, 3430 BB Nieuwegein, The Netherlands

Correspondence to: A. Grefte (a.grefte@tudelft.nl)

Abstract. To guarantee a good water quality at the customers tap, natural organic matter (NOM) should be (partly) removed during drinking water treatment. The objective of this research was to improve the biological stability of the produced water by incorporating anion exchange (IEX) for NOM removal. Different placement positions of IEX in the treatment lane (IEX positioned before coagulation, before ozonation or after slow sand filtration) and two IEX configurations (MIEX® and fluidized IEX (FIX)) were compared on water quality as well as costs. For this purpose the pre-treatment plant at Loenderveen and production plant Weesperkarspel of Waternet were used as a case study. Both, MIEX® and FIX were able to remove NOM (mainly the HS fraction) to a high extent. NOM removal can be done efficiently before ozonation and after slow sand filtration. The biological stability, in terms of assimilable organic carbon, biofilm formation rate and dissolved organic carbon, was improved by incorporating IEX for NOM removal. The operational costs were assumed to be directly dependent of the NOM removal rate and determined the difference between the IEX positions. The total costs for IEX for the three positions were approximately equal (0.0631 € m^{-3}), however the savings on following treatment processes caused a cost reduction for the IEX positions before coagulation and before ozonation compared to IEX positioned after slow sand filtration. IEX positioned before ozonation was most cost effective and improved the biological stability of the treated water.

1 Introduction

Drinking water treatment consists of different steps, depending on the quality of the source water. The presence of natural organic matter (NOM) can cause problems in the treatment, as well as in the distribution of drinking water. NOM can be a source for regrowth in the distribution system, thus decreasing the biological stability of the drinking water (van der Kooij et al., 1982). When the source water contains high NOM concentrations, this should be removed to a high extent during drinking water treatment. NOM can be removed by coagulation, activated carbon filtration, membrane filtration and anion exchange (IEX).

NOM in surface water can be removed by IEX, because the main part of NOM, humic, fulvic and organic acids, is negatively charged (Bolto et al., 2002; Cornelissen et al., 2008). IEX is a promising method for NOM removal, because empty bed contact times can be small and run times of IEX columns can be up to several weeks (van der Helm et al., 2009). IEX is relatively cost effective, easy to operate and a compact installation can be used due to the short contact times (Cornelissen et al., 2009). The efficiency of NOM removal by IEX depends on i.a. NOM concentration, NOM composition, type of IEX resin, empty bed contact time and configuration of the IEX installation. NOM fractions of low and high molecular weight (MW) are known to be removed by IEX (Croue et al., 1999; Bolto et al., 2002; Allpike et al., 2005; Boyer and Singer, 2005; Humbert et al., 2005). Weak base resins do not remove NOM as efficient as strong base resins (Croue et al., 1999) and the bead size of the resin, water retention, capacity and functional groups of the resin (Cornelissen et al., 2008) will also influence the removal efficiency. IEX can be operated as a packed bed or in a fluidized mode, like Magnetic IEX (MIEX®) (Drikas et al., 2002), Fluidized IEX (FIX) (Cornelissen et al., 2009) or suspended IEX (SIX) (Galjaard et al., 2011). When the resin is exhausted, a 10 % sodium chloride solution can be used for regeneration of the resin; the NOM ions are exchanged to chloride ions. The residual or waste of IEX regeneration consists of water, salt (NaCl) and humic substances. The waste can be discharged to the sewer or directly to the waste water treatment plant. However, humic substances are not readily biodegradable and will substantially remain in the effluent of the waste water treatment. Additionally, high salt solutions could give problems in waste water treatment. To limit the residual, the brine can be reused (Schippers et al., 2004). Separating the salt from the humic substances is possible by membrane filtration. In that way the salt can be reused and only the higher concentration of humic substances is discharged to the sewer (Schippers et al., 2004; Kabsch-Korbutowicz et al., 2011).

The objective of this research is to improve the drinking water quality, including biological stability, by incorporating IEX for NOM removal. Different placement positions of IEX in the treatment lane were compared on water quality as well as costs. For this purpose the pre-treatment plant at Loenderveen and production plant Weesperkarspel (WPK) of Waternet, the water cycle company of Amsterdam (NL) and surrounding areas, were used as a case study. The treatment lane consists of coagulation, lake retention, filtration, ozonation, softening, biological activated carbon (BAC) filtration and slow sand filtration (SSF). Assimilable organic carbon (AOC) values below $10\,\mu g\,C\,L^{-1}$ have been derived as a reference value for biostable drinking water during distribution (van der Kooij, 1992), whereas current AOC concentrations are approximately $20\,\mu g\,C\,L^{-1}$ (Baghoth et al., 2009) in the finished water. The biofilm monitor was developed as a tool to determine the biofilm formation characteristics of drinking water (van der Kooij, 1995). A biofilm formation rate (BFR) below $10\,pg\,ATP\,cm^{-2}\,day^{-1}$ reduces the risk of exceeding the Dutch guideline value for Aeromonas in the distribution system to less than 20 % (van der Kooij et al., 1999). Thus, biologically stable water was defined in this study as having an AOC concentration lower than $10\,\mu g\,C\,L^{-1}$ and a BFR below $10\,pg\,ATP\,cm^{-2}\,day^{-1}$. By removing humic substances by IEX before ozonation it is expected that ozonation will form less biodegradable NOM than without IEX. IEX positioned after slow sand filtration is expected to lower the AOC and BFR values by itself. In both cases, a reduction in dissolved organic carbon (DOC) concentration in the finished water from approximately 2.7 to $1\,mg\,C\,L^{-1}$ was aimed for.

First, pilot experiments were conducted with IEX positioned at two locations: (1) halfway the treatment lane, before ozonation and (2) after the treatment lane, before distribution. Experiments were conducted by two IEX configurations, namely FIX and MIEX®. Both IEX pilots were compared on NOM removal by LC-OCD characterization. Second, the effect on the biological stability of the produced water was researched for IEX at these two positions. Third, based on the results three possible positions in the treatment lane (IEX before coagulation was added) were compared on costs per cubic meter of treated water after the complete treatment lane.

2 Materials and methods

2.1 Treatment scheme

For this investigation pre-treatment plant Loenderveen and production plant WPK were used as a case study. The production of WPK is approximately $30\,Mm^3$ per year. At the pre-treatment plant surface water from the Bethune polder is treated by coagulation and sedimentation, followed by natural self-purification in a lake reservoir and rapid sand filtration. The pre-treatment plant partially removes suspended solids, phosphate, heavy metals, and pathogenic micro-organisms. It converts ammonium into nitrate and provides smoothing of peak loads. The pre-treated water is pumped to the production plant at WPK. There the water is treated by ozonation for disinfection, the hardness is reduced by pellet softening followed by removal of organic compounds in BAC filtration. As a final treatment step, the water passes through slow sand filters for the removal of pathogenic micro-organisms and for lowering the AOC concentration. The experiments were conducted at WPK pilot plant, owned and operated by Waternet. The pilot plant of WPK consists of 2 lanes of the same treatment processes and with similar contact times as in the full-scale treatment plant on a scale of approximately 1 : 200 compared to the full-scale treatment plant. The maximum flow in the pilot plant was $7\,m^3\,h^{-1}$ for each lane. The pilot plant was extended with a MIEX® pilot and a FIX pilot in one of the lanes. The MIEX® pilot used in this experiment is the high rate configuration. In this configuration, raw water fed to the base of the reactor vessel is mixed with the MIEX® resin, causing the ion exchange process to occur in a fluidized bed. In the fluidized bed the magnetic particles are attracted to each other to form large agglomerates that form a stable resin suspension. An agitator operating at low velocity keeps the resin/water suspension uniformly mixed. A small stream of resin is withdrawn from the reactor vessel, regenerated twice a week with a 12 % NaCl solution and returned to maintain the ion exchange capacity of the process. A series of plates (or tube settlers) at the top of the reactor vessel separate the resin from the water and treated effluent overflows (OricaWatercare, n.d.). To prevent resin carry-over a magnetic capture device was built in the MIEX® pilot plant. Additional, the MIEX® pilot plant was extended by a filter bag with pore size of $100\,\mu m$ and a settling tank. The influent flow rate of the MIEX® pilot plant was 300 L h^{-1}. The contact time was 3 min. The fluidized bed volume in the contacter was 15 L. The volume ratio of resin to water was 210 ml L^{-1}. Fresh resin was pumped into the contacter with a flow of 12 ml min^{-1}. The resin service was 1984 bed volumes. The MIEX® pilot was placed before ozonation (for 4 weeks) and in one of the lanes after SSF (for 4 months).

The FIX pilot, as described by Cornelissen et al. (2009), was positioned prior to ozonation (for 4 months). Three FIX columns were operated in parallel with a flow of 4 $m^3 h^{-1}$ each and contained Lewatit VP OC 1071 type resin, which is a strong-base gel resin with an acrylic (type 1) structure. This resin was the best performing resin for WPK feed water, according to a selection study on lab scale (Cornelissen et al., 2008). The height of the ion exchange bed was approximately 0.5 m, when fluidized it was approximately 1 m. The surface area was 0.3 m^2, each column was filled with 150 L of resin and the empty bed contact time was approximately 2.3 min. After a run time of 15 000 bed volumes the resin was regenerated with a 10 % NaCl solution.

2.2 Analyses

The water samples were analyzed for general physicochemical characteristics such as DOC, ultra violet absorbance at 254 nm (UV254), pH and temperature, which were determined using standard procedures following Eaton et al. (2005). Liquid chromatography with organic carbon detection (LC-OCD) with ultra violet (UV) and online organic carbon (OC) detection (UVD and OCD, respectively) was used for NOM characterization (Huber et al., 2011). Water samples were analyzed after filtration through 0.45 µm filters. LC-OCD separates chromatographable organic carbon (CDOC) into fractions of different molecular weights. The non-chromatographable organic carbon, which remains on the column, is referred to as hydrophobic organic carbon (HOC). CDOC is fractionated into (a) biopolymers (BP), which is a non-ionic, hydrophilic fraction with a high molecular weight ($\geq$ 10 000 g mol^{-1}); (b) humic substances (HS) (450–1000 g mol^{-1}), that are a heterogeneous fraction of similar chemical composition but varying molecular size and aromaticity, (c) building blocks (BB), a fraction that consists of breakdown products of HS, which is HS-like material of lower molecular weight (300–450 g mol^{-1}); (d) low-molecular weight (LMW) acids $\leq$ 350 g mol^{-1}) and (e) low-molecular weight neutrals, a fraction with a low molecular weight and a low ion density and the fraction is hydrophilic to amphiphilic (Huber and Frimmel, 1996; Huber, 2005; Huber et al., 2011). For data acquisition and data processing of the LC-OCD data a customized software program (FIFFIKUS, DOC-LABOR, Germany) was used. This program integrates the different peaks to determine the concentration of different organic fractions. In Huber et al. (2011) it was explained that a small portion of HS, which is called the LMW-HS, is trapped in the LMW-acids zone. To distinguish between LMW-acids and LMW-HS the UV/OC ratio of HS is used to determine the concentration of LMW-HS, by assuming the same UV/OC ratio for LMW-HS as for HS. The concentration of LMW-acids is calculated by subtracting the concentration of LMW-HS from the total surface of the peak LMW-HS and LMW-acids. In this research the fractions were determined without correction for LMW-HS.

The change in relative signal response from the chromatograms of LC-OCD before and after treatment, is called a differential chromatogram. With a differential chromatogram small qualitative changes in NOM fractions are visualized. A similar differential spectrum analysis was used by Korshin et al. (1999) for light absorption spectra.

The aromaticity per DOC is determined by the specific UV254 absorbance (SUVA = UV254/DOC). SUVA $\geq$ 4 L mg $C^{-1} m^{-1}$ indicates that mainly hydrophobic and especially aromatic material or humic substances are present, while SUVA $\leq$ 2 L mg $C^{-1} m^{-1}$ represents hydrophilic material or non-humic material (Edzwald et al., 1985).

The concentration of AOC was determined, with growth measurements in water samples of 600 ml. Two pure cultures of bacteria were used by applying the simultaneous incubation of strains *Pseudomonas fluorescens* (strain P17), which is capable of utilizing a wide range of low-molecular-weight compounds at very low concentrations and *Spirillum* sp. (strain NOX), which utilizes only carboxylic acids. The AOC concentration was calculated from the obtained maximum colony counts of these strains, using their yield values for acetate (van der Kooij et al., 1982; van der Kooij and Hijnen, 1984; van der Kooij, 1992). AOC was measured in duplicate and the average value was reported.

Biofilm monitors were used to determine the biofilm formation characteristics of drinking water (van der Kooij, 1995). The biofilm monitor consists of a vertically-placed glass column containing glass cylinders on top of each other. Water flows through the column with a flow of 270 L h^{-1}. Every other week two cylinders were collected. The biomass attached to these cylinders was released by sonication and the ATP concentration was determined (van der Kooij et al., 2003).

2.3 Cost comparison and environmental impact

A technical-economical assessment was made using IEX in the WPK treatment. In Table 1 the DOC concentrations after every treatment step, in normal operation (without IEX treatment), are given (Baghoth et al., 2009). In order to obtain a DOC concentration of 1 mg C L^{-1} after treatment, three positions were selected for NOM removal by IEX, namely before coagulation, before ozonation and as polishing step after SSF. For all positions the required DOC concentrations after IEX are given in Table 1. It was assumed that after IEX subsequent treatment processes remove DOC at the same rate as without IEX pre-treatment (coagulation 21 %, reservoir 8 %,

Table 1. The expected variation of dissolved organic carbon (DOC) across the treatment train (based on Baghoth et al., 2009).

DOC concentration ($mg\,C\,L^{-1}$)	Actual	Before coagulation	Before ozonation	After SSF
Raw surface water	9.0	9.0	9.0	9.0
IEX effluent	–	3.5	–	–
Coagulation effluent	7.1	2.7	7.1	7.1
Surface reservoir effluent	6.5	2.5	6.5	6.5
Rapid sand filtration effluent	6.0	2.2	6.0	6.0
IEX effluent	–	–	2.2	–
Ozonation effluent	5.7	2.1	2.1	5.7
Pellet softening effluent	5.4	2.0	2.0	5.4
Biological activated carbon filter effluent	3.0	1.2	1.2	3.0
IEX effluent	–	–	–	1.0
Treated water	2.7	1.0	1.0	1.0

rapid sand filtration 8 %, ozonation 5 %, pellet softening 5 %, biological activated carbon filtration 44 % and SSF 10 %).

A preliminary design for the IEX plant, independent of the location, is given in Table 2. The costs of an IEX installation consists of investment costs, capital costs, maintenance costs, chemicals, disposal costs, energy and staff. Except for chemicals and disposal costs, all costs were considered to be independent of the position of IEX in the treatment lane as well as independent of the removal rate and are given in Table 3. To compare the different positions of IEX in the treatment lane the breakthrough curve of one FIX pilot plant column, operated on WPK feed water, was taken as a starting point (Fig. 1). The design is based on 22 columns, see Table 2. When the filter run time of every individual column is 13 500 BV, then every day one column is regenerated. The effluent DOC concentration of all the columns is the average DOC concentration of the effluent concentrations of the individual columns. The column just regenerated produces water with a lower DOC concentration than the column that runs already for 20 days. For a filter run time of 13 500 BV the average removal will be 60 %, so 40 % of the influent DOC will be found in the effluent. The average DOC concentration was calculated for filter run times of 3000 BV, 6000 BV, 13 500 BV, 25 000 BV and 50 000 BV and interpolated for other filter run times as is shown in Fig. 1. Different NOM removal rates can be obtained by applying different filter run times for 22 columns and thus different regeneration frequencies. The calculations were done with a contact time of 2.3 min for all configurations. The NOM removal rate influences the operational costs, particularly costs for salt consumption and waste disposal. The cost for the discharge of waste were determined by the waste load expressed in pollution equivalents (PE) (van Lier, 2011). The PE is determined by: $PE = \frac{Q}{1000}\cdot(\frac{COD+4\cdot KjN}{54.8})$ in which Q = flow ($m^3\,yr^{-1}$), COD (chemical oxygen demand) = 3·DOC ($mg\,L^{-1}$) (DowChemicalCompany, 2011), KjN (Kjeldahl nitrogen) was assumed to be negligible. The costs per PE are approximately 50 €. The waste is transported by tanker trucks with a volume of 10 m^3.

Table 2. Design of IEX treatment.

Total production	30 $Mm^3\,yr^{-1}$
Empty bed contact time	2.3 min
Reactor	22 columns
Total bed volume	190 m^3 resin
Bed volume per column	8.6 m^3 resin
Flow per column	220 $m^3\,h^{-1}$
Salt solution	10 % NaCl
Regeneration	1.8 BV brine
Rinse	2.4 BV water

The distance to the nearest waste water treatment plant was assumed to be 15 km. A tanker truck with driver costs approximately 70 € h^{-1} and it takes 2 h per tanker truck to be filled, to drive and to discharge the water. With these assumptions, the costs for transportation are 14 € m^{-3}. The costs for salt are 0.1 € kg^{-1}.

The environmental and financial impact of drinking water production plant Weesperkarspel and its pre-treatment plant Loenderveen were determined by Barrios et al. (2008). For calculating cost savings of treatment processes following IEX and for determining the environmental impact of IEX numbers were taken from their study. Barrios et al. (2008) calculated costs for the different treatment steps for 2002, for this research actual costs (2012) were calculated with an inflation of 1.7 % per year.

3 Results and discussion

3.1 NOM removal by MIEX® and FIX

The MIEX® pilot positioned before ozonation caused an average DOC concentration decrease from 5.7 to 2.5 $mg\,C\,L^{-1}$. This is a decrease of 56.1 % and a removal of 3.2 $mg\,C\,L^{-1}$. FIX removed DOC on average from 5.9 to 3.0 $mg\,C\,L^{-1}$ which is a removal of 2.9 $mg\,C\,L^{-1}$ or 49.2 %. The removal of DOC by MIEX® after SSF was on average 57.7 % or 2.2 $mg\,C\,L^{-1}$ (3.8 to 1.6 $mg\,C\,L^{-1}$). The removal rate at both locations was comparable, however, the absolute removal was higher for water before ozonation. This can be explained by the fact that ion exchange from the water to the resin is based on equilibrium reactions, and a higher DOC concentration results in a larger driving force.

The SUVA was lowered by both IEX treatment systems before ozonation from 2.6 to 1.6 $L\,mg\,C^{-1}\,m^{-1}$ (38.4 %), after SSF the decrease was from 1.5 to 0.9 $L\,mg\,C^{-1}\,m^{-1}$ (40.0 %), which is comparable to the decrease before ozonation. The decrease in SUVA means that the aromaticity of the water decreased (Edzwald et al., 1985); NOM with double bonds was removed to a higher extent than NOM with single bonds, and preferentially the higher molecular weight NOM was removed. In Fig. 2 the (differential) OCD chromatograms of the different water types are given. This figure shows that

Table 3. Total fixed costs.

Subject	Amount	Unit		Costs (€ yr^{-1})
Depreciation			Interest (Linear) 5 %	702 000
Civil construction	1 750 000	€	30 yr	102 000
Installation construction	6 600 000	€	20 yr	495 000
Engineering construction	1 000 000	€	30 yr	58 000
Intern process construction	800 000	€	30 yr	47 000
Maintenance				207 000
Civil	7 750 000	€	0.5 %	9000
Installation	6 600 000	€	3 %	198 000
Chemicals				220 000
Resin-Lewatit VP OC 1071	40 000	L yr^{-1}	5.50 € L^{-1}	220 000
Energy	900 000	kWh yr^{-1}	0.09 € kWh^{-1}	81 000
Staff	0.25	FTE	55 000 € yr^{-1} FTE^{-1}	14 000
Total fixed costs				1 224 000

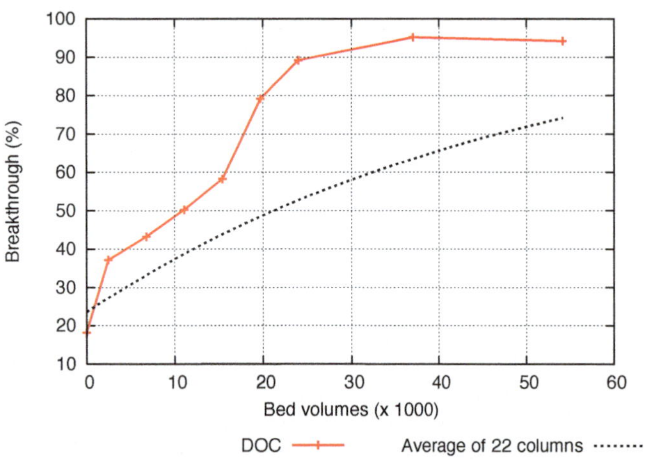

Figure 1. Breakthrough of one FIX column (positioned before ozonation in the WPK treatment plant) and the average of 22 columns for a certain filter run time (x-axis).

specifically the HS were removed from the water, independent of the water source. It also shows that MIEX® removed more of the lower MW NOM and FIX removed more of the HMW-HS. In Fig. 3 the concentrations of the different fractions are given. Although the decrease of the different fractions is higher for water before ozonation, the removal rates are of the same order of magnitude for both water types.

Overall, MIEX® showed a higher removal of NOM (fractions) than FIX. Possible explanations will be the longer contact time of the MIEX® resin compared to the FIX (3 min for MIEX® and 2.3 min for FIX), the smaller resin beads (approximately 150 µm) for the MIEX® resin (Drikas et al., 2002) in comparison to the bead size of 0.4–1.6 mm for Lewatit VP OC 1071 (Lewatit, n.d.) or the different operation conditions (the MIEX® pilot was a continuous process in one column, while every week one FIX column was regenerated and the effluent of three FIX columns were mixed). Figure 3b showed an increase in BP after FIX treatment and an increase in LMW-acids after MIEX treatment at the end of the treatment lane, however the concentrations of both fractions are very low, and the increase rate will not be significant.

3.2 Biological stability

The AOC concentration in the water before ozonation was approximately 7 µg C L^{-1}. After FIX the concentration was lowered to approximately 3.5 µg C L^{-1}, a reduction of 50 %. By extending the treatment lane with FIX columns before ozonation, the ozonation formed less AOC than without FIX treatment (54.5 µg C L^{-1} instead of 115.8 µg C L^{-1}), see Fig. 4. Because of this lower AOC concentration after ozonation, the AOC concentration after SSF was lower as well (14.3 µg C L^{-1} instead of 38.1 µg C L^{-1}). By extending the treatment lane with MIEX® at the end of the treatment, the AOC concentration was decreased from 38.1 to 13.6 µg C L^{-1}. Removing approximately 50 % of DOC concentration before ozonation or 58 % of DOC concentration after SSF resulted in the same AOC concentration after treatment.

The BFR of the water after SSF (before the MIEX® pilot) was 6.6 pg ATP cm^{-2} day^{-1}. Immediately after starting up the experiment the ATP values for determining the BFR were much higher after the MIEX® pilot than after SFF (200 pg ATP cm^{-2} versus 6.6 pg ATP cm^{-2}). The ATP values after MIEX® treatment increased from 200 pg ATP cm^{-2} to 500 pg ATP cm^{-2} and varied for the rest of the experimental

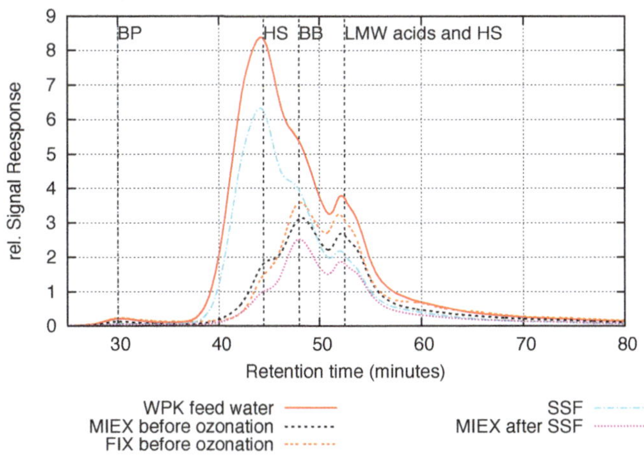

(a) OCD chromatograms

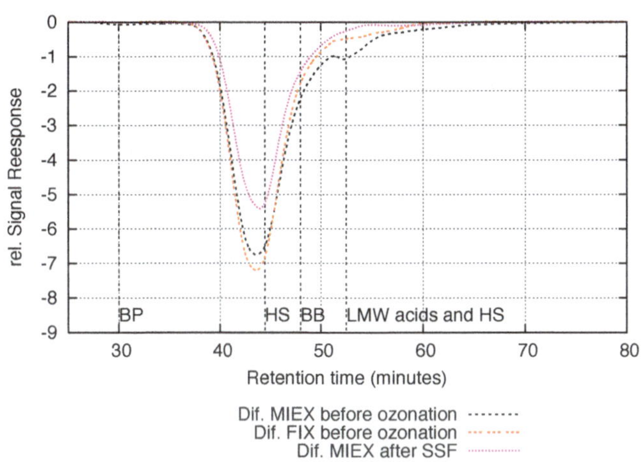

(b) Differential chromatograms

Figure 2. LC-OCD (differential) chromatograms of the different water qualities.

period between these two values, which made it impossible to calculate a BFR for the biofilm monitor positioned after MIEX®. The tube settlers and the magnetic capture device of the MIEX® pilot plant did not remove all spent resin and polluted the biofilm monitor. The MIEX® pilot plant caused an average increase in turbidity from 0.1 NTU to 1.5 NTU. This increase in turbidity after MIEX® treatment was also noticed by Singer and Bilyk (2002); Cornelissen et al. (2010) found an increase in ATP content after FIX. The higher ATP values after MIEX® treatment were probably caused by ATP on the resin that polluted the biofilm monitor. This problem can be solved by adding an extra (magnetite) filter. This filter should remove the resin before the water is transported to the consumers. The BFR of the finished water of the FIX-lane was 2.0 pg ATP cm^{-2} day^{-1} (Grefte et al., 2011), which is an improvement compared to the existing values of 6.6 pg ATP cm^{-2}. The DOC concentration after the treatment lane was lowered from 4.3 to 2.3 mg C L^{-1}, due to FIX pretreatment. MIEX after SSF decreased the DOC concentration

Table 4. Regeneration and waste characteristics of three possible positions for the IEX plant in the treatment lane.

	Before coagulation	Before ozonation	After SSF
DOC removal rate	61 %	62 %	67 %
Regeneration after (BV)	11 100	10 500	6800
Regeneration after (Days)	18.9	17.9	11.6
Regenerations per year	424	448	692
Brine per year (m^3)	6868	7257	11 210
Rinse water per year (m^3)	9158	9677	14 947
Salt per year (kg)	686 880	725 760	1 121 040
DOC (kg C yr^{-1})	165 000	114 000	60 000

from 4.3 to 1.8 mg C L^{-1}. Thus the biological stability of the water after the treatment lane, expressed as DOC, AOC and BFR, improved by extension of the treatment train with FIX. MIEX improved biological stability expressed as DOC and AOC, but results for BFR were obscured by spent resin. Unfortunately the values of DOC and AOC were not below the aimed concentrations of 10 µg C L^{-1} for AOC and 1 mg C L^{-1} for DOC, but the pilot plant (specifically the BAC filters) was not as effective in DOC removal as is the full-scale treatment plant (Grefte et al., 2011), so it is expected that the aimed values will be reached when implementing FIX in the full-scale treatment.

3.3 IEX cost comparison

Three positions were selected for NOM removal by IEX: (1) before coagulation, (2) before ozonation and (3) as a polishing step after SSF. IEX added before coagulation is expected to remove DOC from 9.0 to 3.5 mg C L^{-1} (Table 1), which is a removal rate of 61 %. From Fig. 1 the run time of every column can be determined; for a removal rate of 61 % the run time will be approximately 11 100 BV. The design flow through every column is 220 m^3 h^{-1}, which means that every column should be regenerated every 19 days. When IEX is positioned before coagulation, the DOC removal is 5.5 g C m^{-3}. From 30 Million m^3 treated water 165 000 kg C needs to be discharged. In Table 4 the regeneration and waste characteristics of the three possible positions of the IEX plant are given. The regeneration frequencies of the three positions were calculated based on the expected removal of DOC given in Table 1, combined with the relation between removal rate and run time given in Fig. 1. It is shown that because of the higher regeneration frequency more brine and rinse water will be produced and more salt will be used when IEX is placed after SSF, than when IEX is placed before coagulation or before ozonation. The amount of DOC in the brine is very high for water treated by IEX before coagulation because of the high required (absolute) DOC removal at this position. In Table 5 the estimated operational and fixed costs are given. The highest IEX costs are 0.0656 € m^{-3} when IEX is placed

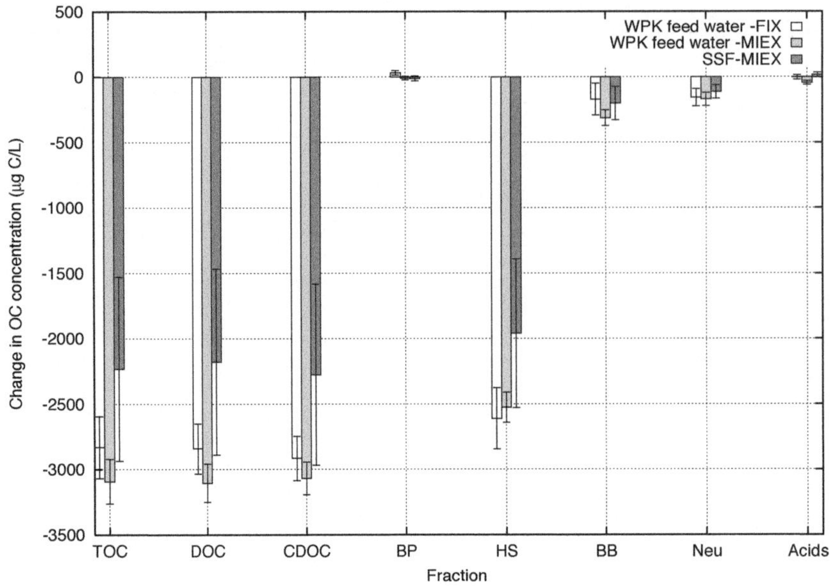

(a) Removal

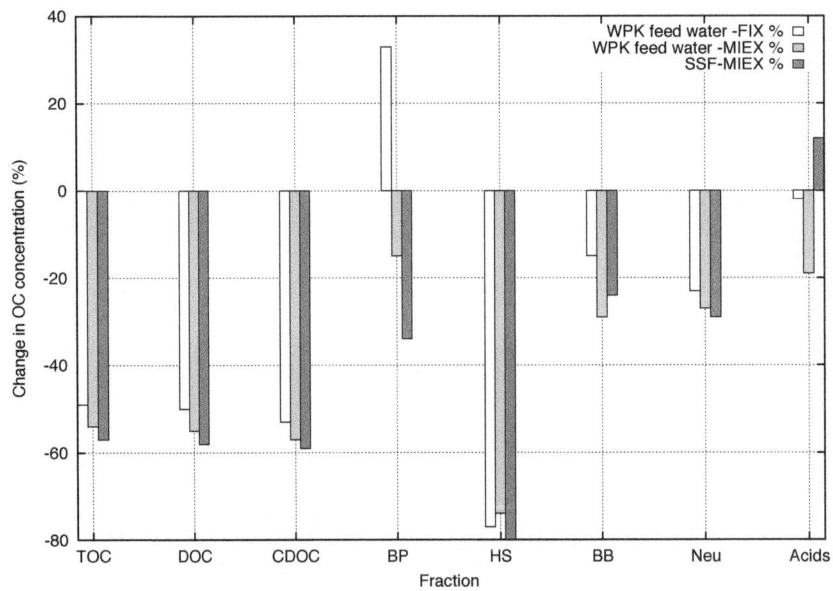

(b) Removal rate

Figure 3. Change in NOM fraction concentrations in the different water types by MIEX® and FIX.

before coagulation. The cheapest option is to place IEX before ozonation, but the difference in costs between these two positions is only 6 %. The costs for waste are 34 % of the total costs. Brine treatment by NF, which means 90 % waste reduction (Schippers et al., 2004), will reduce the price.

3.4 Expected costs savings on subsequent treatment processes

The addition of an IEX installation in the treatment lane will influence subsequent treatment processes in operation and quality as well as costs. Removing NOM before coagulation will save coagulant, even when the main purpose of coagulation is turbidity removal (Singer and Bilyk, 2002). Currently, coagulation removes $1.9\,\mathrm{mg\,C\,L^{-1}}$, when IEX is incorporated before coagulation DOC removal is only $0.75\,\mathrm{mg\,C\,L^{-1}}$. DOC removal will be lowered by 60 %, assuming a linear dependency of the coagulant dose and sludge production on DOC, these can be reduced by 60 % as well (White et al., 1997). This will save $0.0037\,€\,\mathrm{m^{-3}}$ (Barrios et al., 2008). (The purpose of coagulation at Loenderveen is also removal

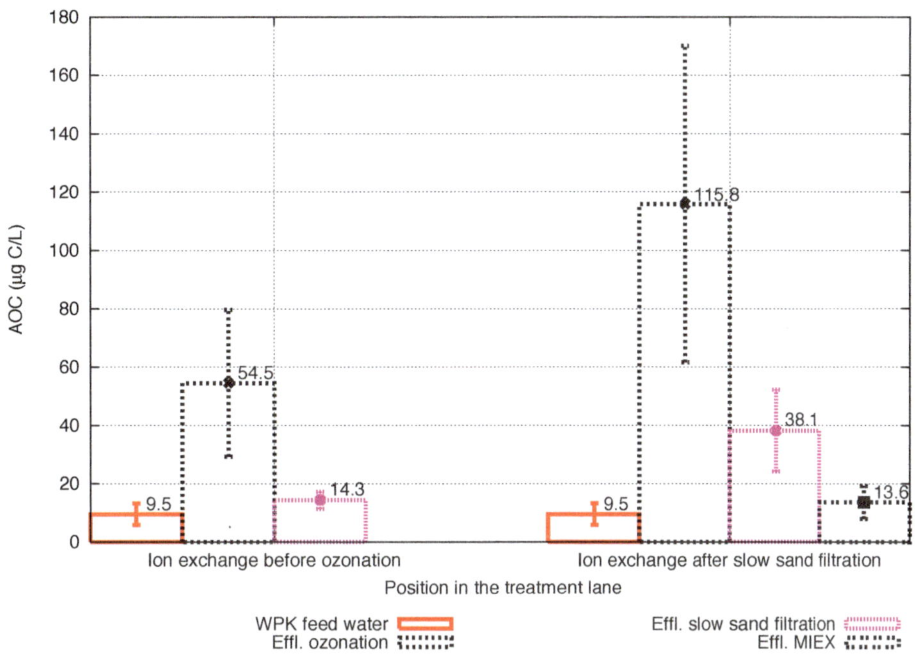

Figure 4. AOC concentrations in the treatment lanes when FIX was positioned before ozonation or MIEX was positioned after SSF.

Table 5. Total costs of IEX at three different positions in the treatment lane.

Costs (€ yr^{-1})	Before coagulation	Before ozonation	After SSF
Waste transport	224 364	237 076	366 198
Waste discharge	451 642	312 044	164 234
Salt	68 688	72 576	112 104
Fixed costs	1 224 000	1 224 000	1 224 000
Total	1 968 694	1 845 696	1 866 536
€ m^{-3}	0.0656	0.0615	0.0622

of phosphate for preventing eutrophication of the lake reservoir. The effect of NOM on phosphate removal is unknown, therefore the reduction in coagulant dose at Loenderveen could be less than 60 %.)

When a linear relation between ozone dosage per DOC concentration and disinfection capacity is adopted, the ozone dosage can be reduced from 2.0 to 0.75 mg L^{-1}. This will save 0.0040 € m^{-3} (Barrios et al., 2008).

Currently, the contact time in the BAC filters is 52 min, approximately 15 out of 26 filters are regenerated per year. The run time of each filter is approximately 1.7 yr. When IEX is incorporated in the treatment lane before BAC, the DOC removal can be reduced by 66 %. Assuming an increase in runtime of 66 %, only 9 columns need to be regenerated per year, resulting in a saving of 0.0089 € m^{-3} (Barrios et al., 2008).

Comparing the net costs (Table 6) shows that IEX before coagulation and IEX before ozonation are the most cost effective options because of the savings on coagulation, ozonation and BAC. In this specific case study IEX before ozonation will be the most cost effective option, because removal of phosphate must be sufficient, the reduction in coagulant and thus the cost savings on coagulation will be less than calculated. IEX positioned after SSF does not influence other treatment steps and therefore the net costs are the highest.

The customer of Waternet pays 1.70 € m^{-3} (VEWIN, 2011) for their tap water. The total production costs of drinking water at Loenderveen and Weesperkarspel treatment plant are 0.1592 € m^{-3} (Barrios et al., 2008). The production costs of IEX consists of replacement of the Lewatit VP OC 1071 resin (220 000 € yr), costs for energy (81 000 € yr), waste transport (237 076 € yr), waste discharge (312 044 € yr) and costs for salt (72 576 € yr) (Tables 3 and 5), this makes the total production costs 0.0308 € m^{-3} for the position before ozonation, without savings on subsequent treatment steps. The increase in production costs is 19 %. The total costs for placing IEX before ozonation are 0.0486 € m^{-3} (see Table 6), the water price for the consumer will increase by 2.9 %

3.5 Environmental impact

Pellet softening is the main contributor to the environmental impact at Waternet (Barrios et al., 2008) due to the use of sodium hydroxide (NaOH). The softening process will not be improved by IEX. The second largest

Table 6. Expected savings of downstream processes because of IEX and the calculated net costs.

Costs (€ m^{-3})	Before coagulation	Before ozonation	After SSF
Costs	0.0656	0.0615	0.0622
Savings			
Coagulation	0.0037	–	–
Ozonation	0.0040	0.0040	–
BAC	0.0089	0.0089	–
Net costs	0.0490	0.0486	0.0622

contributor to the environmental impact is coagulation (2.01×10^{-3} Eco Points m^{-3}) because of the use of ferric chloride (71.3 %) (Barrios et al., 2008). By incorporating IEX before coagulation, the coagulant dose can be reduced by 60 %, which decreases the environmental impact of coagulation to 1.15×10^{-3} Eco Points m^{-3}. Furthermore, regeneration of BAC determines for a large part (59.5 % of 1.17×10^{-3} Eco Points m^{-3}) the height of the environmental impact for BAC (Barrios et al., 2008). The regeneration of BAC columns will be decreased by prior NOM removal by IEX from 15 columns to 9 columns per year, which is a decrease of 40 %. The environmental impact of BAC will decrease to 0.89×10^{-3} Eco Points m^{-3}. Although the reduction of ozone dosage influenced the costs, it will have a negligible effect on the environmental impact (Barrios et al., 2008). It is expected that the environmental impact of an IEX plant would be mainly determined by energy cost, in that way the environmental impact of an IEX treatment will be comparable with the low impact of rapid sand filtration. The environmental impact of rapid sand filtration consists for 80 % of NaOH dose. The expected environmental impact of IEX will be 20 % the environmental impact of rapid sand filtration, so the environmental impact of IEX will be 0.14×10^{-3} Eco Points m^{-3} (Barrios et al., 2008). Extending the treatment lane with IEX before coagulation and ozonation will therefore decrease the (already low) environmental impact of the drinking water treatment. IEX positioned before coagulation will decrease the environmental impact from 8.6×10^{-3} Eco Points m^{-3} to 7.6×10^{-3} Eco Points m^{-3}, the environmental impact for the treatment when IEX is positioned before ozonation will be 8.5×10^{-3} Eco Points m^{-3}.

4 Conclusions

Different placement positions of IEX in the treatment lane (IEX positioned before coagulation, before ozonation or after slow sand filtration) and two IEX configurations (MIEX® and FIX) were compared on water quality as well as costs. For this purpose the pre-treatment plant at Loenderveen and production plant Weesperkarspel of Waternet were used as a case study.

Both, MIEX® and FIX were able to remove NOM (mainly the HS fraction) to a high extent, which improved the water quality. NOM removal can be done efficiently before ozonation and after slow sand filtration, because the position in the treatment lane did not influence the NOM removal percentage.

In this study, biologically stable water was defined in terms of AOC, BFR and DOC. By removing humic substances by FIX before ozonation less biodegradable NOM was formed than without FIX and the biological stability (AOC, BFR and DOC) of the drinking water was improved (Grefte et al., 2011). MIEX® positioned after slow sand filtration caused a decrease in DOC and AOC.

The operational costs were assumed to be directly dependent of the NOM removal rate and determined the difference between the IEX positions. The total costs for IEX for the three positions were approximately equal (0.0631 € m^{-3} on average, Table 6), however the savings on following treatment processes caused a cost reduction for the IEX positions before coagulation and before ozonation compared to IEX positioned after SSF. IEX positioned before ozonation was most cost effective and improved the biological stability of the treated water.

Acknowledgements. This study was part of a project about natural organic matter (NOM) in drinking water distribution networks. Subsidy for the NOM project was provided by Senter Novem. The project is a joint research project of KWR watercycle research institute, UNESCO-IHE, Delft University of Technology, the water supply Companies Vitens and Waternet, water cycle company for Amsterdam and surrounding areas. The MIEX® pilot plant was provided by Orica Watercare, the authors would like to acknowledge the dedicated support from Orica Watercare personnel.

References

Allpike, B., Heitz, A., Joll, C., Kagi, R., Abbt-Braun, G., Frimmel, F., Brinkmann, T., Her, N., and Amy, G.: Size exclusion chromatography to characterize DOC removal in drinking water treatment, Environ. Sci. Technol., 39, 2334–2342, 2005.

Baghoth, S. A., Dignum, M., Grefte, A., Kroesbergen, J., and Amy, G. L.: Characterization of NOM in a drinking water treatment process train with no disinfectant residual, Water Sci. Technol., 9, 379–386, 2009.

Barrios, R., Siebel, M., van der Helm, A., Bosklopper, K., and Gijzen, H.: Environmental and financial life cycle impact assessment of drinking water production at Waternet, J. Clean. Prod., 16, 471–476, 2008.

Bolto, B., Dixon, D., Eldridge, R., King, S., and Linge, K.: Removal of natural organic matter by ion exchange, Water Res., 36, 5057–5065, 2002.

Boyer, T. H. and Singer, P. C.: Bench-scale testing of a magnetic ion exchange resin for removal of disinfection by-product precursors, Water Res., 39, 1265–1276, 2005.

Cornelissen, E., Moreau, N., Siegers, W., Abrahamse, A., Rietveld, L., Grefte, A., Dignum, M., Amy, G., and Wessels, L.: Selection of anionic exchange resins for removal of natural organic matter (NOM) fractions, Water Res., 42, 413–423, 2008.

Cornelissen, E., Beerendonk, E., Nederlof, M., van der Hoek, J., and Wessels, L.: Fluidized ion exchange (FIX) to control NOM fouling in ultrafiltration, Desalination, 236, 334–341, 2009.

Cornelissen, E., Chasseriaud, D., Siegers, W., Beerendonk, E., and van der Kooij, D.: Effect of anionic fluidized ion exchange (FIX) pre-treatment on nanofiltration (NF) membrane fouling, Water Res., 44, 3283–3293, 2010.

Croue, J.-P., Voilleau, D., Bodaire, C., and Legube, B.: Removal of hydrophobic and hydrophilic constituents by anion exchange resin, Water Sci. Technol., 40, 207–214, 1999.

DowChemicalCompany: Dow Water and Process Solutions – Correlation between $KMnO_4$, COD, BOD, DOC and TOC, https://dow-answer.custhelp.com/app/answers/detail/a_id/4219/~/dow-water-and-process-solutions-correlation-between-kmno4,-cod,-bod,-doc-and, 2011.

Drikas, M., Morran, J., Pelekani, C., Hepplewhite, C., and Bursill, D.: Removal of natural organic matter – a fresh approach, Water Sci. Technol., 2, 71–79, 2002.

Eaton, A. D., Clesceri, L. S., Rice, E. W., Greenberg, A. E., and Franson, M. A. H.: Standard methods for the examination of water and wastewater 2005, 21 Edn., American Public Health Association and American Water Works Association and Water Environment Federation, Baltimore, USA, 2005.

Edzwald, J. K., Becker, W. C., and Wattier, K. L.: Surrogate Parameters for Monitoring Organic Matter and THM Precursors, J. AWWA, 77, 122–132, 1985.

Galjaard, G., Martijn, B., Koreman, E., Bogosh, M., and Malley, J.: Performance evaluation SIX-Ceramac in comparison with conventional pre-treatment techniques for Surface Water Treatment, Water Practice & Technology, 6, doi:10.2166/wpt.2011.0066, 2011.

Grefte, A., Dignum, M., Baghoth, S., Cornelissen, E., and Rietveld, L.: Improving the biological stability of drinking water by ion exchange, Water Sci. Technol., 11, 107–112, 2011.

Huber, S. and Frimmel, F.: Size-exclusion chromatography with organic carbon detection (LC-OCD): a fast and reliable method for the characterization of hydrophilic organic matter in natural waters, Vom Wasser, 86, 277–290, 1996.

Huber, S. A.: Origin and Behaviour of Non-Ionic Natural Organic Matter (NOM) in Boiler Feed Water, in: International Water Conference, Orlando, Florida, 2005.

Huber, S. A., Balz, A., Abert, M., and Pronk, W.: Characterisation of aquatic humic and non-humic matter with size-exclusion chromatography – organic carbon detection – organic nitrogen detection (LC-OCD-OND), Water Res., 45, 879–885, 2011.

Humbert, H., Gallard, H., Suty, H., and Croue, J.-P.: Performance of selected anion exchange resins for the treatment of a high DOC content surface water, Water Res., 39, 1699–1708, 2005.

Kabsch-Korbutowicz, M., Wisniewski, J., Lakomska, S., and Urbanowska, A.: Application of UF, NF and ED in natural organic matter removal from ion-exchange spent regenerant brine, Desalination, 280, 428–431, 2011.

Korshin, G., Benjamin, M., and Li, C.-W.: Use of differential spectroscopy to evaluate the structure and reactivity of humics, Water Sci. Technol., 40, 9–16, 1999.

Lewatit: Product information Lewatit VP OC 1071 http://www.lewatit.com, n.d.

OricaWatercare: MIEX Treatment Systems-High Rate Configuration, http://www.miexresin.com/files/High%20Rate%20Brochure%20V0307.pdf, n.d.

Schippers, D., Kooi, M., Sjoerdsma, P., and de Bruijn, F.: Colour removal by ion exchange and reuse of regenerant by means of nanofiltration, Water Sci. Technol., 4, 57–64, 2004.

Singer, P. C. and Bilyk, K.: Enhanced coagulation using a magnetic ion exchange resin, Water Res., 36, 4009–4022, 2002.

van der Helm, A. W. C., Grefte, A., Baars, E. T., Rietveld, L. C., van Dijk, J. C., and Amy, G. L.: Effects of natural organic matter (NOM) character and removal on ozonation for maximizing disinfection with minimum bromate and AOC formation, J. Water Supply Res. T., 58, 373–385, 2009.

van der Kooij, D.: Assimilable organic carbon as an indicator of bacterial regrowth, J. Am. Water Works Ass., 84, 57–65, 1992.

van der Kooij, D. and Hijnen, W.: Substrate utilisation by an oxalate-consuming *Spirillum* species in relation to its growth in ozonated water, Appl. Environ. Microb., 47, 551–559, 1984.

van der Kooij, D., Visser, A., and Hijnen, W.: Determining the concentration of easily assimilable organic carbon in drinking water, J. Am. Water Works Ass., 74, 540–545, 1982.

van der Kooij, D., Veenendaal, H. R., Baars-Loristand, C., van der Klift, H. W., and Drost, Y. C.: Biofilm formation on surfaces of glass and Teflon exposed to treated water, Water Res., 29, 1655–1662, 1995.

van der Kooij, D., van Lieverloo, J., Schellart, J., and Hiemstra, P.: Maintaining quality without a disinfectant residual, J. Am. Water Works Ass., 91, 55–64, 1999.

van der Kooij, D., Vrouwenvelder, J. S., and Veenendaal, H. R.: Elucidation and control of biofilm formation processes in water treatment and distribution using the unified biofilm approach, Water Sci. Technol., 47, 83–90, 2003.

van Lier, J.: Lecture Notes Wastewater Treatment, Sanitary Engineering (CT3420), TU Delft, 2011.

VEWIN: Tarievenoverzicht drinkwater per 1 januari 2012, http://www.vewin.nl/SiteCollectionDocuments/Publicaties/tarievenoverzicht_2012.pdf, 2012.

White, M., Thompson, J., Harrington, G., and Singer, P.: Evaluating criteria for enhanced coagulation compliance, J. Am. Water Works Ass., 89, 64–77, 1997.

5

Application of DVC-FISH method in tracking *Escherichia coli* in drinking water distribution networks

L. Mezule, S. Larsson, and T. Juhna

Department of Water Engineering and Technology, Riga Technical University, Azenes 16/20-263, Riga, Latvia

Correspondence to: L. Mezule (linda.mezule@rtu.lv)

Abstract. Sporadic detection of live (viable) *Escherichia coli* in drinking water and biofilm with molecular methods but not with standard plate counts has raised concerns about the reliability of this indicator in the surveillance of drinking water safety. The aim of this study was to determine spatial distribution of different viability forms of *E. coli* in a drinking water distribution system which complies with European Drinking Water Directive (98/83/EC). For two years coupons (two week old) and pre-concentrated (100 times with ultrafilters) water samples were collected after treatment plants and from four sites in the distribution network at several distances. The samples were analyzed for total, viable (able to divide as DVC-FISH positive) and cultivable *E. coli*. The results showed that low numbers of *E. coli* enters the distribution sytem from the treatment plants and tend to accumulate in the biofilm of water distribution system. Almost all of the samples contained metabolically active *E. coli* in the range of 1 to 50 cells per litre or cm^2 which represented approximately 53 % of all *E. coli* detected. The amount of viable *E. coli* significantly increased into the network irrespective of the season.

The study has shown that DVC-FISH method in combination with water pre-concentration and biofilm sampling allows to better understand the behaviour of *E. coli* in water distribution networks, thus, it provides new evidences for water safety control.

1 Introduction

Over the last century, monitoring of *Escherichia coli* by using standard plate count methods in water samples taken from several locations of water supply system (grab sampling) has allowed to increase the safety of public water supply system and significantly reduce the mortality caused by waterborne outbreaks in the developed countries. However, this strategy has not been always successful in detecting outbreaks, including several major ones (Lee et al., 2002). This approach has not also been able to explain sporadic contamination cases of drinking water (Hunter, 1997). There are several reasons for this bias including the following ones: (i) due to different rates of transport in aquatic environment and resistance to disinfection, the presence of *E. coli* does not always correlate with other pathogens (Ashbolt et al., 2001), (ii) some *E. coli* cells could be injured or stressed and unable to reproduce on growth media used for their detection (Scheusner et al., 1971) and (iii) the probability of detecting the contamination with grab sampling is low due to the small size of samples and limited sampling locations in the networks (van Lieverloo et al., 2007). The significance of the latter two factors in the context of understanding the fate of *E. coli* is only marginally studied.

Although it is still disputable whether so called viable but not cultivable (VBNC) state of bacteria actually exist (Bogosian et al., 1998), there are convincing evidences that not all alive (viable) *E. coli* are able to form colonies in the conditions used for standard plate counting (growth media, temperature, cultivation time) (Na et al., 2006). The importance of this bacterial subpopulation for public health is stressed by numerous molecular studies showing that they still retain ability of gene expression (Asakura et al., 2007; Oliver, 2010) including the ones responsible for pathogenicity (Liu et al., 2010). In order to better understand the fate of *E. coli*, the application of molecular methods are needed which allow to detect microorganisms in different metabolic states including VBNC. Detection of bacteria which are able to elongate using, so called, direct viable count (DVC) (Kogure et al., 1979) in combination with fluorescence in situ hybridization (FISH) is one of the tools allowing to study VBNC bacteria.

In previous studies, to increase the probability of detecting *E. coli*, the composition of the biofilm was analysed (Juhna et al., 2007) The results showed that ca. 0.1 % from total microbial population in the biofilm consist of *E. coli* even though no indicators of faecal contamination were found by routine sampling and analyses of water. In the present study the sampling of biofilm and large volumes of water (concentrates) from the networks over the period of two years was performed and total, cultivable and potentially dividing (detected as DVC-FISH) cells were analysed to obtain comprehensive data about the distribution of *E. coli* in water distribution networks. With this approach it was expected to detect even a low number of *E. coli* and, hence, better understand the fate of these faecal origin bacteria in the networks. The study was carried out in a large water distribution network (Riga, Latvia) meeting drinking water quality standards with respect to microbiology.

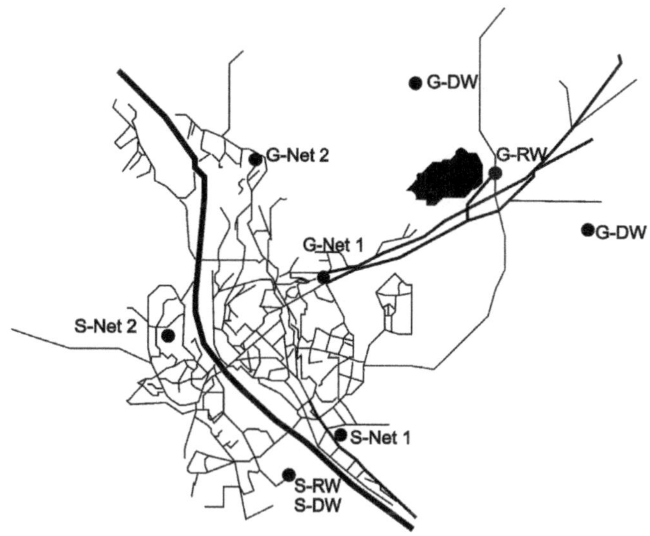

Figure 1. Riga drinking water distribution network with 9 biofilm and water sampling sites including water before (G-RW – raw lake water; S-RW – raw surface water) and after treatment (S-DW – surface water after final chlorination) and, G-DW (infiltrated groundwater after final chlorination) and in the network S-NET1, S-NET2, G-NET1 and G-NET2. Numbers 1 and 2 indicate the residence time after treatment – "1" having shorter residence time.

2 Materials and methods

2.1 Site location and sampling

The total length of Riga drinking water supply system is 1374 km with a demand of 1500 L s^{-1}. The 76 % of all pipe length consists of cast and ductile iron, 19 % of steel and the rest of reinforced concrete and polyethylene (Rubulis et al., 2010). The left bank is supplied with two-stage treated surface water, whereas the right bank is supplied with groundwater treated by artificial groundwater recharge through infiltration. In both cases final chlorination is applied. Detailed parameters of the water can be found elsewhere (Tihomirova et al., 2010).

A total of 24 biofilm samples and 40 water samples were collected from eight sites at water treatment stations and four sites in Riga drinking water supply system (Latvia) (Fig. 1). Points were located at surface and groundwater treatment plants before and during treatment and just after final chlorination (S-RW, S-DW, G-RW, G-DW and G-DW') and at four places in the network (S-NET1, S-NET2, G-NET1, G-NET2). Samples were collected over a period of two years and included all four seasons.

To collect the biofilm, a stainless steel biofilm collector containing five 3 mm diameter stainless steel coupons inserted into 25 to 40 mm pipes was attached to the distribution system and kept for two weeks to allow the biofilm formation. After two weeks the collectors were removed and immediately transported to the laboratory. Within two hours the biofilm was removed by ultrasonication for 2 min at 20 µA and 22 KHz. A total of 16–40 mL biofilm suspension in sterile distilled water was obtained and further analysed. All samples were analysed in double.

A large volume of water samples (more than 100 L) were concentrated to approximately 1 L using tangential ultrafiltration method. Total concentration time varied from 12–24 h. The apparatus used for concentration was similar to the one developed within TECHNEAU project. The estimated recovery rate for the concentration of drinking water was 81 ± 33 % (Veenendaal and Brouwer-Hanzens, 2007). The obtained concentrate was collected into sterile glass bottles, transported to the laboratory and analysed within two hours. All samples were analysed in triple.

2.2 Total bacterial numbers (TBN)

A known volume of sample (~ 0.1 mL, adjusted to obtain 15–100 cells per microscope field of view) was filtered onto 25 mm diameter, 0.2 µm pore-size filters (Anodisc; Whatman plc), fixed with 3–4 % (v/v) formaldehyde and stained with 10 µg mL^{-1} DAPI (4',6-diamidino-2-phenylindole, Merck) for 15 min. Cell numbers were determined by epifluorescence microscopy by counting 20 random fields of view (Ex. 340/380 nm; Em. > 425 nm, dichromatic mirror 565 nm, Leica DM, LB).

Table 1. Total cultivable, FISH positive and DVC positive (viable) *Escherichia coli* in the biofilm and water samples. Values are the average for four seasons with standard deviation specified as a maximum and minimum interval.

Sampling site	Sample type*	Cultivable *E. coli*, cfu cm^{-2} or cfu L^{-1}**	FISH positive *E. coli*, cells cm^{-2} or cells L^{-1}	DVC positive *E. coli*, cells cm^{-2} or cells L^{-1}
S-RW	W	26.74	n/d	$(2.52 \pm 0.17) \times 10^3$
	B	0	44.21 ± 34.72	1.99 ± 2.54
S-DW	W	0	n/d	0
	B	0	8.40 ± 3.92	2.95 ± 5.14
S-NET1	W	0	n/d	0.93 ± 1.61
	B	0	111.58 ± 57.87	45.20 ± 30.25
S-NET2	W	0.13	n/d	9.30 ± 4.13
G-RW	W	30.24	n/d	$(2.99 \pm 0.94) \times 10^3$
	B	0	9.39 ± 1.67	0.66 ± 0.85
G-DW	W	0	n/d	1.01 ± 1.75
G-DW'	W	0	n/d	0
	B	0	14.44 ± 4.94	3.54 ± 4.79
G-NET1	W	0.06	n/d	0.94 ± 1.62
	B	0	70.62 ± 19.97	84.75 ± 57.37
G-NET2	W	0.006 ± 0.01	103.97 ± 20.05	52.41 ± 58.72

Values are the average for four seasons with standard deviation specified as a maximum and minimum interval.
* W – water, B – biofilm
** For biofilm samples the results are represented as cfu or cells per cm^2, for water samples – cfu or cells per Liter of water.
N/D – not determined

2.3 ATP measurements

ATP was determined according to a modified method described by Vital et al. (2008). In brief, 500 µL of pre-warmed sample were mixed with BacTiter-GloTM System (Promega, USA) and measured with luminometer. Results were expressed as ng ATP per cm^2 of biofilm sample. For all samples total ATP and free ATP was measured and then cell ATP calculated. To obtain free ATP the sample was filtered through 0.1 µm pore size syringe filters (Sartorious, Minisart®, Germany).

2.4 Culture based methods

To obtain heterotrophic plate counts (HPC) samples were serially diluted in sterile distilled water and then inoculated onto R2A agar plates by spread plate technique. All plates were incubated in dark at 22 °C for 7 days. Results were expressed as colony forming units (CFU) per cm^2 (biofilm) or ml (water) of sample. For cultivable *E. coli* samples were analysed by membrane filtration method in a certified reference laboratory according to EN ISO 9308-1:2000 (ISO, 2001).

2.5 DVC-FISH

Cell viability-potential to divide, was determined by modified DVC method by Kogure et al. (1979) and combined with FISH. The validity of the FISH method used for detection of *E. coli* in this study has been proven earlier where it was compared with other molecular methods. In brief, samples were mixed with equal amount of Tryptone Soya broth (Oxoid Ltd., UK) and 10 µg mL^{-1} nalidixic acid and incubated for 6 h at 30 °C. After the incubation, samples were fixed with 3–4 % (*v*/*v*) formaldehyde for 20 min. Then the samples were filtered onto 25 mm diameter, 0.2 µm pore-size filters (Anodisc; Whatman plc) washed with sterile distilled water, removed from filtration device and air-dried. Then, 20 µL of PNA hybridization mix consisting of hybridization buffer (50 mM Tris-HCl, 10 % *w*/*v* 50 % dextran sulphate, 0.1 mM of NaCl, 30 % *v*/*v* formamide, 30 % *v*/*v* tetra-sodium pyrophosphate, 0.2 % *w*/*v* polyvinylpyrrolidone, 0.2 % *w*/*v* Ficoll 400, 5 mM Na_2 EDTA, 0.1 % *v*/*v* Triton X-100) and 200 nM fluorescently labelled PNA probe, was applied to the filters and covered with cover glass. The PNA probe (TCA ATG AGC AAA GGT) (Perry-O'Keefe et al., 2001) was labelled with CY3 (Ex: 550, Em: 570) and flanked with solubility enhancers (Applied Biosystems, Foster City, CA, USA). The samples were incubated at 57 °C for 60 min in a tight vessel in dark. After hybridization the filters were washed with plenty of sterile distilled water, air-dried and visualized with epifluorescence microscopy (Ex. 535 ± 25 nm; Em. 610±37 nm, dichromatic mirror 565 nm, Leica DM LB). The detection limits were calculated based on volume of analysed sample, repetitions made and microscope fields of view examined. In this study filter sectors were scanned for positive events, enabling to overview 1/5 of the filter surface (Mezule, 2012). The calculated limits for *E. coli* was 6 cells per mL in concentrated water or 4 cells per cm^2 in the biofilm.

Table 2. Total bacterial numbers (TBN) and heterotrophic plate counts (HPC) in the biofilm samples.

Sampling site	TBN, cells $cm^{-2} \times 10^6$	HPC, cfu $cm^{-2} \times 10^4$	% HPC
S-DW	5.64 ± 7.62	4.04 ± 5.33	0.72
S-NET 1	3.17 ± 3.88	8.40 ± 11.89	2.65
S-NET 2	17.13 ± 24.90	41.51 ± 21.62	2.42
G-DW	1.12 ± 0.48	0.78 ± 1.44	0.7
G-NET 1	7.71 ± 10.35	3.14 ± 2.71	0.41
G-NET 2	28.97 ± 18.14	178.50 ± 49.82	6.16

Values are the average for all biofilm samples tested during four seasons with standard deviation specified as a maximum and minimum interval.

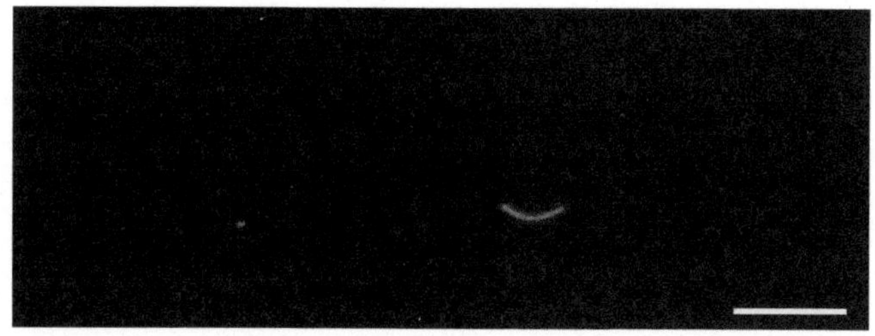

Figure 2. G-DW biofilm sample obtained in winter season with FISH positive *E. coli* (left) and DVC-FISH positive *E. coli* (right) Bar 10 μm.

3 Results and discussion

Previous studies have shown that *E. coli* is harboured in drinking water biofilm in a nonculturable form (Juhna et al., 2007). In this study we sampled both water and biofilm at several distances away from water treatment plants to understand if the number these bacteria is changing with increase of water residence time. Water supply system (Riga, Latvia) which meets European Drinking Water Directive (98/83/EC) with respect to microbial parameters was selected for study. Analyses of raw water samples (river and groundwater) with standard methods (based on sample cultivation on growth media) showed that water sources contained less than four cultivable *E. coli* per 100 mL of sample (Table 1), which is an indication of relatively clean surface water. In contrast, DVC-FISH analyses showed that viable *E. coli* cell concentration was about 10 times higher. This is in agreement with previous studies about raw waters (Garcia-Armisen and Servais, 2004) which showed that only a fraction of alive *E. coli* are detected with traditional methods in raw water sources used for drinking water production.

Analyses of drinking water after treatment and in the distribution network showed that none of water samples exceeded the standard value of zero cultivable *E. coli* in 100 mL. The applied sample concentration method allowed to decrease the detection limit when compared to traditional sampling methodology. About 30 % of water samples contained 0.02 to 1.2 cells in 10 L. All positive samples were obtained further in the networks (S-NET2, G-NET1 and G-NET2) rather than directly after water treatment. Hence, cultivable *E. coli* cells were present in drinking water, however, the concentration was very low to be detected with traditional sampling and analytical methodology. The analyses of water samples showed that the lowest amount of FISH positive cells was obtained from water samples (G-DW and S-DW) directly after the treatment plants (Table 1) and this number tended to increase further in the network. The same trend was observed also for viable *E. coli* (DVC-FISH positive) cells, where their concentration increased from few cells to more than 50 cells per litre as water proceeded into the networks.

The analyses of biofilm samples showed that at the treatment plant (S-DW) as much as 44 *E. coli* cells per cm^2 were present. This particular measurement could be biased by high background fluorescence intensity from humic substances present in the water (Tihomirova et al., 2010). However, the interference from autofluorescent background was overcome by analysing the viable cells with DVC-FISH as in this case the bacterial cells appeared very bright and visible due to their increased size (Fig. 2). The amount of viable *E. coli* in the biofilm at the treatment plant was 1.99 cells cm^{-2} (S-DW) and 0.66 cells cm^{-2} (G-DW) indicating that faecal bacteria sporadically can pass though the treatment plant and are harboured within the biofilm. A significant increase ($p < 0.03$) in DVC-FISH counts was observed when treated (both surface and ground) water was compared to (NET) sites in the networks. In total the results of FISH analyses showed that each cm^2 of two week old biofilm contain a minimum of 5 *E. coli* cells (Table 1) and all 24 biofilm and 25 water samples were positive for *E. coli*. From all FISH positive *E. coli* in the biofilm, on average 53 % of cells showed the ability to divide.

Along with cell distribution trends, the influence on seasonality was analysed and showed that there is no significant difference ($p > 0.05$) in sampling time for DVC-FISH. Again, irrespective of the season, the highest amount of DVC positive *E. coli* was observed for points further in the network. For total *E. coli* counts significantly different results were obtained in 2 sites – S-DW ($p < 0.02$) and S-NET1 ($p < 0.05$) where in winter and spring months, respectively, higher counts were obtained, indicating that warmer outside temperatures are not linked to the occurrence of *E. coli* in the network, especially when the water temperature in the system is around 13 °C throughout the year.

The findings of faecal bacteria in the drinking water is not so surprising since previous research has shown that *E. coli* can survive in the drinking water biofilters (Li et al., 2006) and even multiply in the biofilm (Camper et al., 1991; Fass et al., 1996; Williams and Braun-Howland, 2003). Our study confirmed that the biofilm act as a source for accumulation and possible growth of *E. coli* in drinking water distribution networks. This should be taken into account when studying the fate of these bacteria from the source to the tap.

The highest amount of *E. coli* was found in the samples collected further away from both water treatment plants (Table 1). There was no correlation between cultivable and viable *E. coli* numbers in drinking water, unlike observed for raw water samples (Garcia-Armisen and Servais, 2004). Similarly the biofilm in locations further away in the network contained more *E. coli* cells with a higher percentage of DVC positive ones. The high numbers of *E. coli* could be due to their resistance to chlorine, low chlorine levels at all in the network or low protozoan grazing (Sibille et al., 1998). Their release from the biofilm by detachment (Parsek and Singh, 2003) can have a serious effect on human health, especially if VBNC forms of bacteria, including *E. coli* O157:H7 are present (Liu et al., 2009). Moreover, lately discussions about the existence of naturalized *E. coli*, although no pathogenicity was proven, raise concerns about secondary contamination of water supply systems due to pathogenic bacteria regrowth (Jjemba et al., 2010).

The particular water supply system is known to support bacterial growth due to the high level of labile organic carbon in drinking water (Tihomirova et al., 2010). Thus, biofilm formation was also studied. The analysis of coupon samples inserted in Riga drinking water distribution network showed that extensive biofilm formation occurred on surfaces just at the outlet from the treatment plant, reaching the concentration of more than 10^6 bacterial cells per cm^2 of the biofilm (Table 2).

The amount of total bacteria, heterotrophic plate counts and *E. coli* in the biofilm tended to increase with the water residence time in the network. A positive correlation ($R^2 > 0.95$) between the total bacterial numbers and heterotrophic plate counts in the biofilm was obtained. For groundwater supply the percent of cultivable bacteria in the biofilm increased from $< 1\,\%$ just after treatment to more than $6\,\%$ at G-NET2 site (Table 2), indicating on the formation of more favourable conditions for microbial colonization and growth. Similar increase in the distribution network was also observed for water samples, where a correlation of $R^2 = 0.8869$ was obtained between HPC and TBN in five water samples from one sampling site. But, similarly as in previous studies, no correlation between total bacterial counts or HPC in biofilm and water was observed (Flemming, 2002). In water samples from both treatment plants the bacterial numbers tended to increase with the water residence time in the network. This has been observed previously (Juhna et al., 2007) and could be connected to decrease of disinfectant in the network in addition to increase of bacterial resistance and survival capacity (Gilbert and Brown, 1995; LeChevallier and Kwok-Keung, 2004; Saby et al., 2005). Thus, more favourable conditions for biofilm formation and maturing are created.

The results of ATP showed a high variation between the samples (from 1.1 till 2518 ng ATP mL^{-1}) and between the different seasons of samples from one location (7.35 till 2518 ng ATP mL^{-1}). The correlation between ATP and HPC or TBC was weak, which is similar to previous observations (Delahaye et al., 2003). Comparison of DVC positive *E. coli* with total bacterial counts showed a linear correlation ($R^2 = 0.9637$) in the biofilm, thus, indicating on the role of biofilm community on pathogen accumulation. Most of the studies that have examined the presence of *E. coli* in biofilms have used culture-based assays (Castonguay et al., 2006; Hu et al., 2005) which have limitations, including the duration of incubation, antagonistic organism interference, lack of specificity, and poor detection of slow-growing or non-dividing microorganisms (Rompre et al., 2002). Plate count methods also result in some inaccuracy since the cells can be clumped together and intertwined with other biofilm components. It must be emphasized that methods using microbial growth will not detect non-dividing cells at all. Therefore, the number of *E. coli* in the drinking water distribution network could be, and likely is, underestimated. The presence of *E. coli* was inadequately indicated by the traditional culture-based methods in the present study, a finding in agreement with previous studies showing that cultivation-independent detection methods detect at least 10 times more cells (Bjergbaek and Roslev, 2005).

This study has shown that analyses with culture based methods does necessary describe transport of *E. coli* in drinking water distribution system. We have shown that low number of these bacteria can pass though the treatment plant into the networks and accumulate on the surfaces of the pipes in a biofilm. Since no major outbreaks were detected in the water supply system studied, the link between a low number of *E. coli* and infections from drinking water cannot be ruled out. However, the findings raise questions about the margin of safety provided by conventional drinking water microbial quality control methods. Further studies should be directed to address this issue.

4 Conclusions

The combination of biofilm sampling, and water sample preconcentration and analyses of different viability forms revealed new information about behaviour of *E. coli* in drinking water systems. Despite that both treatment processes are generally regarded as safe, some sporadic breakthrough of *E. coli* occur which lead to small numbers of *E. coli* missed in drinking water when analysed with traditional sampling methodology. Total and viable amount of *E. coli* tend to accumulate in the networks were it is less exposed to disinfectant. Thus, biofilm formation in the drinking water distribution networks increases the risk (at least during the first two weeks) of accumulation of viable but non cultivable *E. coli* and their presence does not necessary directly links to recent faecal contamination.

Acknowledgements. This work has been undertaken as a part of the research project "Technology enabled universal access to safe water – TECHNEAU" (Nr. 018320) which is supported by the European Union within the 6th Framework Programme. There hereby follows a disclaimer stating that the authors are solely responsible for the work. It does not represent the opinion of the Community and the Community is not responsible for any use that might be made of data appearing herein.

References

Asakura, H., Ishiwa, A., Arakawa, E., Makino, S., Okada, Y., Yamamoto, S., and Igimi, S.: Gene expression profile of *Vibrio cholerae* in the cold stress-induced viable but non-culturable state, Environ. Microbiol., 9, 869–879, 2007.

Ashbolt, N. J., Grabow, W. O. K., and Snozzi, M.: World health organization (WHO). Water quality: Guidelines, standards and health, in: Indicators of microbial water quality, edited by: Fewtrell, L. and Bartram, J., IWA Publishing, London, 2001.

Bjergbaek, L. A. and Roslev, P.: Formation of nonculturable *Escherichia coli* in drinking water, J. Appl. Microbiol., 99, 1090–1098, 2005.

Bogosian, G., Morris, P. J. L., and O'Neil, J. P.: A mixed culture recovery method indicates that enteric bacteria do not enter the viable but nonculturable state, Appl. Environ. Microbiol., 64, 1736–1742, 1998.

Camper, A., McFeters, G., Characklis, W., and Jones, W.: Growth kinetics of coliform bacteria under conditions relevant to drinking water distribution systems, Appl. Environ. Microbiol., 57, 2233–2239, 1991.

Castonguay, M. H., van der Schaaf, S., Koester, W., Krooneman, J., van der Meer, W., Harmsen, H., and Landini, P.: Biofilm formation by *Escherichia coli* is stimulated by synergistic interactions and co-adhesion mechanisms with adherence-proficient bacteria, Res. Microbiol., 157, 471–478, 2006.

Delahaye, E., Welte, B., Levi, Y., Leblon, G., and Montiel, A.: An ATP-based method for monitoring the microbiological drinking water quality in a distribution network, Water Res., 37, 3689–3696, 2003.

Fass, S., Dincher, M. L., Rasoner, D. J., Gatel, D., and Block, J.-C.: Fate of *Escherichia coli* experimentally injected in a drinking water distribution pilot system, Water Res., 30, 2215–2221, 1996.

Flemming, H. C.: Biofouling in water systems – cases, causes and countermeasures, Appl. Microbiol. Biot., 59, 629–640, 2002.

Garcia-Armisen, T. and Servais, P.: Enumeration of viable *E. coli* in rivers and wastewaters by fluorescent in situ hybridization, J. Microbiol. Meth., 58, 269–279, 2004.

Gilbert, P. and Brown, M.: Phenotypic plasticity and mechanisms of protection of bacterial biofilms from antimicrobial agents, in: Microbial biofilms, edited by: Lappin-Scott, H. E. C., and Costerton, J. W., Cambridge University Press, 118–132, 1995.

Hu, J., Yu, B., Feng, Y., Tan, X., Ong, S., Ng, W., and Hoe, W.: Investigation into biofilms in a local drinking water distribution system, in: Biofilms, Cambridge University Press, Cambridge, 19–25, 2005.

Hunter, P. R.: Waterborne disease: Epidemiology and ecology, John Wiley and Sons, West Sussex, 1997.

ISO 9308-1:2000: Water quality – detection and enumeration of *Escherichia coli* and coliform bacteria – Part 1: Membrane filtration method, International Organisation for Standartization, 2001.

Jjemba, P. K., Weinrich, L. A, Cheng, W., Giraldo, E., and LeChevallier, M. W.: Regrowth of potential opportunistic pathogens and algae in reclaimed-water distribution systems, Appl. Environ. Microbiol., 76, 4169–4178, 2010.

Juhna, T., Birzniece, D., Larsson, S., Zulenkovs, D., Sharipo, A., Azevedo, N. F., Ménard-Szczebara, F., Castagnet, S., Féliers, C., and Keevil, C. W.: Detection of *Escherichia coli* in biofilms from pipe samples and coupons in drinking water distribution networks, Appl. Environ. Microbiol., 73, 7456–7464, 2007.

Kogure, K., Simidu, U., and Taga, N.: A tentative direct microscopic method for counting living marine bacteria, Can. J. Microbiol., 25, 415–420, 1979.

LeChevallier, M. W. and Kwok-Keung, A.: Water treatment and pathogen control: Process efficiency in achieving safe drinking water, IWA Publishing on behalf of the World Health Organization, London, 2004.

Lee, S. H., Levy, D. A., Craun, G. F., Beach, M. J., and Calderon, R. L.: Surveillance for waterborne-disease outbreaks – United States, 1999–2000, MMWR Surveill Summ, 51, 1–47, 2002.

Li, J., McLellan, S., and Ogawa, S.: Accumulation and fate of green fluorescent labeled *Escherichia coli* in laboratory-scale drinking water biofilters, Water Res., 40, 3023–3028, 2006.

Liu, Y., Wang, C., Tyrrell, G., Hrudey, S. E., and Li, X. F.: Induction of *Escherichia coli* O157:H7 into the viable but non-culturable state by chloraminated water and river water, and subsequent resuscitation, Environ. Microbiol. Rep., 2, 155–161, 2009.

Liu, Y., Wang, C., Tyrrell, G., and Li, X. F.: Production of Shiga-like toxins in viable but nonculturable Escherichia coli O157:H7, Water Res., 44, 711–718, 2010.

Mezule, L.: Significance of nonculturable Escherichia coli in drinking water: experimental and pilot studies in large drinking water systems, Lambert Academic Publishing, 2012.

Na, S. H., Miyanaga, K., Unno, H., and Tanji, Y.: The survival response of *Escherichia coli* K12 in a natural environment, Appl. Microbiol. Biot., 72, 386–392, 2006.

Oliver, J. D.: Recent findings on the viable but nonculturable state in pathogenic bacteria, FEMS Microbiol. Rev., 34, 415–425, 2010.

Parsek, M. and Singh, P.: Bacterial biofilms: An emerging link to disease pathogenesis, Annu. Rev. Microbiol., 57, 677–701, 2003.

Perry-O'Keefe, H., Stender, H., Broomer, A., Oliveira, K., Coull, J., and Hyldig-Nielsen, J.: Filter-based PNA in situ hybridization for rapid detection, identification and enumeration of specific micro-organisms, J. Appl. Microbiol., 90, 180–189, 2001.

Rompre, A., Servais, P., Baudart, J., de-Roubin, M., and Laurent, P.: Detection and enumeration of coliforms in drinking water: Current methods and emerging approaches, J. Microbiol. Meth., 49, 31–54, 2002.

Rubulis J., Dejus S., and Meksa, R.: Online measurement usage for predicting water age from tracer tests to validate a hydraulic model, 12th Annual Water Distribution Systems Analysis Conference, Arizona, Tucson (USA), 2010.

Saby, S., Vidal, A., and Suty, H.: Resistance of Legionella to disinfection in hot water distribution systems, Water Sci. Technol., 52, 15–28, 2005.

Scheusner, D. L., Busta, F. F., and Speck, M. L.: Inhibition of injured *Escherichia coli* by several selective agents, Appl. Microbiol., 21, 46–49, 1971.

Sibille, I., Sime-Ngando, T., Mathieu, L., and Block, J. C.: Protozoan bacterivory and *Escherichia coli* survival in drinking water distribution systems, Appl. Environ. Microbiol., 64, 197–202, 1998.

Tihomirova, K., Rubulis, J., and Juhna, T.: Changes of nom fractions during conventional drinking water treatment process in Riga, Latvia, Wa. Sci. Technol., 10, 157–163, 2010.

van Lieverloo, J. H., Mesman, G. A., Bakker, G. L., Baggelaar, P. K., Hamed, A., and Medema, G.: Probability of detecting and quantifying faecal contaminations of drinking water by periodically sampling for *E. coli*: A simulation model study, Water Res., 41, 4299–4308, 2007.

Veenendaal, H. R. and Brouwer-Hanzens, A. J.: A method for the concentration of microbes in large volumes of water, in: D 3.2.4, TECHNEAU, 1–30, 2007.

Vital, M., Hammes, F., and Egli, T.: *Escherichia coli* O157 can grow in natural freshwater at low carbon concentrations, Environ. Microbiol., 10, 2387–2396, 2008.

Williams, M. M. and Braun-Howland, E. B.: Growth of *Escherichia coli* in model distribution system biofilms exposed to hypochlorous acid or monochloramine, Appl. Environ. Microbiol., 69, 5463–5471, 2003.

6

Status of organochlorine pesticides in the drinking water well-field located in the Delhi region of the flood plains of river Yamuna

P. K. Mutiyar[1], A. K. Mittal[1], and A. Pekdeger[2]

[1]Department of Civil Engineering, Indian Institute of Technology Delhi, 110016 New Delhi, India
[2]Institute of Geological Sciences, Free University of Berlin, 12249 Berlin, Germany

Abstract. This study presents the occurrence of pesticides in a well-field located in Yamuna flood plain of Delhi region. Ground water sampling campaigns were carried out during pre-monsoon and post-monsoon periods covering 21 borewells and 5 Ranney wells. Major 17 organochlorine pesticides (OCP's) along with other water quality parameters were monitored during this period. Pesticide concentrations were determined using GC-ECD, while GC-MS was used for confirmatory purposes. OCP's groups like $\sum$HCH, $\sum$DDT, endosulfan and aldrin were observed in this well-field. Concentration of OCPs from Ranney well exceeded the limit ($1\ \mu g\ l^{-1}$) prescribed by the Bureau of Indian Standards (BIS) in pre-monsoon season, though OCP levels in borewells were within BIS limits. However, these levels were very close to the World Health Organisation (WHO) and European Union (EU) limit of for pesticides ($0.5\ \mu g\ l^{-1}$) in many samples. Borewell produced better quality water compared to the water from Ranney wells. Although, the level of OCP's was slightly lower than prescribed limit of national regulatory agency but such low doses may cause long-term damage to human populations if such water is consumed for longer durations. At low doses OCP's acts as endocrine disrupting agent and cause metabolic disorders in local population.

1 Introduction

Every year hundreds of new molecules are produced by the industry in addition to the production of existing ones. These chemicals either in the original form or as residues enter different components of the ecosystem. Several of these synthetic organics are persistent compounds. Important groups include pesticides, polycyclic aromatic hydrocarbons (PAHs) and pharmaceutical and personal care products (PPCPs). Pesticide contamination in drinking water resources is widely recorded in many parts of the world. Majority of pesticides enter the environment through non-point sources, following normal spraying in agricultural field and further surface run-off systems. Incidences of occurrence of pesticide residues in ground water may be of prime concern because majority of rural and urban populations in the developing world depend upon ground water resources for potable purposes.

The organochlorine groups of pesticides are non-biodegradable and thereby these compounds can be concentrated through food chains and produce a significant magnification of the original concentration at the end of the chain. Once contaminated, (entered) these pesticides remain there for long. It is also supported by long half life of community like DDT's half life is 30 yr. Other chlorinated pesticides such as aldrin, endosulfan, HCH, lindane, endrin and dieldrin remain stable in water for many years after their use. Due to long residence time of these substances, there is a great interest in examining their presence in the environment. Reports showing their ubiquitous presence have made the situation worse as their residues are found in water (Hassan et al., 1996), ground water (Kaushik et al., 2011; Mudiam et al., 2011), food commodities (Mukherjee and Gopal, 1996), dairy milk (John et al., 2001), bovine milk (Sharma et al., 2007), edible oil (Bajpai et al., 2007), animal feed (Mukherjee and Gopal, 1996), mother's milk (Kumar et al., 2006a), human blood (Waliszewski et al., 2001) and in skin (Dua et al., 1998). Organochlorine pesticides (OCP's) were detected even in the most remote parts of the pristine environment like ice core from Mt. Everest (Wang et al., 2008), Antarctic marine ecosystem (Geisz et al., 2008), Alpine glaciers (Villa et al., 2003; Bizzotto et al., 2009) and the breast milk of Inuit communities of Canada (Barrie et al., 1992; Blais et al., 2007) where these pesticides have never been used, but reached due to global circulation pattern and cold deposition on mountains. In India, OCP's have been reported in drinking water sources at: Bhopal (Dikshith et al., 1990), Hyderabad (Shukla et al., 2006), Jaipur (Bakore et al., 2004), Kanpur (Sankararamakrishnan et al., 2005) and Ahemedabad (Jani et al., 1991). The total pesticides consumption in Delhi area is continuously decreasing due to conversion of agricultural land to residential and industrial land. The total pesticides consumption in Delhi area was 100 MT 1990 which reduced to 53 MT in 2005 (www.ncipm.org.in). Mukherjee and Gopal (2002) reported higher level ($\mu g\,ml^{-1}$) of OCP's contamination in ground water samples taken from different zones of Delhi. But now we expected to have lower levels of pesticide residues in the ground water due to decreased pesticide consumption. It is evident from the literature that no comprehensive report is available on pesticide residues in Yamuna flood areas of Delhi region which is the main watershed zone for harvesting water for urban supply of Delhi NCT. The present study is focused on the special Palla-Burari well-field. The study area is strategically important because this area is also used by the local water supply organization as a potable water source. It has around 80 borewells and 5 Ranney wells to extract the good quality water from the Yamuna river flood plains to meet the growing water demand of Delhi. About 15 % of the public water supply of Delhi is contributed by this well-field, making the area hotspot as this area contributes a significant fraction of drinking water to 16.8 million people of Delhi. Since, this well-field has both the borewells and Ranney wells, so it is also contemplated to study effect of type of well on the water quality.

2 Materials and methods

2.1 Description of sampling site

This site is located along the upstream of urbanised parts of Delhi comprising, North West and North district of Delhi, on the flood plain of the western bank of the river Yamuna. The total sampling area of well-field is around 40 km^2, on the western bank of river Yamuna, starting from outskirts of Delhi (un-urbanised area) to the border of neighbouring state of Haryana. The flood plains have an embankment on both sides at a distance of 500–2000 m from the actual course of river Yamuna to protect the populated rural area from flood events. The flood plain is of younger alluvium and the sediments have been deposited upon a series of several hundreds of meters of older alluvium, which predominately composed of fine sand, silt and clay (Shekhar and Prasad, 2009). The well-field has relatively plain terrain within the embankment and the ground water level is about 4–6 m b.g.l. (below ground level). The well-field contains two types of wells, Ranney wells and borewells. Ranney wells comprise a central concrete caisson excavated to a target depth (shallow) at which well screens project laterally outward in a radial pattern and are designed to induce infiltration. The site is on the bank of river Yamuna, thus can support Ranney wells by inducing infiltration through river-bank filtration from river Yamuna. The well-field also contains borewells which pumps the water from relatively deeper aquifer. Yield of borewells ranges from 100–280 m^3 h^{-1} in the area (Kumar et al., 2006b). The distance of wells sampled, from the feeding water source Yamuna river varies from 50–1000 m. The site is strategically important to compare the water quality from shallower and deeper aquifers, as Ranney wells water sample represent ground water quality in shallow aquifer while borewell water sample represent the deep aquifer water quality. Thus water samples were collected from borewells and Ranney wells, which are located between latitude 28°46′56.69″ N, longitude 77°11′58.62″ E and latitude 28°50′34.94″ N, longitude 77°12′34.91″ E. Sampling sites are shown in Fig. 1.

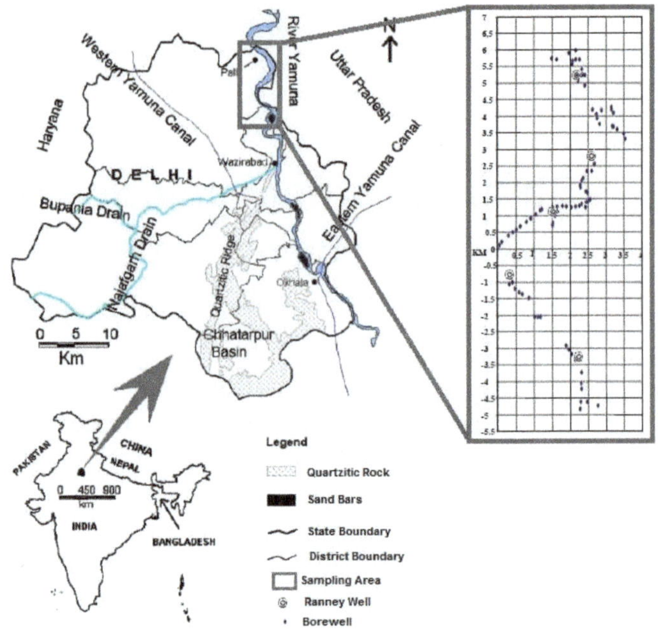

Figure 1. Map showing location of the study area (NCT Delhi) with the grids selected for the OCP analysis (map: modified from Shekhar and Prasad, 2009).

2.2 Water sampling

Sampling area was divided in a grid of 500 × 500 m, to reduce the total number of samples for OCP's analyses and to make a representative sampling as the sampling points were located in close proximity. Samples were taken from all the functional borewells and Ranney wells. One sample from each grid was chosen for OCP analysis in order to obtain complete representation of the well-field. Average number of borewells in a grid of 2.5 km^2 was around 5. Thus 21 borewells and all 5 Ranney wells were selected for OCP analysis. The average depth of the Ranney wells ($N = 5$) was 25.56 m while the average depth of the borewells ($N = 80$) was 44.07 m. Samples were collected in high density polyethylene (HDPE) amber coloured bottles with Teflon lined caps. Prior to sampling, pumps were run for at-least 10 min to clear the casing of standing water and to bring in fresh water from the aquifer. The sampling bottles were rinsed with well water and were carefully filled to overflowing, without passing air bubbles through the samples or trapping air bubbles in sealed bottles. Field blanks were prepared with distilled water and were analysed only when pesticide residues were detected in primary samples. Two samples were taken from each site during each sampling campaign. One sample was used to determine physico-chemical parameters while the second sample was used for OCP analysis. The sample collection bottles were transported in cool-box with ice packs and subsequently stored in a refrigerator at 4 °C until further analysis.

2.3 Reagents and standards

Analytical grade (AR) chemicals (Merck, Germany) were used throughout the study without any further purification. To prepare reagents and calibration standards, double glass distilled water was used. The glass-wares were washed with dilute nitric acid (1.15 N) followed by several portions of distilled water. EPA 502 Pesticide Standard Mix (49690-U) was procured from Sigma-Aldrich USA. The working standards of pesticides were prepared by diluting EPA pesticide mixture standard in *n*-hexane and were stored at −20 °C.

2.4 Physico-chemical parameters

Samples were analysed for different physico-chemical parameters: pH, electrical conductivity (EC), alkalinity, hardness, major anions (Cl^-, F^-, NO_3^-, SO_4^{-2}) and major cations (Na^+, K^+, Ca^{+2}, Mg^{+2}) as per APHA (1998). EC and pH were measured using a portable EC and pH meter, respectively.

2.4.1 Extraction of pesticides residues

Method prescribed by APHA (1998) with some modifications was used for the extraction of OCP residues from the water samples. A liquid liquid extraction (LLE) method, using *n*-hexane as solvent, was used for extraction of pesticide residues. The water samples were prefiltered using 0.45 µm glass-fiber filter to remove any suspended impurity and were extracted without any pH adjustment. For extraction, one litre water sample was taken into a separating flask. It was mixed with 30 g of NaCl and 50 ml of *n*-hexane. Sample was shacked properly and hexane layer was separated. This process was repeated thrice with same water sample and hexane portions were pooled and transferred to a Buchi Condenser flask through a column of anhydrous sodium sulfate and florosil mixture to remove any remaining water molecules. The extract was condensed to around 2 ml on Rotavac and further reduced to 1 ml under gentle N_2 stream. The concentrated extract was transferred to air-tight GC vials and stored at −20 °C until analysis.

2.4.2 Qualitative and quantitative analysis

Analysis of pesticide residue was carried out on a Thermo® Trace GC, microprocessor controlled gas chromatograph, equipped with electron capture detector (ECD), having Nickel (63) foil as the electron source and auto sampler. The column specifications and operating conditions are given in Table 1. Confirmation of the identity of the OCP was done on a Mass Spectrometer (MS), coupled with Trace GC Ultra. The column type, column condition, temperature programming and injector temperature were kept the same for GC-ECD and GC-MS analysis. MS transfer line temperature and ionization source temperature were kept 280 °C and 200 °C, respectively (Table 1). The samples were injected by a Thermo AS 3000 autosampler by setting injection volume of 1 µl in splitless mode for each sample. Analysis was done in full scan mode, range of 50–650 units with 70-eV electron impact (EI) mode and specific ions were monitored for confirmation. The instrument was operated by Xcaliur software from Thermo Finnigan. To ensure the quality of extraction and detection procedure, 5 different concentrations of each OCP standards were mixed with distilled water, extracted by the same method, and recovery was measured. Table 2 presents retention time (RT), recovery efficiency (RE), method detection limit (MDL) and ions monitored for confirmation of different pesticides for this method. Important physico-chemical properties of investigated OCPs are given in Table 3.

3 Result and discussion

Physico-chemical quality of the ground water of the area is presented in Table 4. Water quality of groundwater at selected locations for major parameters: pH, electrical conductivity (EC), alkalinity, hardness, major anions (Cl^-, F^-, NO_3^-, SO_4^{-2}) and major cations (Na^+, K^+, Ca^{+2}, Mg^{+2}) was within BIS limits (IS:10500), except for EC and hardness for Ranney wells (1934.2 $\mu S\,cm^{-1}$ and 653.6 $mg\,l^{-1}$, respectively) in pre-monsoon season which is slightly higher than prescribed limits. However, the EC and hardness decreased during the post-monsoon season. This could be attributed to the recharge of aquifer by good quality rain water and high flood level in river Yamuna during monsoon season. Lorenzen et al. (2010) reported a steep gradient from surface water

Table 1. Operating conditions used for the operation of GC-ECD and GC-MS.

	GC-ECD	GC-MS
Column used	DB-5, fused silica capillary column (30 m × 0.25 mm ID, film thickness 0.25 μm)	DB-5, fused silica capillary column (30 m × 0.25 mm ID, film thickness 0.25 μm)
Injector temperature	250 °C	250 °C
Oven programming	90 to 150 °C at 15 °C min^{-1}, 150 to 220 °C at 3 °C min^{-1} and 220 to 270 °C at 5 °C min^{-1}	90 to 150 °C at 15 °C min^{-1}, 150 to 220 °C at 3 °C min^{-1} and 220 to 270 °C at 5 °C min^{-1}
Detector temperature	280 °C	–
Carrier gas	Helium at 1.5 ml min^{-1}	Helium at 1.5 ml min^{-1}
Makeup gas	Nitrogen at 40 ml min^{-1}	
MS transfer line temperature	–	280 °C
Ionisation source temperature	–	200 °C

Table 2. Standardisation of OCP's compounds using GC-ECD and GC-MS.

Compound	RT (min)	Recovery (%)	R^2	MDL[a] (ng l^{-1})	Selected Ions (m/z)
α-HCH (H1)	12.28	71.28	0.9994	0.01	
β-HCH (H2)	13.52	79.42	0.9984	0.01	
γ-HCH (H3)	13.74	70.99	0.999	0.01	
δ-HCH (H4)	14.97	78.56	0.9985	0.01	**181**, 183, 109, 288*
Heptachlor (He1)	17.22	87.53	0.9956	0.01	**100**, 272, 274, 370*
Aldrin (A1)	18.99	146.78	0.9988	0.01	**66**, 263, 79, 362*
Hepta-Epoxide (He2)	21.21	59.55	0.9987	0.01	**81**, 353, 355, 386*
α-Endo (E1)	23.15	107.97	0.9993	0.01	**195**, 197, 241, 404*
Dieldrin (A2)	24.6	123.87	0.9993	0.01	**79**, 81, 263, 378*
4,4′-DDE (D1)	24.8	86.16	0.999	0.01	**246**, 318, 316*
Endrin (A3)	25.79	87.70	0.9983	0.01	**81**, 79, 263, 378*
β-Endo (E2)	26.4	90.85	0.9978	0.01	**195**, 197, 241, 404*
4,4′-DDD (D2)	27.18	87.47	0.9972	0.01	**235**, 237, 165, 318*
Endrin-aldehyde (A4)	27.47	133.12	0.9951	0.01	**67**, 345, 250, 378*
Endo-Sulfate (E3)	28.81	85.80	0.9948	0.01	272, **387**, 420*
4,4′-DDT (D3)	29.24	99.22	0.9895	0.01	**235**, 237, 352*
Methoxychlor (M1)	32.28	95.48	0.9868	0.01	**227**, 228, 344*

MDL = methods detection limit
The ions were used for confirmation purpose only; the bold ions represent the base peak ion while the ion with a asterisk shows the molecular peak ion.

to groundwater and between shallow and deeper groundwater in this area, which is likely to be a consequence of heavy pumping in the well-field leading to high recharge rates from the river and a strong vertical flow component. Water quality improved in the post-monsoon season for all the parameters (Table 4), except for NO_3, which has slightly increased in the post-monsoon season. It may be due to recharge of water from surface run-off from agricultural field. The borewell water quality was much superior to Ranney well water and was less prone to seasonal change (Table 4).

Both, borewell (TW-1 to TW-21, grid wise) and Ranney well (RW-1 to RW-5) samples were analyzed for the presence of OCP's (Fig. 1, grid). One sample from each grid (500 × 500 m) of the well-field was taken. One sample may represent the whole grid since OCP's contamination may not vary within the grid. In these grids, OCP's are contributed by the anthropogenic activities only. Data revealed the trace levels of OCP's in the ground water. Tables 5 and 6 show levels of OCPs detected. In pre-monsoon, only 2 samples (TW-6, TW-9) out of 21 samples did not contain any of 17

Table 3. Physico-chemical properties of investigated organochlorine pesticides.

OCP	CAS No	Molecular weight	Molecular formula	Density ($g\,cm^{-3}$)	Melting Point (°C)	Boiling Point (°C)	Water Solubility ($mg\,l^{-1}$)	Log K_{OC}	Log K_{OW}
α-HCH	319-84-6	290.8	$C_6H_6Cl_6$	1.87	159	288	0.02–2	3.3–3.6	3.8–5.8
β-HCH	319-85-7	290.8	$C_6H_6Cl_6$	1.89	314	60	0.05–0.2	3.6	3.78
γ-HCH	58-89-9	290.8	$C_6H_6Cl_6$	1.85	112.5	323.4	0.7–5.59	3.57	3.6–3.7
δ-HCH	319-86-8	290.8	$C_6H_6Cl_6$	NA	141.5	60	0.2–259	3.8	4.14
Heptachlor	76-44-8	373.3	$C_{10}H_5Cl_7$	1.58	135–145	95-96	0.002–0.18	4.4	5.2–6.1
Aldrin	309-00-2	364.9	$C_{12}H_8Cl_6$	1.7	105	385	0.002–0.078	4.68	5.7–7.4
Hepta-Epoxide	1024-57-3	389.3	$C_{10}H_5Cl_7O$	1.58	161	425	0.02–0.275	4.48	4.6–5.4
α-Endo	959-98-8	406.9	$C_9H_6Cl_6O_3S$	1.94	108	NA	0.05–1.73	4.2	3.8–4.9
Dieldrin	60-57-1	380.9	$C_{12}H_8Cl_6O$	1.75	176	385	0.002–0.19	5.28	4.5–6.2
DDE	72-55-9	318	$C_{14}H_8Cl_4$	1.51	89	316	0.065	5.04	5.9–6.9
Endrin	72-20-8	380.9	$C_{12}H_8Cl_6O$	1.7	240	416	0.026–0.25	3.97	4.7–5.2
β-Endo	33213-65-9	406.9	$C_9H_6Cl_6O_3S$	1.7	106	NA	0.2–1.6	4.2	3.6–4.8
DDD	72-54-8	321	$C_{14}H_{10}Cl_4$	1.385	59.8	376	0.016–0.09	4.9	4.7–6.3
Endrin Aldehyde	7421-93-4	382.9	$C_{12}H_8Cl_6O$	NA	163	340	0.016–0.26	6.33	6.44
Endo-Sulfate	1031-07-8	422.9	$C_9H_6Cl_6O_4S$	NA	181	481	0.001–0.1	5.9	6.1
DDT	50-29-3	354.4	$C_{14}H_9Cl_5$	1.6	109	260	0.003	5.2	5.4–6.9
Methoxychlor	72-43-5	345.6	$C_{16}H_{15}Cl_3O_2$	1.39	86–96	436	0.001–0.045	4.89	5.66

NA = Not available

Table 4. Water quality of ground water of the Palla well field.

Parameters	Pre-monsoon Season Borewell			Pre-monsoon Season Ranney Well			Post-monsoon season Borewell			Post-monsoon season Ranney Well			IS Standards
	Ranges	Avg.	SD	Ranges	Avg.	SD	Ranges	Avg.	SD	Ranges	Avg.	SD	
pH	6.7–7.5	7.1	0.2	6.7–7.3	6.9	0.2	7.2–7.9	7.5	0.2	7.2–7.4	7.3	0.1	6.5–8.5
EC ($\mu S\,cm^{-1}$)	87–1770	725.3	309.4	825–4030	1934.2	1451.7	440–1570	728.6	248.5	847–2330	1210.6	470	1500
Sulfate ($mg\,l^{-1}$)	43.6–187.7	83.8	42.4	59.7–255.8	160.5	80.8	7–199.8	54.	43.9	39.2–255.7	93.9	94.7	200–400
Cl^- ($mg\,l^{-1}$)	38–232.3	86.4	38.3	72.8–1063.5	369.3	425.7	32–503	129.2	113.5	64–950	292.6	414.3	250–1000
Nitrate ($mg\,l^{-1}$)	0.4–0.9	0.6	0.1	0.4–7.9	2.2	3.2	2.1–30	6.9	5.4	3.6–7.9	3.5	1.6	<45
F^- ($mg\,l^{-1}$)	0.3–1.8	0.8	0.4	0.2–0.6	0.4	0.2	0.1–1.7	0.6	0.3	0.2–0.4	0.3	0.1	1–1.5
Na^+ ($mg\,l^{-1}$)	28–122	54.1	20.3	44–491.5	223.1	227.0	7–199.8	70.7	43.9	35.4–491.5	119.6	192.2	–
K^+ ($mg\,l^{-1}$)	2–31	3.7	4.8	3–8.1	5.0	2.0	1.5–8.4	7.3	1.2	4.7–8.1	9.6	1.3	–
Hardness ($mg\,l^{-1}$)	184–792	353.7	178.8	252–960	653.6	308.5	180–584	260.1	73	337–484	400	53.3	300–600
Alkalinity ($mg\,l^{-1}$)	176–356	241.8	51.0	244–320	290.4	34.6	180–388	259.2	56.8	284–336	300	21.8	200–600

targeted OCP's (Fig. 2). Rest of the samples contained 2 or more pesticides. Residues of all 17 targeted OCP's were found in 3 samples only. The most frequently occurring pesticides in this sampling campaign were aldrin, followed by ∑HCH, endosulfan and degradation product of DDT. The total numbers of pesticide occurrences were more in Ranney well, with an average of 11 OCP's per sample compared to the 7 OCP's in the borewell. None of the Ranney well sample was free from all the pesticides (Fig. 3). Concentrations of ∑HCH in the borewell water varied from not detected (ND) to $0.269\,\mu g\,l^{-1}$, with maximum of β-HCH being $0.201\,\mu g\,l^{-1}$ as individual HCH. The aldrin residues were detected in the range of ND to $0.047\,\mu g\,l^{-1}$, while dieldrin, endrin and endrin aldehyde were found in same range ND to $0.094\,\mu g\,l^{-1}$ but the frequency of occurrences was 27 % as compare to aldrin showing 76 % occurrences. During the sampling, it was noticed that the area has problem of mites and root worms. High occurrence of aldrin could be due to the use of pesticide to control the mites in the past. In case of Ranney wells, the most frequently occurring contaminants were ∑HCH, aldrin, DDE (frequency = 100 %), followed by β-endosulfan and methoxychlore with 80 % occurrence frequency. The maximum residue of single pesticide detected were similar to borewell water i.e. β-HCH, ($0.206\,\mu g\,l^{-1}$). The maximum concentration of endosulfan was $0.095\,\mu g\,l^{-1}$ for β-endosulfan in all the samples. During post-monsoon sampling campaign, more grids reported no pesticides (Fig. 4) and lower occurrences frequency where only 1 sample showed the presences of all the targeted OCP (Table 7) while in pre-monsoon periods 3 samples showed presence of all OCP's (Table 5). All the banned pesticides, viz ∑HCH, aldrin, heptachlor, DDT, DDD and DDE

Table 5. Organochlorine residues ($\mu g\, l^{-1}$) in ground water (bore well) of Palla area during pre-monsoon seasons.

	α-HCH	β-HCH	γ-HCH	δ-HCH	Hepta chlor	Aldrin	Hepta-Epoxide	α-Endo	Di-eldrin	DDE	Endrin	β-Endo	DDD	Endrin-aldehyde	Endo-Sulfate	DDT	Methoxy Chlor
TW-1	ND	0.004	ND	ND	ND	0.012	ND	0.024	ND	0.017	ND	ND	ND	ND	ND	0.008	ND
TW-2	0.017	0.045	0.057	0.018	0.018	0.047	ND	ND	ND	ND	ND	ND	ND	ND	ND	0.035	ND
TW-3	0.011	0.024	0.036	0.007	0.007	0.029	ND	ND	ND	ND	ND	ND	ND	ND	ND	0.015	ND
TW-4	ND	ND	0.012	ND	ND	0.010	ND	ND	ND	ND	ND	0.013	ND	ND	ND	ND	ND
TW-5	ND	ND	ND	ND	ND	0.022	ND	ND	ND	ND	ND	0.009	ND	ND	ND	ND	0.041
TW-6	ND	ND	ND	ND	ND	ND	ND	ND	ND	ND	ND	ND	ND	ND	ND	ND	ND
TW-7	0.014	0.079	0.055	0.078	0.022	0.009	0.019	0.020	0.066	0.066	0.041	0.024	0.064	0.052	0.074	0.082	0.089
TW-8	ND	ND	ND	ND	ND	ND	ND	ND	ND	ND	0.058	ND	0.030	ND	ND	ND	ND
TW-9	ND	ND	ND	ND	ND	ND	ND	ND	ND	ND	ND	ND	ND	ND	ND	ND	ND
TW-10	0.012	0.018	0.028	0.012	ND	0.033	ND	ND	0.009	0.009	ND	ND	ND	0.044	ND	ND	ND
TW-11	0.015	0.089	0.055	0.078	0.017	0.009	0.021	0.020	0.075	0.033	0.048	0.045	0.033	0.047	0.044	0.052	0.066
TW-12	0.005	0.014	0.013	ND	ND	0.015	ND	ND	ND	0.009	ND	ND	0.011	ND	ND	ND	ND
TW-13	0.004	0.008	0.013	0.005	0.022	0.016	0.012	ND	ND	0.014	ND	0.030	0.007	ND	ND	0.018	0.050
TW-14	0.015	0.201	0.041	0.012	0.017	0.015	0.009	0.015	ND	0.010	ND	0.020	0.013	0.059	ND	ND	0.034
TW-15	0.018	0.033	0.061	0.022	0.019	0.042	ND	ND	ND	0.010	0.018	0.050	0.012	0.069	ND	ND	ND
TW-16	ND	0.009	ND	ND	ND	ND	ND	ND	ND	0.024	ND	ND	ND	ND	ND	0.014	ND
TW-17	ND	ND	ND	ND	ND	ND	ND	0.012	ND	ND	ND	0.017	ND	ND	ND	ND	ND
TW-18	0.007	0.017	ND	0.012	0.022	0.013	0.012	ND	ND	0.008	ND	0.029	ND	ND	ND	ND	0.075
TW-19	0.008	0.016	0.025	0.010	0.016	0.016	0.008	ND	ND	0.011	ND	ND	0.014	ND	ND	0.037	0.034
TW-20	ND	ND	ND	ND	ND	0.014	ND	ND	0.015	ND	0.021	0.016	ND	ND	ND	ND	ND
TW-21	0.012	0.034	0.048	0.015	0.018	0.030	0.009	0.019	0.018	0.013	0.021	0.094	0.016	0.053	0.020	0.035	0.035

Table 6. Organochlorine residues ($\mu g\, l^{-1}$) in ground water (Ranny well) of Palla area during pre-monsoon seasons.

	α-HCH	β-HCH	γ-HCH	δ-HCH	Hepta chlor	Aldrin	Hepta-Epoxide	α-Endo	Di-eldrin	DDE	Endrin	β-Endo	DDD	Endrin-aldehyde	Endo-Sulfate	DDT	Methoxy Chlor
RW-1	0.003	0.012	0.016	0.008	ND	0.013	ND	ND	ND	0.009	ND	0.015	ND	ND	ND	ND	ND
RW-2	0.006	0.017	0.027	0.011	ND	0.030	ND	ND	0.010	0.009	ND	0.020	ND	0.035	ND	ND	0.033
RW-3	0.012	0.022	0.035	0.015	0.015	0.021	0.008	ND	ND	0.009	ND	0.022	0.012	ND	ND	ND	0.034
RW-4	0.014	0.206	0.032	0.012	0.017	0.015	0.009	0.015	ND	0.011	ND	ND	0.014	ND	ND	0.037	0.035
RW-5	0.016	0.099	0.069	0.100	0.027	0.030	0.041	0.067	0.115	0.073	0.083	0.095	0.073	0.087	0.066	0.072	0.070

were below their regulatory concentrations in the borewells. α-endosulfan was present only in one sample ($0.034\, \mu g\, l^{-1}$), which shows the wise use of highly controversial pesticide, endosulfan in Yamuna catchment area. In case of Ranney well samples, the trends of occurrences were similar to the pre-monsoon sampling campaign; but the levels were lower (Table 8, Fig. 5). The student's t-test was carried out for pre-monsoon and post-monsoon samples taking all the samples as independent variables. It showed that there was no significant difference in total pesticide concentration in both the season ($P > 0.05$). Thus, it can be inferred that during whole year people receiving water supply from this well-field are at high risk of pesticide exposure. Though, levels in post-monsoon season were lower than safe limit, but were not statistically different from pre-monsoon season. However, concentration of individual pesticides for both the seasons was statistically different ($P < 0.01$) for borewells and Ranney wells. This indicates that concentrations of individual pesticides decreased in post-monsoon season. It can be attributed to the dilution due to recharge of the aquifer during monsoon. Figures 6 and 7 indicate the total pesticides concentration in individual samples. The water quality of borewells was superior to the Ranney well as none of the borewell sample was found to contain the OCP residues level more than BIS permissible limit (Figs. 6 and 7). As per BIS drinking water standard, the water should be free from all pesticides, but in case of non availability of other sources, the limit is $1\, \mu g\, l^{-1}$ of total pesticides. Water, other than packaged mineral water, is not an important commodity in international trade, so there is no initiative in place to harmonize these limits at the international level. It varies from place to place, depending on the awareness level and economy of the country. The safe limit for pesticide residues are different in different parts of the world: WHO and EU have a safe limit of $0.5\, \mu g\, l^{-1}$ as compare to BIS limit of $1\, \mu g\, l^{-1}$. Recent reports on the total OCP levels in groundwater of neighbouring areas of Palla (Delhi) have reported much higher concentrations, $2.184\, \mu g\, l^{-1}$ in Haryana (Kaushik et al., 2011) and up to $4\, \mu g\, l^{-1}$ in Lucknow city (Mudiam et al., 2011). This well-field is being recharged with the bank filtrate of river Yamuna (Lorenzen et al., 2010). Both external flux through bank filtrate and local use of pesticides can contribute to the pesti-

Table 7. Organochlorine residues ($\mu g\,l^{-1}$) in ground water (bore well) of Palla area during post-monsoon season.

	α-HCH	β-HCH	γ-HCH	δ-HCH	Hepta chlor	Aldrin	Hepta-Epoxide	α-Endo	Di-eldrin	DDE	Endrin	β-Endo	DDD	Endrin-aldehyde	Endo-Sulfate	DDT	Methoxy Chlor
TW-1	0.004	0.009	ND	ND	ND	0.024	ND	ND	ND	ND	ND	ND	ND	ND	ND	ND	ND
TW-2	ND	ND	ND	ND	ND	ND	ND	ND	ND	ND	ND	ND	ND	0.040	ND	ND	ND
TW-3	ND	ND	ND	ND	ND	0.017	ND	ND	ND	ND	ND	ND	ND	0.039	ND	ND	ND
TW-4	ND	ND	0.007	ND	ND	ND	ND	ND	ND	ND	ND	ND	ND	ND	ND	ND	ND
TW-5	ND	ND	ND	ND	ND	0.019	ND	ND	ND	ND	ND	ND	ND	ND	ND	ND	ND
TW-6	ND	ND	ND	ND	ND	ND	ND	ND	ND	ND	ND	ND	ND	ND	ND	ND	ND
TW-7	0.005	0.037	0.012	0.039	0.032	0.034	0.026	0.034	0.058	0.046	0.046	0.041	0.055	0.069	0.057	0.093	0.094
TW-8	ND	ND	ND	ND	ND	ND	ND	ND	ND	ND	0.054	ND	0.024	ND	ND	ND	ND
TW-9	ND	ND	ND	ND	ND	0.024	ND	ND	ND	ND	ND	ND	ND	ND	ND	ND	ND
TW-10	0.003	ND	ND	ND	ND	0.011	ND	ND	ND	ND	ND	ND	ND	ND	ND	ND	ND
TW-11	ND	ND	ND	ND	ND	0.014	ND	ND	ND	ND	ND	ND	ND	ND	ND	ND	ND
TW-12	ND	ND	ND	ND	ND	ND	ND	ND	0.009	0.018	ND	ND	ND	ND	ND	ND	ND
TW-13	ND	ND	ND	ND	ND	0.018	ND	ND	ND	0.018	ND	ND	0.022	ND	ND	ND	0.066
TW-14	ND	ND	ND	ND	ND	ND	ND	ND	ND	ND	0.029	ND	ND	0.083	ND	ND	ND
TW-15	ND	0.013	0.009	0.014	0.028	0.025	0.017	ND	0.016	0.021	ND	ND	0.025	ND	0.038	0.069	0.069
TW-16	ND	0.008	ND	ND	ND	ND	ND	ND	ND	ND	ND	ND	ND	ND	ND	ND	ND
TW-17	ND	ND	ND	ND	ND	ND	ND	ND	ND	ND	ND	0.031	ND	ND	ND	ND	ND
TW-18	ND	ND	ND	ND	0.028	0.016	0.016	ND	ND	0.018	ND	0.060	ND	ND	ND	ND	0.066
TW-19	0.005	ND	0.014	0.013	ND	0.018	0.016	ND	0.030	0.019	ND	ND	0.024	ND	0.036	0.068	0.068
TW-20	ND	ND	ND	ND	ND	ND	ND	ND	0.010	ND	0.042	ND	ND	ND	ND	ND	ND
TW-21	ND	ND	ND	ND	ND	0.028	ND	ND	ND	ND	ND	0.031	ND	ND	ND	ND	ND

Table 8. Organochlorine residues ($\mu g\,l^{-1}$) in ground water (Ranny well) of Palla area during post-monsoon seasons.

	α-HCH	β-HCH	γ-HCH	δ-HCH	Hepta chlor	Aldrin	Hepta-Epoxide	α-Endo	Di-eldrin	DDE	Endrin	β-Endo	DDD	Endrin-aldehyde	Endo-Sulfate	DDT	Methoxy Chlor
RW-1	0.006	ND	ND	0.013	ND	0.018	0.016	ND	0.010	0.019	ND	0.018	0.024	ND	0.036	0.068	0.068
RW-2	0.005	0.011	0.037	0.022	ND	0.018	ND	ND	ND	ND	ND	0.023	ND	0.015	ND	ND	ND
RW-3	0.005	ND	0.024	ND	ND	0.021	ND	0.024	ND	ND	ND	0.022	0.022	ND	ND	ND	ND
RW-4	0.009	0.015	0.019	ND	ND	ND	ND	ND	ND	0.013	ND	ND	ND	ND	ND	ND	0.068
RW-5	0.008	0.015	0.018	0.017	ND	0.018	0.019	0.044	0.021	0.023	0.030	0.032	0.031	0.018	0.038	0.074	0.070

cides in the aquifer. The presence of α-HCH and γ-HCH reveals the use of technical HCH (55–80 % α-HCH and 8–15 % γ-HCH plus other compounds) and lindane (more than 99 % γ-HCH) in the Yamuna catchment area. Most of the developed countries have banned the use of technical HCH in the 1980s, but in India, it is still under production for export and restricted use, other than agriculture. The α-HCH/γ-HCH ratio within the technical formulation ranges from 4–15 and 0.2–1 in lindane (Ridal et al., 1996). The low ratios of α-HCH/γ-HCH (Table 9) in the ground water samples indicate that lindane may also be an important source of HCHs in Yamuna river watershed. Among the DDTs, 4, 4-DDD and DDE were found to be the predominant compound (Tables 5, 6, 7 and 8). The DDE/DDTs and DDD/DDTs ratios can be used to assess whether the DDT input is of recent origin or from the past (Kannan et al., 1997). Since DDE/DDTs and DDD/DDTs ratios in all the samples were much lower than unity (Table 9), it appears DDT inputs is of recent origin from the Yamuna river catchment area. Total annual production of DDT in India is still more than 3000 t. (UNEP/POPS, 2008). It is used for export and malaria control programme in India, which supports the findings of recent inputs of DDT.

Table 9. Concentration ratios of selected OCP's compounds in the ground water according to the sampling source and seasons.

Sampling sites		α-HCH/γ-HCH	DDE/DDTs	DDD/DDTs	DDE/DDD
Borewell	Pre-monsoon	0.245	0.518	0.269	0.931
	Post-monsoon	0.347	0.692	0.327	1.334
Ranney well	Pre-monsoon	0.292	0.319	0.319	1.137
	Post-monsoon	0.329	0.274	0.2	0.709

The DDE/DDD ratio is indicative of the prevalent redox conditions in the area. While, in oxic conditions, DDE is the main metabolite of DDT (Wolfe et al., 1977), in anoxic conditions, DDD is the major degradation product (Zoro et al., 1974). Here, the DDE/DDD ratio varies from 0.709–1.334, more than unity is indicative of the less reducing conditions or an extended degradation during the long- range transport of DDT. As the major share of the ground water is replenished by river Yamuna and the river is in losing condition (Lorenzen et al., 2010) in this area, the background level of OCP present in river water could also reach into

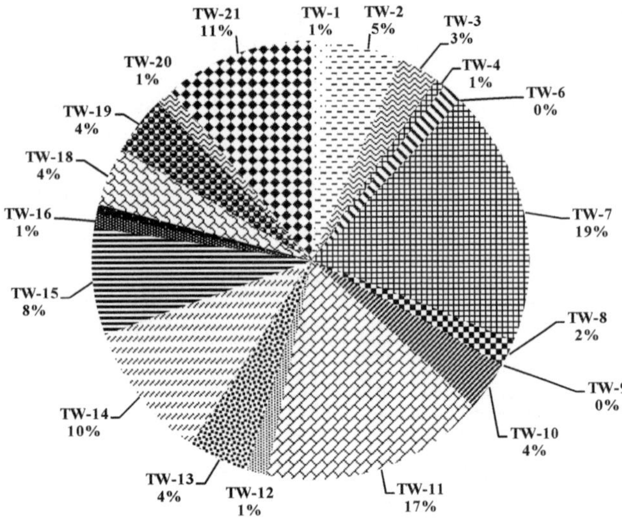

Figure 2. Relative abundance of OCP in different borewell samples (Pre-monsoon).

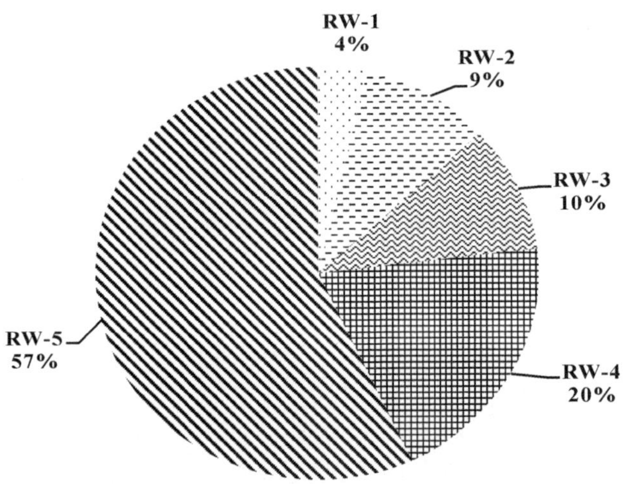

Figure 3. Relative abundance of OCP in different Ranney well samples (Pre-monsoon).

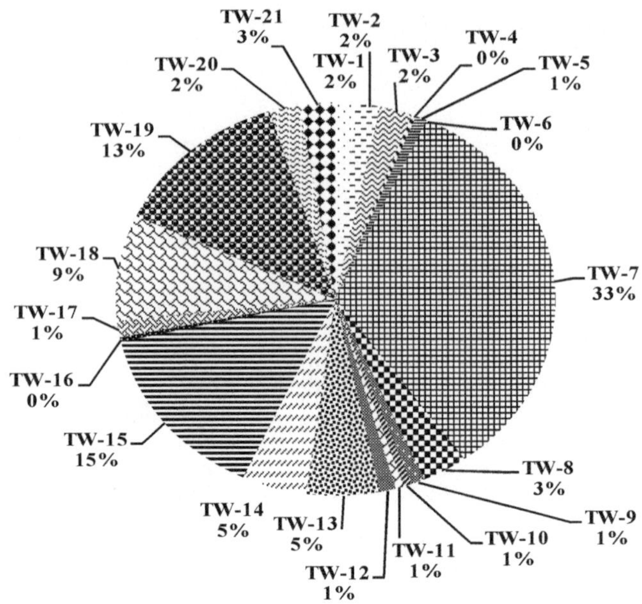

Figure 4. Relative abundance of OCP in different borewell samples (Post-monsoon).

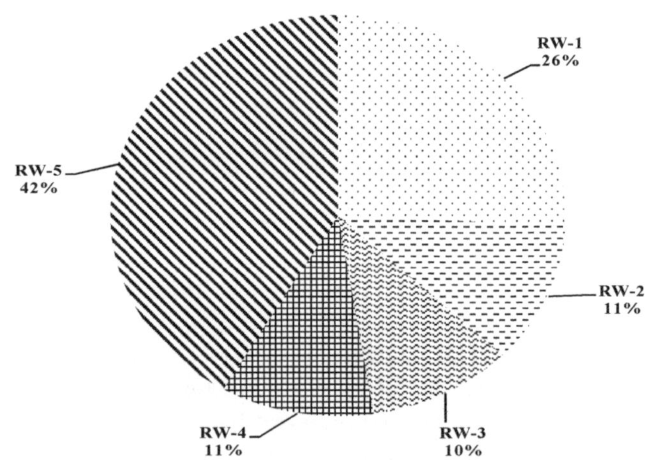

Figure 5. Relative abundance of OCP in different Ranney well samples (Post-monsoon).

the aquifers. Kaushik et al. (2008), monitored the levels of OCP's in the river Yamuna and reported $\sum$HCH 0.012–0.593 $\mu g\,l^{-1}$, and $\sum DDT = 0.066–0.723\,\mu g\,l^{-1}$ in surface water of river Yamuna, in upstream of this well-field. Levels of these pesticides in the aquifer of this well-field were much lesser. It may be because of dilution or adsorption of organic compounds by soil strata during recharge. The high affinity to soil organic matter (high K_{OC}) make these pesticides strongly bound to soil organic matter and thus they are less leached in the aquifer while recharging the well-field. The levels of OCP residues were higher for Ranney wells than borewells (Tables 5, 6, 7 and 8) in all the cases. Ranney well have much higher hydraulic gradient as compared to borewells. Thus, higher gradient may result in higher desorption of pesticides from solid phase to aquifer or lower transfer of pesticides from aquifer to soil particles.

4 Conclusions

There is wide prevalence of OCPs throughout the well-field though at low concentrations. During monsoon high flood levels results in recharge of the well-field. But, there is no statistical difference between pre-monsoon and post-monsoon periods. The borewell water quality was superior to the Ranney well. So it is advantageous to develop a well-field with borewells rather using Ranney wells. Although the levels of pesticides were lower than the local regulatory

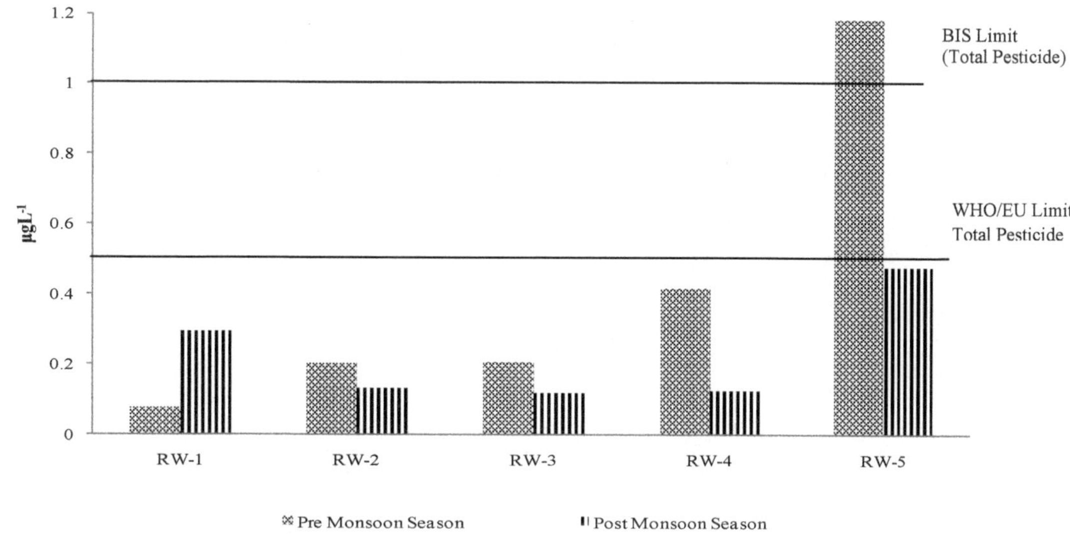

Figure 6. Comparison of Ranney well water quality with respect to various limits.

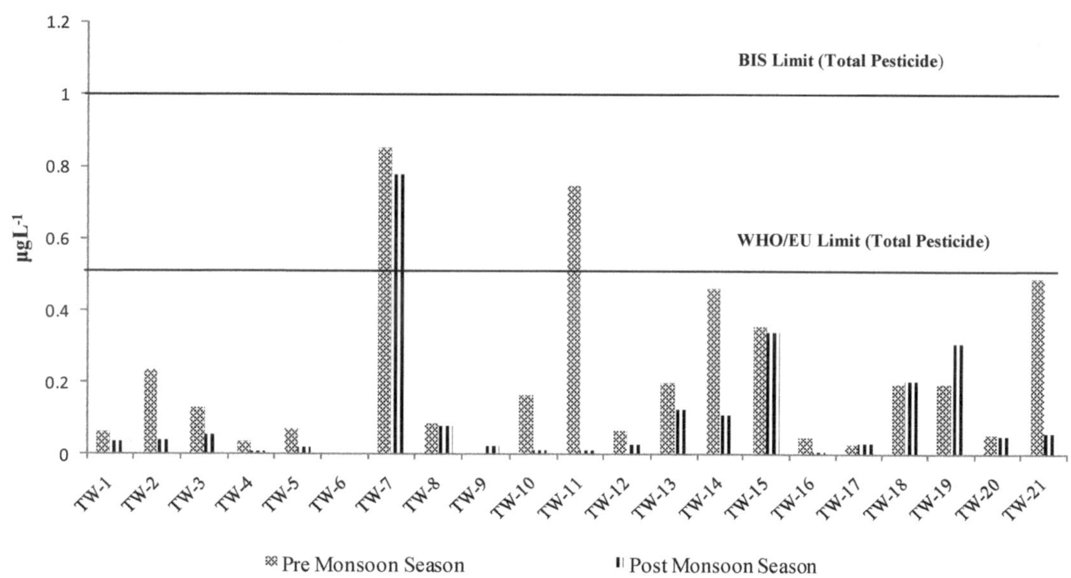

Figure 7. Comparison of borewell water quality with respect to various limits.

limits but regular consumption of such contaminated water may pose serious health hazard to the people who depend for the water on this well-field. In order to rule out any possibility of health risk by OCPs, regular monitoring of pesticide residues in river water and ground water is required. Delhi is a rapid growing urban centre, so there is a great political and social pressure to utilize this well-field to supply water for the city of Delhi since this well-field produces 60–76 million m^3 day^{-1} potable water in all the seasons. There is need for comprehensive investigations to monitor the breakdown products of the pesticides in order to get better idea of the overall risk due to well contamination (since breakdown products are often toxic), and the mechanism causing wells to be contaminated.

Acknowledgements. One of the authors (Pravin K. Mutiyar) is thankful to University Grants Commission (UGC), New Delhi (India) for providing financial assistance in the form of Junior Research Fellowship (JRF) to conduct this work.

References

American Public Health Association (APHA): Standard Methods for the Examination of Water and Wastewater, 20th edn., USA, 1998.

Bajpai, A., Shukla, P., Dixit, B., and Banerji, R.: Concentrations of organochlorine insecticides in edible oils from different regions of India, Chemosphere, 67, 1403–1407, 2007.

Bakore, N., John, P., and Bhatnagar, P.: Organochlorine pesticide residues in wheat and drinking water samples from Jaipur, Rajasthan, India, Environ. Monit. Assess., 98, 381–389, 2004.

Barrie, L. A., Gregor, D., Hargrave, B., Lake, R., Muir, D., Shearer, R., Tracey, B., and Bidleman, T.: Arctic contaminants: Sources, occurrence and pathways, Sci. Total Environ., 122, 1–74, 1992.

Bizzotto, E., Villa, S., Vaj, C., and Vighi, M.: Comparison of glacial and non-glacial-fed streams to evaluate the loading of persistent organic pollutants through seasonal snow/ice melt, Chemosphere, 74, 924–930, 2009.

Blais, J. M., Macdonald, R. W., Mackay, D., Webster, E., Harvey, C., and Smol, J. P.: Biologically mediated transport of contaminants to aquatic systems, Environ. Sci. Technol., 41, 1075–1084, 2007.

Dikshith, T., Raizada, R., Kumar, S., Srivastava, M., Kulshrestha, S., and Adholia, U.: Residues of DDT and HCH in major sources of drinking water in Bhopal, India, Bull. Environ. Contam. Toxicol., 45, 389–393, 1990.

Dua, V., Pant, C., and Sharma, V.: HCH and DDT in surface extractable skin lipid as a measure of human exposure in India, Bull. Environ. Contam. Toxicol., 60, 238–244, 1998.

Geisz, H. N., Dickhut, R. M., Cochran, M. A., Fraser, W. R., and Ducklow, H. W.: Melting glaciers: A probable source of DDT to the Antarctic marine ecosystem, Environ. Sci. Technol., 42, 3958–3962, 2008.

Hassan, A. A. M., Benfenati, E., and Fanelli, R.: Detection and quanification of trihalomethanes in drinking water from Alexandria, Egypt, Bull. Environ. Contam. Toxicol., 56, 397–404, 1996.

Jani, J., Raiyani, C., Mistry, J., Patel, J., Desai, N., and Kashyap, S.: Residues of organochlorine pesticides and polycyclic aromatic hydrocarbons in drinking water of Ahmedabad city, India, Bull. Environ. Contam. Toxicol., 47, 381–385, 1991.

John, P., Bakore, N., and Bhatnagar, P.: Assessment of organochlorine pesticide residue levels in dairy milk and buffalo milk from Jaipur city, Rajasthan, India, Environ. Int., 26, 231–236, 2001.

Kannan, K., Tanabe, S., Giesy, J. P., and Tatsukawa, R.: Organochlorine pesticides and polychlorinated biphenyls in foodstuffs from asian and oceanic countries, Rev. Environ. Contam. Toxicol., 152, 1–56, 1997.

Kaushik, C., Sharma, H., Jain, S., Dawra, J., and Kaushik, A.: Pesticide residues in river Yamuna and its canals in Haryana and Delhi, India, Environ. Monit. Assess., 144, 329–340, 2008.

Kaushik, C., Sharma, H., and Kaushik, A.: Organochlorine pesticide residues in drinking water in the rural areas of Haryana, India, Environ. Monit. Assess., doi:10.1007/s10661-011-1950-9, in press, 2011.

Kumar, A., Dayal, P., Shukla, G., Singh, G., and Joseph, P.: DDT and HCH residue load in mother's breast milk: A survey of lactating mother's from remote villages in Agra region, Environ. Int., 32, 248–251, 2006a.

Kumar, M., Ramanathan, A., Rao, M., and Kumar, B.: Identification and evaluation of hydrogeochemical processes in the groundwater environment of Delhi, India, Environ. Geol., 50, 1025–1039, 2006b.

Lorenzen, G., Sprenger, C., Taute, T., Pekdeger, A., Mittal, A., and Massmann, G.: Assessment of the potential for bank filtration in a water-stressed megacity (Delhi, India), Environ. Earth Sci., 61, 1419–1433, 2010.

Mudiam, M. K. R., Pathak, S., Gopal, K., and Murthy, R.: Studies on urban drinking water quality in a tropical zone, Environ. Monit. Assess., doi:10.1007/s10661-011-1980-3, in press, 2011.

Mukherjee, I. and Gopal, M.: Insecticide residues in baby food, animal feed, and vegetables by gas liquid chromatography, Bull. Environ. Contam. Toxicol., 56, 381–388, 1996.

Mukherjee, I. and Gopal, M.: Organochlorine insecticide residues in drinking and ground water in and around Delhi, Environ. Monit. Assess., 76, 185–193, 2002.

Ridal, J. J., Kerman, B., Durham, L., and Fox, M. E.: Seasonality of air-water fluxes of hexachlorocyclohexanes in Lake Ontario, Environ. Sci. Technol., 30, 852–858, 1996.

Sankararamakrishnan, N., Kumar Sharma, A., and Sanghi, R.: Organochlorine and organophosphorous pesticide residues in ground water and surface waters of Kanpur, Uttar Pradesh, India, Environ. Int., 31, 113–120, 2005.

Sharma, H., Kaushik, A., and Kaushik, C.: Pesticide residues in bovine milk from a predominantly agricultural state of Haryana, India, Environ. Monit. Assess., 129, 349–357, 2007.

Shekhar, S. and Prasad, R.: The groundwater in the Yamuna flood plain of Delhi (India) and the management options, Hydrogeol. J., 17, 1557–1560, 2009.

Shukla, G., Kumar, A., Bhanti, M., Joseph, P., and Taneja, A.: Organochlorine pesticide contamination of ground water in the city of Hyderabad, Environ. Int., 32, 244–247, 2006.

UNEP/POPS: Global status of DDT and its alternatives for use in vector control to prevent disease, 2008.

Villa, S., Vighi, M., Maggi, V., Finizio, A., and Bolzacchini, E.: Historical trends of organochlorine pesticides in an Alpine glacier, J. Atmos. Chem., 46, 295–311, 2003.

Waliszewski, S., Aguirre, A., Infanzon, R., Silva, C., and Siliceo, J.: Organochlorine pesticide levels in maternal adipose tissue, maternal blood serum, umbilical blood serum, and milk from inhabitants of Veracruz, Mexico, Arch. Environ. Contam. Toxicol., 40, 432–438, 2001.

Wang, X., Xu, B., Kang, S., Cong, Z., and Yao, T.: The historical residue trends of DDT, hexachlorocyclohexanes and polycyclic aromatic hydrocarbons in an ice core from Mt. Everest, central Himalayas, China, Atmos. Environ., 42, 6699–6709, 2008.

Wolfe, N. L., Zepp, R. G., Paris, D. F., Baughman, G. L., and Hollis, R. C.: Methoxychlor and DDT degradation in water: Rates and products, Environ. Sci. Technol., 11, 1077–1081, 1977.

Zoro, J. A., Hunter, J. M., Eglinton, G., and Ware, G. C.: Degradation of p, p[prime]-DDT in reducing environments, Nature, 247, 235–237, 1974.

7

Nanofiltration for water and wastewater treatment

H. K. Shon[1], S. Phuntsho[1], D. S. Chaudhary[1], S. Vigneswaran[1], and J. Cho[2]

[1]School of Civil and Environmental Engineering, University of Technology, Sydney (UTS), P.O. Box 129, Broadway, NSW 2007, Australia

[2]Department of Civil and Environmental Engineering, Yonsei University, Yonsei-ro 50, Seodaemun-gu, Seoul 120-749, Korea

Correspondence to: H. K. Shon (hokyong.shon-1@uts.edu.au)

Abstract. The application of membrane technology in water and wastewater treatment is increasing due to stringent water quality standards. Nanofiltration (NF) is one of the widely used membrane processes for water and wastewater treatment in addition to other applications such as desalination. NF has replaced reverse osmosis (RO) membranes in many applications due to lower energy consumption and higher flux rates. This paper briefly reviews the application of NF for water and wastewater treatment including fundamentals, mechanisms, fouling challenges and their controls.

1 Introduction

Membrane filtration is a pressure driven process in which membrane acts as selective barriers to restrict the passage of pollutants such as organics, nutrients, turbidity, microorganisms, inorganic metal ions and other oxygen depleting pollutants, and allows relatively clear water to pass through (Mulder, 1997). With technological advances and the ever-increasing stringency of water quality criteria, membrane processes are becoming a more attractive solution to the challenge of quality water, and water reuse (Shannon et al., 2008). Several studies have been done on the application of microfiltration/ultrafiltration for wastewater treatment and reuse (Vigneswaran et al., 1991; Seo et al., 1996, 1997; Snoeyink et al., 2000; Visvanathan et al., 2000; Ben Aim and Semmens, 2001; Kim et al., 2001; Matsui et al., 2001a, b).

The membrane process has been classified into four broad categories as depending on their pore sizes as: microfiltration (MF), ultrafiltration (UF), nanofiltration (NF) and reverse osmosis (RO) membranes. This paper briefly reviews the application of NF for water and wastewater treatment including for water reuse. This article begins with the brief fundamentals of membrane process followed by discussion on the mechanisms of NF process and its few basic models. The article also covers the challenges of NF fouling and their control mechanisms adopted to mitigate fouling. Finally the article concludes with a brief summary.

2 Fundamentals of membrane process

There are many types of membrane processes in use. RO membrane is essentially non-porous, and it preferentially passes liquid and retains most of the solutes including ions. The RO is characterized by high operating pressure (20 to 100 bar). NF has pore size 1–5 nm and it can retains ions, and low molecular weight organics. It has significantly higher water permeability than that of RO membrane and operates at lower pressure (typically 7 to 30 bar). Similarly, UF membrane has pore size typically 5 to 20 nm and retains fine colloids, macromolecules, and microorganism. The UF operates with pressure range of 1 to 10 bar. The other membrane processes that are used in liquid separation process are microfiltration (MF), electrodialysis (ED), liquid membrane (LM), pervaporation (PV), vapour permeation (VP), and gas permeation (GP). The types of membrane processes, the particle size typically removed by the membrane, and the driving force of the processes are illustrated in Fig. 1. The major

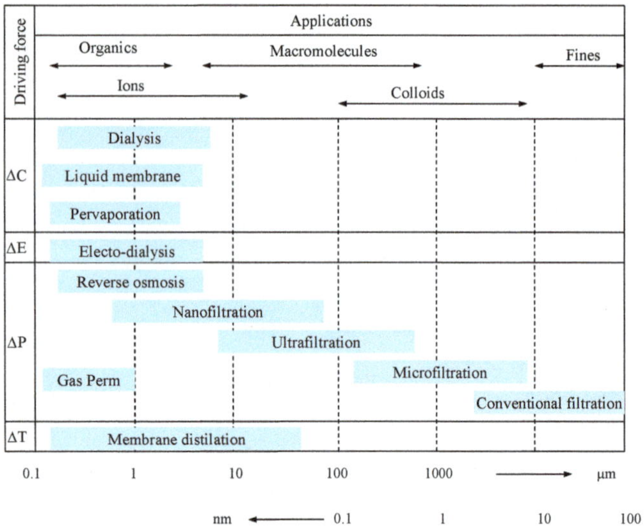

Figure 1. Effective range of membrane processes.

difference between these membrane processes is shown in Table 1.

3 Nanofiltration

NF is the most recently developed pressure-driven membrane process for liquid-phase separations. NF has replaced reverse osmosis (RO) in many applications due to lower energy consumption and higher flux rates (Cadotte et al., 1988; Gozalvez et al., 2002). The properties of NF membranes lie between those of non-porous RO membranes (where transport is governed by a solution-diffusion mechanism) and porous ultrafiltration (UF) membranes (where separation is usually assumed to be due to size exclusion and, in some cases, charge effects). Commercial NF membranes possess a fixed charge developed by dissociation of surface groups such as sulphonated or carboxyl acids. The properties of NF membranes, therefore, allow ions to be separated by a combination of the size and electrical effects of UF and the ion interaction mechanisms of RO (Bowen and Welfoot, 2002).

The NF membrane is the relatively newly introduced technology in wastewater treatment system. The size of pores in NF membranes (nominally $\sim$ 1 nm) is such that even small uncharged solutes are highly rejected while the surface electrostatic properties allow monovalent ions to be reasonably well transmitted with multivalent ions mostly retained. These characteristics make NF membranes extremely useful in the fractionation and selective removal of solutes from complex process streams. The development of NF technology as a viable process over recent years has led to a remarkable increase in its application in a number of industries such as treatment of pulp-bleaching effluents from the textile industry, separation of pharmaceuticals from fermentation broths, demineralization in the dairy industry, and metal recovery from wastewater and virus removal (Bowen et al., 2002).

NF is one of the promising technologies for the treatment of natural organic matter and inorganic pollutants in surface water. Since the surface water has low osmotic pressure, a low-pressure operation of NF is possible. There is a high rejection of organic substances such as disinfection-by-products precursors by the NF process. In the NF of surface waters, natural organic compounds, which have relatively large molecules compared to membrane pore size, could be removed by sieving mechanism, whereas the inorganic salts by the charge effect of the membranes and ions (Thanuttamavong et al., 2001, 2002). The past studies on NF are summarized in Table 2 (Ernst et al., 2000; Xu and Lebrun, 1999; Tsuru et al., 2000; Seidel and Elimelech, 2002; Van der Bruggen et al., 2002; Lee et al., 2002; Choi et al., 2002; Trebouet et al., 2001).

4 Separation mechanisms in NF

Since NF membrane exhibits properties between those of ultrafiltration (UF) and reverse osmosis (RO), both charge and size of particle play important role in NF rejection mechanism. Simpson et al. (1987) has described NF as a charged UF system whereas Rohe et al. (1990) has referred it as low pressure RO system. However, NF has advantages of lower operating pressure compared to RO, and higher organic rejection compared to UF. For the colloids and large molecules, physical sieving would be the dominant rejection mechanism whereas for the ions and lower molecular weight substances, solution diffusion mechanism and charge effect of membrane play the major role in separation process. Macoun (1998) presented the NF rejection mechanisms into following five steps.

- Wetted surface – water associates with the membrane through hydrogen bonding and the molecules which form the hydrogen bonding with the membrane can be transported.

- Preferential sorption/Capillary rejection – membrane is heterogeneous and microporous, and electrostatic repulsion occurs due to different electrostatic constants of solution and membrane.

- Solution diffusion – membrane is homogeneous and non-porous, and solute and solvent dissolve in the active layer of the membrane and the transport of the solvent occurs due to the diffusion through the layer.

- Charged capillary – electric double layer in the pores determines rejection. Ions of same charge as that of membrane are attracted and counter-ions are rejected due to the streaming potential.

Table 1. Difference between MF, UF, NF, and RO.

Particular	MF	UF	NF	RO
Membrane	Porous isotropic	Porous asymmetric	Finely porous asymmetric/composite	Nonporous asymmetric/composite
Pore size	50 nm–1 μm	5–20 nm	1–5 nm	–
Transfer mechanism	Sieving and adsorptive mechanisms (the solutes migrate by convection)	Sieving and preferential adsorption	Sieving/electrostatic hydration/diffusive	Diffusive (solutes migrate by diffusion mechanism)
Law governing transfer	Darcy's law	Darcy's law	Fick's law	Fick's law
Typical solution treatment	Solution with solid particles	Solution with colloids and/or macromolecules	Ions, small molecules	Ions, small molecules
Typical pure water flux ($L\,m^{-2}\,h$)	500–10 000	100–2000	20–200	10–100
Pressure requirement (atoms)	0.5–5	1–10	7–30	20–100

Table 2. Past studies of nanofiltration.

Membrane	Pretreatment/Feed solution	Remarks
– Four types of flat sheet membranes: DK5 (polysulfone + polypiperazinamide), PES10 (polyethersulfone), C5F (cellulose), and MP 36 (modified PVDFK) with MWCO 200, 1000, 5000, and 1000, respectively	PAC adsorption, Tertiary wastewater effluent for ground water recharge	Cut-off characteristics of membrane play important role to remove smaller molecular weight substances
– Two types of flat sheet membranes: NF70 (negatively charged), NF45 (hydrophilic, pore size 2–5 nm)	No pretreatment, Electrolytic solutions (NaCl, Na_2SO_4, and polyethylene glycol)	The transport of solute through NF depends on sieving mechanism and surface force interaction.
– Three inorganic stirred cell membranes with MCO 200, 600, and 2000.	No pretreatment, different electrolytic solutions	The rejection of solutes decreased but the permeate volume increased with an increase in temperature.
– Thin film composite, flat sheet membranes	No pretreatment, NOM solutions	NOM fouling of NF membranes is governed by the combined effects of initial permeate flux or applied pressure, crossflow velocity and divalent (calcium) ion concentration.
– Flat sheet membrane NF70	No pretreatment, 14 chemical solutions	Flux decline is caused by the molecules that fill the pores of the membrane and adsorption of molecules on the membrane surfaces which is enhanced by the hydrophobicity of the solutes.
– Polyamide TFC flat sheet membrane with MWCO 250 Dalton	No pretreatment, different NOM solutions	Mass transport is more affected by difference in NOM structure than solution chemistry, and is dominated by diffusion.
– Cellulose acetate hollow fibre submerged membrane, aerated.	No pretreatment, synthetic wastewater	Low pressure NF bioreactor can be used for long-term without fatal fouling and cleaning.
– MPT-20 and MPT-31 made of polyster with polyacrilonitrile support and polypropylene with polysulphone support, respectively, MWC 450 Da and negatively charged	Ferric chloride coagulation – filtration (10 μm polypropylene mesh), Landfill leachate	Presence of Fe^{3+} ions may change the surface charges, ionic force of solutions and the structure of membrane surface, and thus may reduce the organic retention capacity of the membrane
– Polyamide, negatively charged, 200 Da MWCO	Millipore filtered (10 μm and 1 μm) feed, Secondary sewage effluent	Fouling due to colloids (such polysaccharides or proteins) are more severe than the hybrophobic and transphilic fractions of organics in the sewage effluent.

– Finely porous – membrane is a dense material punctured by pores. Transport is determined by partitioning between bulk and pore fluid.

The characteristics of NF membranes lies between the non-pores reverse osmosis membranes (where the rejection is due to solution-diffusion mechanism) and porous UF membranes (where the rejection is by size exclusion and electrostatic charge effects). Thus, the rejection of uncharged molecules is dominated by size exclusion, while that of ionic species is influenced by both size exclusion and electrostatic interactions. Electrostatic characteristics of NF membranes have been known as playing an important role in rejection anions, namely, negative zeta potential on the membrane surface varies with different pH and concentration of an electrolyte solution (Choi et al., 2002) .

5 Mathematical modelling of nanofiltration process

NF is a complex phenomenon. The NF membranes exhibit properties between those of RO membranes and UF membranes, and hence the solution-diffusion mechanism, the size exclusion, and charge effects need to be considered in modeling the governing phenomenon of NF process. The basic equation to describe the transport of ions/solutes through the membranes is given by the extended Nernst-Planck equation (Eq. 1).

$$J = D_p \frac{dc}{dx} - \frac{zcD_p}{RT} F \frac{d\psi}{dx} + K_c c V \quad (1)$$

where, J = Ion flux based on membrane area (mol m^{-2} s^{-1}), D_p = Hindered diffusivity (m^2 s^{-1}), c = Ion concentration in the membrane (mol m^{-3}), x = distance from the membrane (m), z = Valence of ion, R = Gas constant (J mol^{-1} K^{-1}), T = Absolute temperature (K), F = Faraday constant (C mol^{-1}), K_c = Hindrance factor for conversion, ψ = Potential difference, and V = Solvent velocity (m s^{-1}).

The terms on the right hand side of the equation represent transport of solutes due to diffusion, electric gradient, and convention respectively. Thus the equation can predict solute rejection as a function of feed concentration, ion charge, convection across the membrane, and solute diffusion (Braghetta, 1995). It can be used to calculate the effective pore size (which does not necessarily mean that pores exist), and to determine the thickness and effective charge of the membrane (Bowen and Mukhtar, 1996).

The mass transport through a membrane could have as many as five steps such as (i) diffusion from the water phase to the surface of the membrane, (ii) selective portioning into the membrane phase, (iii) selective transport (diffusion) through the membrane, (iv) desorption from the permeate side of the membrane, and (iv) diffusion away from the membrane and into the bulk fluid of the extracting phase. Among these transfer operations, steps (i), (iii), and (v) may control the rate of mass transfer as the slowest step. Mass transport through NF can be described as diffusion-controlled processes. The different mechanisms and models used to describe the transport of solutes through a semi-permeable NF membrane are Donnan equilibrium, extended Nernst-Planck, hindered transport, and irreversible non-equilibrium thermodynamics model. The use of the extended Nernst-Planck model in conjunction with the Donnan equilibrium condition suggests the possibility of characterizing the effective membrane pore size and effective charge density to predict the separation of mixtures of electrolytes at the membrane/solution interface. Secondly, the non-equilibrium thermodynamic model provides a real description of ion transport through membranes even though the membrane is treated as a black box and the model gives little insight into the physico-chemical processes involved in solute and solvent transport across a membrane. Thus, the thermodynamic model can be accepted especially when the parameters used in the model are experimentally measured. The steric-hinderance pore (SHP) model is the modification of pore model where the gradient of mechanical pressure across a membrane is taken into account. Summary of the previous study on NF modelling is given in Table 3 (Lee and Lee, 2000; Yoon et al., 2002; Thanuttamavong et al., 2001; Ratanatamskul et al., 1998).

6 Membrane fouling in the NF process

Like any other membrane processes, NF is also susceptible to membrane fouling. Membrane fouling is one of the significant challenges in any membrane process and therefore understanding the fouling mechanism and identifying a suitable control option is one of the essential components of the membrane applications (Hong and Elimelech, 1997; Mulder, 1997; Lee et al., 2010; Phuntsho et al., 2011b). The solutions to fouling issues require a multipronged approach involving the membrane properties, operational conditions, feed characteristics, etc. (Chapman et al., 2002; Shon et al., 2005, 2009; Phuntsho et al., 2011a). The NF membrane fouling could be due to inorganic precipitation or scaling, colloidal fouling, organic adsorption and/or biofouling. While biofouling is important in long term, most likely, biofouling occurs only after organic or inorganic or colloidal fouling. Since interactions between solutes and the membranes are poorly understood, it is possible that effects like charge interactions, bridging, and hydrophobic interactions may play an important role in NF fouling. Normally, membranes with larger pores exhibit a greater flux decline as filtration proceeds because of internal clogging. However, flux decline is not necessarily due to fouling. Other phenomena such as concentration polarization or osmotic pressure or membrane compaction can appear as fouling during the NF process.

Reiss and Taylor (1994) compared three parameters, silt density index (SDI), modified fouling index (MFI), and the

Table 3. Summary of the mathematical modeling work on Nanofiltration

Modeling	Types	Remarks
$\frac{dm \cdot s}{dt} = -\frac{d(\frac{J_b}{J_w})}{dt}$	Cross flow	$\frac{dm \cdot s}{dt}$: the degree of concentration polarization
$\frac{dm \cdot c}{dt} = \frac{d(\frac{J_b}{J_t})}{dt}$		$\frac{dm \cdot c}{dt}$: deposition probability and the bulk crystallization rate
$R = 1 - \frac{[1/(t_2-t_1)\int_{t_1}^{t_2} L_s(C_m-C_p)dt]}{[(f/(t_2-t_1)\int_{t_1}^{t_2} L_v \Delta P_{eff} dt]}$	Stirred cell	The measured rejection matches the model results quite well
$J_s = \omega\Delta\pi + (1-\sigma)C_{avg}J_v$ (non-equilibrium thermodynamic equation)	Cross flow	Solute flux is governed by convection for a relatively large pore size membrane
$P = H_D S_D D_S(\frac{A_k}{\Delta x})$ $H_D = 1$ $S_D = (1-\eta)^2$ $\eta = r_s/r_p$ (steric hindrance pore model)	Cross flow	$\frac{A_k}{\Delta x}$: the ratio of porosity to membrane thickness (value: $4.4 \times 10^3 \sim 7.8 \times 10^4$ for NF) P: solute permeability
$J_i = -u_i RT\frac{dc_i}{dx} - Z_i c_i u_i F\frac{d\Psi}{dx} + c_i J_v$ (the extended Nernt-Planck equation)	Cross flow	The difference in rejection between chloride and nitrate was well explained by introducing a new parameter
$\sum_i Z_i C_i = 0$ outside membrane $\sum_i Z_i C_i + \phi X = 0$ inside membrane $(\frac{c_i}{C_i})^{1/Z_i} = \exp(\frac{-F\Delta\Psi D}{RT})$ (Donnan equilibrium)		

linear correlation of the mass transfer coefficient (MTC) to investigate the NF fouling. However, no correlation between these parameters was obtained indicating that the simple filtration laws might not be valid for NF process. DiGiano et al. (1994) found that the organic compounds with molecular weight higher than 30 kDa was responsible for NF fouling. They further noticed the change in fouling mechanism after 20 h operation of NF, possibly due to the interactions of the hydrophobic and hydrophilic fractions of organics. Thorsen et al. (1999) recommended the use of highly hydrophilic NF membranes with pore size of 1–2 nm and low operating pressure to reduce fouling. They found that hydrophilic membranes were more fouling resistant irrespective of the pore size of the membranes. Membrane fouling would be very severe in positively charged membranes which can attract the negatively charged organics easily (Nystrom et al., 1995). Inorganic ions such as calcium, phosphorus, aluminium and iron etc. were found to enhance the membrane fouling during water treatment process (Baker et al., 1995). Hong and Elimelech (1997) showed that membrane fouling by NOM was increased in the presence of calcium ions, at lower pH, and higher ionic strength. They further noted that permeation drag and electrostatic double layer repulsion controlled the membrane fouling. Chellam et al. (1997) found that colloidal materials could cause more fouling than organic in NF. The NF membrane fouling can occur due to the following reasons: (i) biological fouling which is the growth of biological species on the membrane surface, (ii) colloidal fouling which results in a loss of permeate flux through the membrane, (iii) organic fouling due to the deposition of organic substances, and (iv) scaling which is defined as the formation of mineral deposits precipitating from the feed stream to the membrane surface (Duranceau, 2001).

6.1 Membrane fouling control

Membrane fouling is normally controlled either by operating the system within the critical flux range or adding chemicals (especially to prevent inorganic scaling and fouling), and/or by pretreatment. Pretreatment is emerging as the most promising solution to control the fouling as it is simple and easy to implement. Gusses et al. (1997) and Glucina et al. (1997) found that conventionally used filter media was not sufficient to reduce the fouling of NF, and suggested a combination of coagulation, ozonation and biofiltration as a better alternative to reduce the NF fouling. Normally for coagulation as a pretreatment, iron or aluminium sulphate are commonly used. However, Nystrom et al. (1995) observed that when humic acid was filtered alone, it was retained to 100 %, but when filtered together with $FeCl_3$, humic acid retention decreased. Thus, it requires to be wise on the use of physico-chemical pretreatments before a NF module. Levenstein et al. (1996) found that addition of a polyelectrolyte enhanced the ion rejection in NF. Activated carbon adsorption is a very effective pretreatment process. Many researchers have used activated carbon adsorption as a pretreatment to membrane processes (Kim et al., 2001; Matsui et al., 2001a, b; Vigneswaran et al., 2003). Since the initial decrease in the permeate flux is mainly due to rapid, irreversible adsorption of organic substances on the membrane surface (Ben Aim et al., 1993), providing a pre-treatment such as adsorption or flocculation of organics before passing the feed solution through the membrane is very effective solution to the membrane fouling problem (Chapman et al., 2002). Backwashing, backflushing or chemical cleaning are some other options to reduce the NF fouling.

6.2 Importance of pre-treatments prior to NF process

Pretreatment of the feed to NF is one of the important considerations to protect the membrane and to improve the performance of NF. Protection refers usually to the prevention of fouling, but also includes the protection against mechanical and chemical damage. A high solids load can damage the membrane surface mechanically and restrict the flow in the filtration system. Meanwhile, oxidation agent, e.g. chlorine and ozone, are harmful to many membrane materials.

NF can be used in the tertiary wastewater especially to remove persisting organic pollutants. In order to improve the filtration flux of NF and extend the operation of NF without extensive organic fouling, effective pretreatment is necessary. In recent years, high rate flocculation and magnetic ion exchange resin have been tried to remove hydrophobic and hydrophilic organics respectively. This can greatly reduce the organic fouling on the NF membranes.

7 Summary

NF that is the widely used membrane process for water and wastewater treatment in addition to other applications such as desalination where its application is increasing plays an important role to partially replace RO, which reduces energy and operational costs. The fundamentals of membrane process in general and the mechanisms of the NF process in particular with some of its basic models were discussed and the issues and challenges of the membrane fouling with NF applications have also been identified including the pretreatment options to mitigate the membrane fouling with the NF process. For the future, NF on behalf of RO will be preferentially considered if it meets water quality requirements.

Acknowledgements. This research was funded by Australian Research Council (ARC). This work was supported in part by the National Research Foundation of Korea (NRF) grand funded by the Korea government (MEST) (No. 2012047029, TOC).

References

Baker, J., Stephenson, T., Dard, S., and Cote, P.: Characterisation of fouling of nanofiltration membranes used treat surface waters, Environ. Technol., 16, 977–985, 1995.

Ben Aim, R. and Semmens, M. J.: Membrane bioreactors for wastewater treatment and reuse: a success story, Proceedings and key-note speech of the IWA-Asian Environmental Technology Conference, Singapore, IWA, 2001.

Ben Aim, R., Liu, M. G., and Vigneswaran, S.: Recent development of membrane processes for water and wastewater treatment, Water Sci. Technol., 27, 141–149, 1993.

Bowen, W. R. and Mukhtar, H.: Characterisation and prediction of separation performance on nanofiltration membranes, J. Membrane Sci., 112, 263–274, 1996.

Bowen, W. R. and Welfoot, J. S.: Modelling the performance of membrane nanofiltration critical assessment and model development, Chem. Eng. Sci., 57, 1121–1137, 2002.

Bowen, W. R., Welfoot, J. S., and Williams, P. M.: Linearized transport model for nanofiltration: development and assessment, AIChE J., 48, 760–772, 2002.

Braghetta, A.: The influence of solution chemistry and operating conditions on nanofiltration of charged and uncharged organic macromolecules, Dissertation, University of North Carolina, USA, 1995.

Cadotte, J., Forester, R., Kim, M., Petersen, R., and Stocker, T.: Nanofiltration membranes broaden the use of membrane separation technology, Desalination, 70, 77–88, 1988.

Chapman, H., Vigneswaran, S., Ngo, H. H., Dyer, S., and Ben Aim, R.: Pre-flocculation of secondary treated wastewater in enhancing the performance of microfiltration, Desalination, 146, 367–372, 2002.

Chellam, S., Jacangelo, J. G., Bonacquisti, T. P., and Schauer, B. A.: Effect of pretreatment for surface water nanofiltration, J. Am. Water Works Assoc., 89, 77–88, 1997.

Choi, J. H., Cockko, S., Fukushi, K., and Yamamoto, K.: A novel application of a submerged nanofiltration membrane bioreactor (NF MBR) for wastewater treatment, Desalination, 146, 5 413–420, 2002.

DiGiano, F. A., Braghetta, A., Nilson, J., and Utne, B.: Fouling of nanofiltration membranes by natural organic matter, National Conference on Environmental Engineering, American Society of Civil Engineers, 1994.

Duranceau, S. J.: Membrane practices for water treatment, Denver, American Water Works Association, 2001.

Ernst, M., Sachse, A., Steinberg, C. E. W., and Jekel, M.: Characterization of the DOC in nanofiltration permeates of at tertiary effluent, Water Res., 34, 2879–2886, 2000.

Glucina, K., Laine, J. M., and Robert, C.: Intergrated, multi-objective membrane systems for surface water treatment, Proceedings AWWA Membrane Technology Conference, New Orleans, USA, 1997.

Gozalvez, J. M., Lora, J., Mendoza, J. A., and Sancho, M.: Modelling of a low-pressure reverse osmosis system with concentrate recirculation to obtain high recovery levels, Desalination, 144, 341–345, 2002.

Gusses, A. M., Allgeier, S. C., Speth, T. F., and Summers, R. S.: Evaluation of surface water pretreatment processes using rapid bench-scale membrane test, Proceedings AWWA Membrane

Technology Conference, New Orleans, USA, 1997.
Hong, S. and Elimelech, M.: Chemical and physical aspects of natural organic matter (NOM) fouling of nanofiltration membranes, J. Membr. Sci., 132, 159–181, 1997.
Kim, H. S., Katayama, H., Takizawa, S., and Ohgaki, S.: Removal of coliphage Qb and organic matter from synthetic secondary effluent by powdered activated carbon-microfiltration (PACMF) process, Proceedings of IWA Specialized Conference on Membrane Technology, Israel, 211–219, 2001.
Lee, E., Shon, H. K., and Cho, J.: Biofouling characteristics using flow field-flow fractionation: Effect of bacteria and membrane properties, Bioresource Technol., 101, 1487–1493, 2010.
Lee, S. and Lee, C.-H.: Effect of operating conditions on $CaSO_4$ scale formation mechanism in nanofiltration for water softening, Water Res., 34, 3854–3866, 2000.
Lee, S., Shim, Y., Kim, I. S., Sohn, J., Yim, S. K., and Cho, J.: Determination of mass transport characteristics for natural organic matter (NOM) in ultrafiltration (UF) and nanofiltration (NF) membranes, Water Sci. Technol., 2, 151–160, 2002.
Levenstein, R., Hasson, D., and Semiat, R.: Utilization of the Donnan effect for improving electrolyte separation with nanofiltration membranes, J. Membr. Sci., 116, 77–92, 1996.
Macoun, R. G.: The mechanisms of ionic rejection in nanofiltration, Chemical Engineering, Ph.D. thesis, University of New South Wales, Sydney, Australia, 1998.
Matsui, Y., Colas, F., and Yuasa, A.: Removal of a synthetic organic chemical by PAC-UF systems – II: Model application, Water Res., 35, 464–470, 2001a.
Matsui, Y., Yuasa, A., and Ariga, K.: Removal of a synthetic organic chemical by PAC-UF systems – I: theory and modeling, Water Res., 35, 455–463, 2001b.
Mulder, M.: Basic principles in membrane technology, Kluwer Academic Publishers, 1997.
Nystrom, M., Kaipia, L., and Luques, S.: Fouling and retention of nanofiltration membranes, J. Membr. Sci., 98, 249–262, 1995.
Phuntsho, S., Listowski, A., Shon, H. K., Le-Clech, P., and Vigneswaran, S.: Membrane autopsy of a 10 year old hollow fibre membrane from Sydney Olympic Park water reclamation plant, Desalination, 271, 241–247, 2011a.
Phuntsho, S., Shon, H. K., Vigneswaran, S., and Cho, J.: Assessing membrane fouling potential of humic acid using flow field-flow fractionation, J. Membr. Sci., 373, 64–73, 2011b.
Ratanatamskul, C., Urase, T., and Yamamoto, K.: Description of behaviour in rejection of pollutants in ultra low pressure nanofiltration, Water Sci. Technol., 38, 453–462, 1998.
Reiss, C. R. and Taylor, J. S.: Prediction of nanofilter productivity decline using fouling indices, Proceedings AWWA Membrane Technology Conference, Nevada, USA, 1994.
Rohe, D. L., Blanton, T. C., and Marinas, B. J.: Drinking water treatment by nanofiltration, National conference on Environmental Engineering, 1990.
Seidel, A. and Elimelech, M.: Coupling between chemical and physical interactions in natural organic matter (NOM) fouling of nanofiltration membranes: implications for fouling control, J. Membr. Sci., 203, 245–255, 2002.
Seo, G. T., Suzuki, Y., and Ohgaki, S.: Bilogical powdered activated carbon (BPAC) microfiltration for wastewater reclamation and reuse, Desalination, 106, 39–45, 1996.
Seo, G. T., Ohgaki, S., and Suzuki, Y.: Sorption characteristics of biological powdered activated carbon in BPAC-MF (biological activated carbon-microfiltration) system for refractory organic removal, Water Sci. Technol., 35, 163–170, 1997.
Shannon, M. A., Bohn, P. W., Elimelech, M., Georgiadis, J. G., Marinas, B. J., and Mayes, A. M.: Science and technology for water purification in the coming decades, Nature, 452, 301–310, 2008.
Shon, H. K., Vigneswaran, S., Aim, R. B., Ngo, H. H., Kim, I. S., and Cho, J.: Influence of Flocculation and Adsorption as Pretreatment on the Fouling of Ultrafiltration and Nanofiltration Membranes: Application with Biologically Treated Sewage Effluent, Environ. Sci. Technol., 39, 3864–3871, 2005.
Shon, H. K., Vigneswaran, S., Zareie, M. H., Aim, R. B., Lee, E., Lee, J., Cho, J., and Kim, I. S.: Physico-chemical pretreatment to seawater reverse osmosis (SWRO): organic characterization and membrane autopsy, Desalination, 236, 282–290, 2009.
Simpson, A. E., Kerr, C. A., and Buckley, C. A.: The effect of pH on the nanofiltration of the carbonate system in solution, Desalination, 64, 305–319, 1987.
Snoeyink, V. L., Campos, C., and Marinas, B. J.: Design and performance of powdered activated carbon/ultrafiltration systems, Water Sci. Technol., 42, 1–10, 2000.
Thanuttamavong, M., Oh, J. I., Yamamoto, K., and Urase, T.: Comparison between rejection characteristics of natural organic matter and inorganic salts in ultra low pressure nanofiltration for drinking water production, Water Sci. Technol., 1, 77–90, 2001.
Thanuttamavong, M., Yamamoto, K., Oh, J. I., Choo, K. H., and Choi, S. J.: Rejection characteristics of organic and inorganic pollutants by ultra low-pressure nanofiltration of surface water for drinking water treatment, Desalination, 145, 257–264, 2002.
Thorsen, T.: Membrane filtration of humic substances – State of the art, Water Sci. Technol., 40, 105–112, 1999.
Trebouet, D., Schlumpf, J. P., Jaouen, P., and Quemeneur, F.: Stabilized landfill leachate treatment by combined physicochemical-nanofiltration processes, Water Res., 35, 2935–2942, 25 2001.
Tsuru, T., Izumi, S., Yoshioka, T., and Asaeda, M.: Temperature effect on transport performance by inorganic nanofiltration membranes, AlChE J., 46, 565–574, 2000.
Van der Bruggen, B., Braeken, L., and Vandecasteele, C.: Evaluation of parameters describing flux decline in nanofiltration of aqueous solutions containing organic compounds, Desalination, 147, 281–288, 2002.
Vigneswaran, S., Vigneswaran, B., and Ben Aim, R.: Application of microfiltration for water and wastewater treatment, Environmental Sanitation Reviews, 31 June, 1991.
Vigneswaran, S., Chaudhary, D. S., Ngo, H. H., Shim, W. G., and Moon, H.: Application of a PAC-Membrane Hybrid System for Removal of Organics from Secondary Sewage Effluent: Experiments and Modelling, Separ. Sci. Technol., 38, 2183–2199, 2003.
Visvanathan, C., Ben Aim, R., and Parameshwaran, K.: Membrane separation bioreactors for wastewater treatment. Critical reviews, Environ. Sci. Technol., 30, 1–48, 2000.
Xu, Y. and Lebrun, R. E.: Comparison of nanofiltration properties of two membranes using electrolyte and non-electrolyte solutes, Desalination, 122, 95–106, 1999.
Yoon, Y., Amy, G., Cho, J., Her, N., and Pellegrino, J.: Transport of perchlorate (ClO_4) through NF and Membranes, Desalination, 147, 11–17, 2002.

8

Fluoride in the drinking water of Pakistan and the possible risk of crippling fluorosis

M. A. Tahir and H. Rasheed

Pakistan Council of Research in Water Resources, Khayaban-e-Johar, H-8/1, Islamabad, Pakistan

Correspondence to: M. A. Tahir (pcrwr2005@yahoo.com)

Abstract. To explore the possibility of fluoride toxicity, 747 water samples were collected from surface water and groundwater sources of 16 major cities of Pakistan, adopting a uniform sampling design with distribution of samples: Lahore (79), Kasur (46), Faisalabad (30), Khushab (50), Chakwal (51), Mianwali (30), Jhelum (53), Bahawalpur (60), Karachi (60), Mirpur Khas (55), Peshawar (38), Risalpur (35), Quetta (81), Ziarat (21), Loralai (21), and Mastung (37). Comparison of analytical findings with WHO Guidelines of Drinking Water for Fluoride (i.e., 1.5 ppm) has concluded that 16 % of the monitored water sources have fluoride concentration beyond the permissible safe limit of 1.5 $mg L^{-1}$ falling in the concentration range of 1.6–25 $mg L^{-1}$. The highest fluoride contamination (22 %) is detected in the Balochistan province followed by 19 % in Punjab province. Comparatively higher fluoride levels of > 20 % in the groundwater sources like hand pumps supported the possibility of increased groundwater contamination as excessive fluoride concentrations are expected to come from calcium-poor aquifers and in areas where fluoride-bearing minerals are common or where cation exchange of sodium for calcium occurs. Field observations have also indicated the prevalence of fluoride-associated health implications in the study areas with excessive fluoride in water sources. Findings of this study have provided bidirectional vision for the epidemiological investigations as well as to mitigate the issues in the affected vicinities of fluoride-rich areas.

1 Introduction

Per capita water availability in Pakistan has decreased from 5000 m^3 per annum in 1951 to 1100, which is just above the internationally recognized scarcity rate. It is projected that water availability will be less than 700 m^3 per capita by 2025 (Pak-SCEA, 2006). Recent estimations of the availability and use of groundwater of an acceptable quality have also indicated the heavy over-exploitation of the water resources resulting in the deteriorated quality as well as quantity of the groundwater (Pakistan Water Partnership, 2000). Findings of the several studies conducted in the country to have exact evaluation of the drinking water quality have disclosed the presence of a few potentially toxic substances in the groundwater with higher concentration accelerated by the human activities. The major reasons of poor water quality may be the untreated disposal of municipal and industrial effluents, excessive use of fertilizers and insecticides. In addition, out of the total, 40 % of diseases prevalent in the country are waterborne and 20–40 % hospitalizations are due to such waterborne diseases (PCRWR, 2008). Therefore, considering the demand of the time, Pakistan Council of Research in Water Resources (PCRWR) launched a National Water Quality Monitoring Programme (2001–2006) in the country, and the findings of this mega-water quality monitoring program for 23 major cities of Pakistan have revealed the prevalence of four major water quality tribulations such as bacteriological contamination (68 %), arsenic (24 %), nitrate (13 %) and fluoride (5 %) in the surface or groundwater sources of Pakistan, specifically in the cities like Kasur, Loralai, Quetta, Bahawalpur, Karachi, Faisalabad and Ziarat (Kahlown et al., 2008).

Fluoride is an important water quality parameter and has beneficial effects on teeth at low concentrations in drinking-water; however, excessive exposure to fluoride in drinking-water, or in combination with exposure to fluoride from other sources, can give rise to a number of adverse effects, which may range from mild dental fluorosis to crippling skeletal fluorosis as the level and period of exposure increases. Many countries in the world, especially Iraq, Iran, Syria, Turkey, India, Algeria, Morocco, Southern Parts of the USA and former USSR, have also reported the higher fluoride concentration in their groundwater (Edmunds and Smedley, 1996 and Mangla, 1991).

Khan et al. (2002) have investigated the presence of fluoride in drinking water of Lahore city within safe limits; however, adjoining areas like Mangamandi with higher fluoride concentrations in the water sources have also reported health effects like dental and skeletal fluorosis among the natives due to excessive fluoride intake through drinking water (Khan et al., 2002). Similar case was reported in the groundwater of Mastung valley in Kalat division of Balochistan province (A. Khan, unpublished data, 1999).

Considering the potential toxicity of higher fluoride concentration appeared as dental or skeletal fluorosis, Pakistan Council of Research in Water Resources with financial assistance from United Nation's Children Education Funds (UNICEF) carried out a detailed investigation on prevalence of fluoride in drinking water sources of 16 major suspected cities of Pakistan, and the main objectives of the study were to authenticate the reported levels of fluoride and to review the possible associated health implications to be disseminated to implementing agencies in order to provide guidelines to the affected communities for locating alternate safe water sources.

2 Materials and methods

Seven hundred and forty-seven (747) water samples were collected from various surface or groundwater sources such as hand pumps, tube wells, wells, nullahs, springs, dams, bores and water supply of 16 major cities of the country on the basis of grid size of 0.25 km^2 for small cities, 9 km^2 for medium cities and 16 km^2 for big cities as given in Table 1.

Following the grids on the city maps, one sample per grid was taken maintaining a distance of 0.5 or 1 km between two monitoring points. Replicating every fifth sample for quality control purposes, 298 samples were collected additionally to check the reliability and accuracy of the sampling and analytical procedures.

Following the standard methods, all the samples were collected in 0.5 L polystyrene bottles and transported to laboratory in the least possible time (APHA-AWWA-WEF, 1992). Field information regarding sampling was recorded on the sampling pro forma. All the samples were analyzed for fluoride (mg L^-) using SPANDS method recommended by standard methods based on the comparatively higher sensitivity and efficiency of the method (APHA-AWWA-WEF, 1992). A sample of blank, known fluoride standard and previously analyzed sample were analyzed after every ten samples to check the reproducibility of method, and their results were found in the range of ± 5 % deviation compared to the actual analysis.

Table 1. Sampling criteria and no. of samples collected from four provinces of country.

Sr. No.	Province	City	Grid Size (km^2)	No. of Samples
1	Punjab	Lahore	3	79
		Kasur	0.5	46
		Faisalabad	3	30
		Khushab	0.5	50
		Chakwal	0.5	51
		Mianwali	0.5	30
		Jhelum	0.5	53
		Bahawalpur	3	60
2	Sindh	Karachi	4	60
		Mirpur Khas	0.5	55
3	North West Frontier Province (NWFP)	Peshawar	3	38
		Risalpur	0.5	35
4	Balochistan	Quetta	1.5	81
		Ziarat	0.5	21
		Loralai	0.5	21
		Mastung	0.5	37
		Grand Total	–	747

3 Results and discussion

Analytical findings were compared with WHO Drinking Water Guideline value for fluoride, which also considers the impact of climatic conditions, volume of water intake as well as fluoride intake from other sources (WHO, 2004). Thus, considering the annual average of maximum daily air temperature (84 °F) of Pakistan, WHO guideline value for fluoride in the drinking water of Pakistan is considered as 1.5 mg L^-. City-wise status of fluoride as percentage contamination, out of total monitored sites of each city, is graphically reflected in Fig. 1. Mianwali city has shown the highest contamination level of 37 % in Punjab province, Mirpur Khas city (3.6 %) in Sindh province, Quetta city (14 %) in Balochistan province and Risalpur city (11 %) in North-West Frontier Province (NWFP).

Overall, only 16 % samples out of total 747 were found to be unsafe, whereas the remaining 84 % were declared to be safe. This overall distribution of fluoride for 747 locations monitored in selected 16 cities is shown as Fig. 2.

Table 2 provides details of quantitative summaries of each water quality variable, namely the minimum, maximum, median (middle value of the ordered data), mean, the first and third quartiles, $Q1$ and $Q3$, respectively (which are the 25th and 75th percentile values of the ordered data), and the total number of observations recorded (Count). Comparing

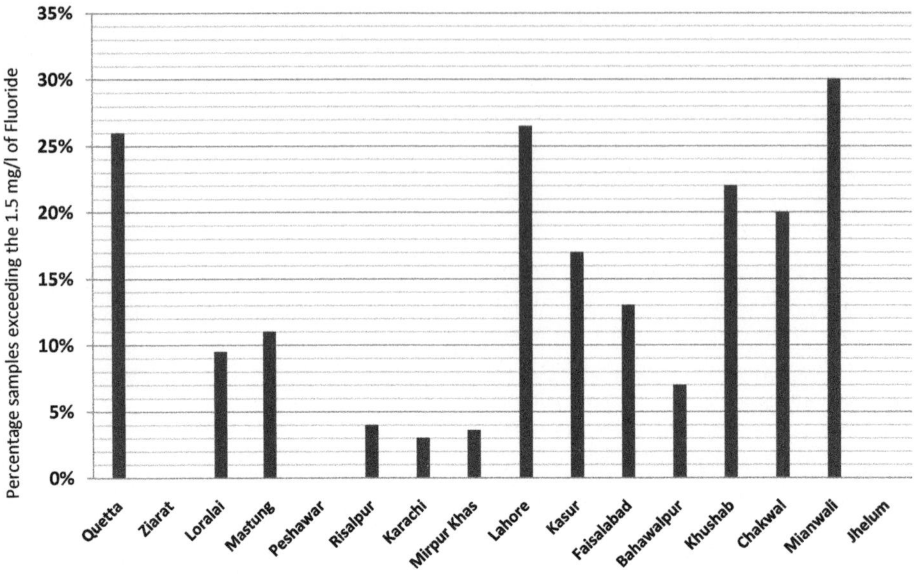

Figure 1. Distribution of fluoride in 16 cities of four provinces.

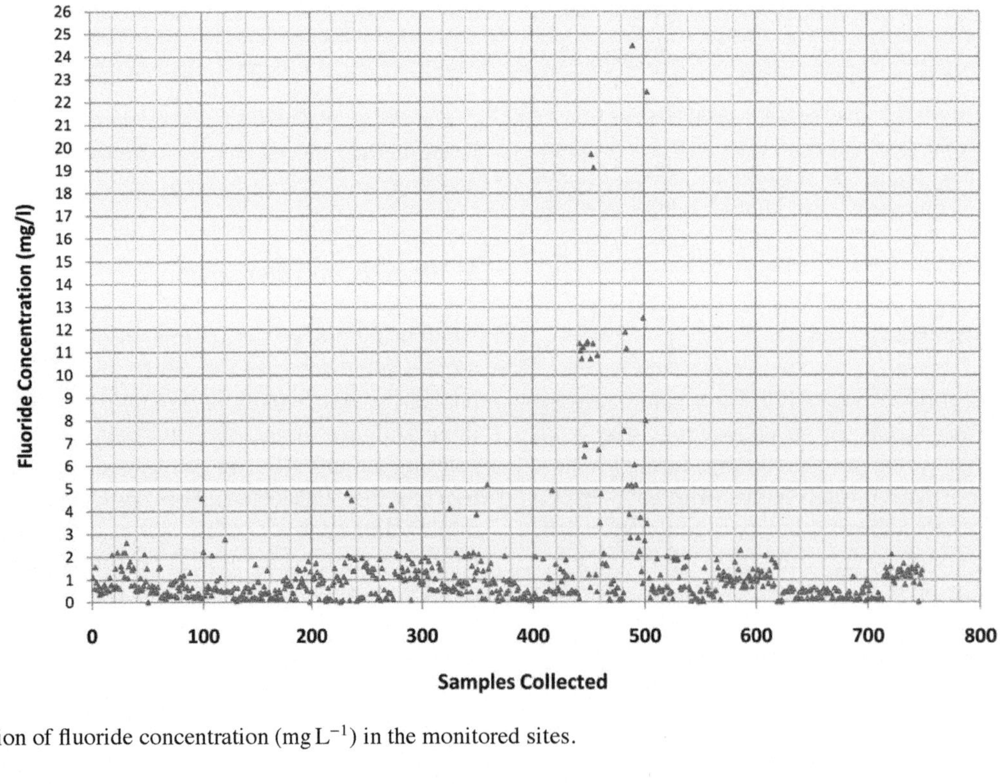

Figure 2. Variation of fluoride concentration (mg L^{-1}) in the monitored sites.

means and medians, we see that the means of fluoride are substantially greater than their medians (except for Jhelum city). The distributions of the data for these variables are highly skewed to the right.

An overall comparison of four provinces indicated a higher percentage contamination in Balochistan province followed by Punjab province. The province-wise distribution is also given in Figs. 3–6 to provide an idea of range of fluoride concentration within each province.

The results of this perspective survey have finally led to recognition of the fact that fluoride is a considerable issue in a few places of Pakistan. Observations of the sampling team have also disclosed the possible signs and symptoms including skeletal and dental fluorosis among the few individuals of the communities associated most probably with the excessive

Table 2. Basic statistics of fluoride in water sources of 16 cities.

Sr. No.	Cities	Count	Mean	SD	Min	*Q*1	Median	*Q*3	Max
1	Chakwal	51	1.02	0.60	0.00	0.59	0.76	1.49	2.63
2	Bahawalpur	60	0.67	0.67	0.15	0.30	0.53	0.74	4.58
3	Karachi	60	0.48	0.44	0.05	0.17	0.43	0.54	2.77
4	Mirpur Khas	55	0.74	0.48	0.00	0.28	0.81	1.06	1.80
5	Khushab	50	1.09	1.10	0.04	0.16	1.05	1.50	4.78
6	Mianwali	30	1.37	0.47	0.08	1.06	1.36	1.77	2.14
7	Kasur	46	1.11	0.84	0.35	0.54	0.81	1.47	4.10
8	Faisalabad	30	0.89	0.97	0.04	0.36	0.79	1.00	5.16
9	Lahore	79	2.62	4.46	0.05	0.26	0.51	1.76	19.70
10	Quetta	81	2.47	4.23	0.03	0.45	0.94	2.11	24.48
11	Ziarat	21	0.39	0.33	0.01	0.14	0.34	0.47	1.51
12	Loralai	21	1.08	0.38	0.08	0.88	1.04	1.18	1.86
13	Mastung	37	1.13	0.41	0.63	0.82	1.08	1.35	2.27
14	Jhelum	53	0.37	0.18	0.05	0.16	0.41	0.50	0.64
15	Peshawar	38	0.29	0.29	0.05	0.10	0.15	0.40	1.10
16	Risalpur	35	1.27	0.27	0.78	1.12	1.30	1.42	2.09

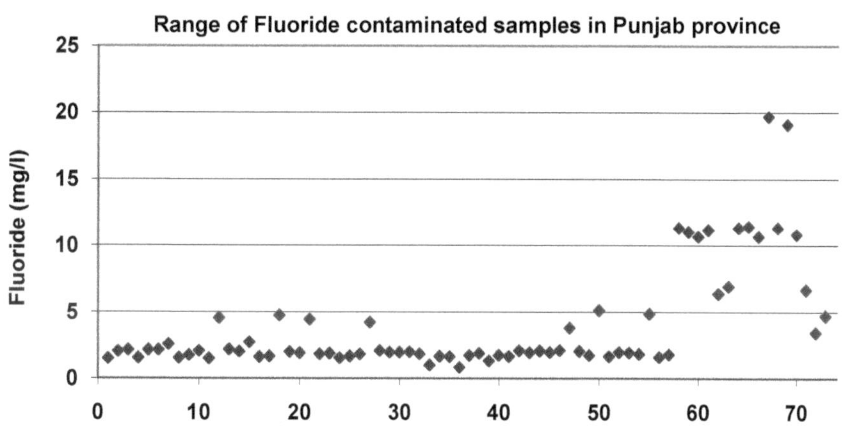

Figure 3. Range of fluoride-contaminated samples (> 1.5 mg L^{-1}) in Punjab province.

fluoride level in their drinking water. In the same way, deaths of the native people at locations with excessive fluoride due to an unknown chronic disease symptomized by bone and joint deformities have also been reported by teams of World Health Organization (WHO) and other NGOs in the year 2000. Experts found prominent signs of "dental fluorosis" in the native areas, where the possibility of chronic fluoride toxicity was confirmed further by the epidemiological and laboratory investigation in the problem areas by a joint team of Pakistan Council of Research in Water Resources (PCRWR), World Health Organization (WHO), National Institute of Health (NIH), and Federal Ministry of Health and Sindh Provincial Health Department in the year 2005.

Fluoride is unlikely to produce crippling skeletal fluorosis leading to osteosclerosis, ligamentous and tendinous calcification, hypersensitivity and other immunological effects (USNRC, 1993). In addition, fluoride has an effect on the kidneys, muscular and nervous system, and there are some reports about the relationship between erythrocyte abnormalities and serum fluoride levels with excessive fluoride intake and therefore leading to death after prolonged illness. Prolonged exposure to 10–20 mg fluoride/person/day for 10–20 yr can lead to crippling skeletal fluorosis, leading to osteosclerosis, ligamentous and tendinous calcification and extreme bone deformities (USNRC, 1993). Health impacts from long-term use of various ranges of fluoridated water have been summarized as the following:

< 0.5 mg L^{-1}	Dental caries;
0.5–1.5 mg L^{-1}	Promotes dental health;
1.5–4 mg L^{-1}	Dental fluorosis;
> 4 mg L^{-1}	Dental, skeletal fluorosis;
> 10 mg L^{-1}	Crippling fluorosis.

Fluoride in drinking water increases the risk of hip fractures in women as evidence has suggested that fluoride may be associated with some gender-dependent mechanisms

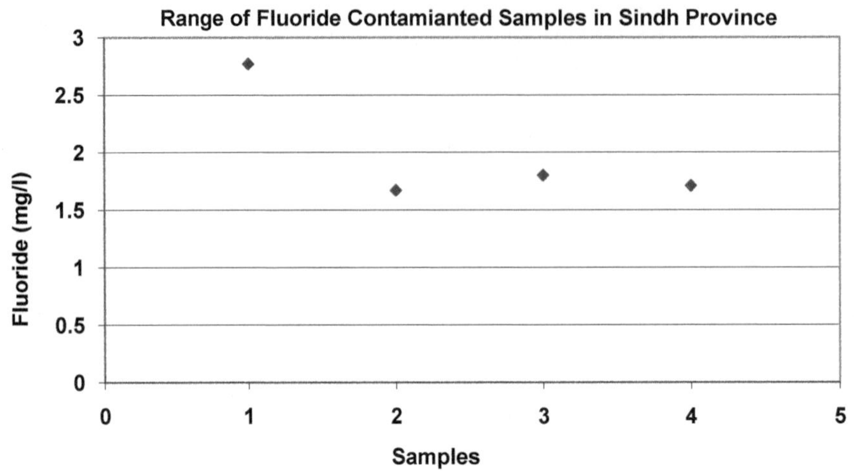

Figure 4. Range of fluoride-contaminated samples (> 1.5 mg L^{-1}) in Sindh province.

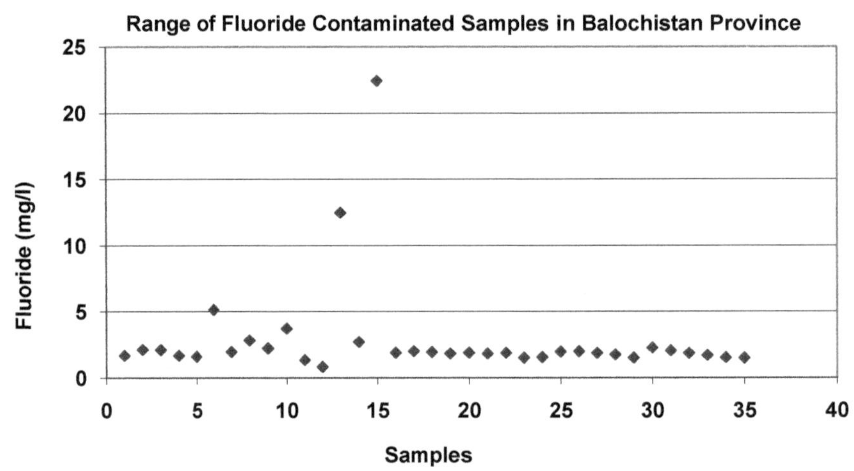

Figure 5. Range of fluoride-contaminated samples (> 1.5 mg L^{-1}) in Balochistan province.

or risk factors for hip fractures (Kurttio et al., 1999). A large number of epidemiological studies showed that fluoride is carcinogenic, a bone seeker and is associated with hip fractures and brittle bones (Yelena et al., 2001). It has also been found that the crippling malady of fluorosis affects not only the bones and teeth, but every tissue and organ of the body, leading to death after prolonged illness (Susheela et al., 1992). USNRC has concluded that fluoride is unlikely to produce hypersensitivity and other immunological effects. Fluoride is mainly excreted from the kidneys (USNRC, 1993). Thus, it is reasonable that those with impaired renal function might be at greater risk of fluoride toxicity than others. However, there is much evidence that the amount of fluoride absorbed and retained by the body is mainly dependent on nutrition, especially for children. The study mentioned has clearly indicated that most of the water sources of the 16 cities of the country are found free from the presence of excessive fluoride and therefore from the possibility of potential fluoride toxicity. In total, 16 locations have fluoride level > 10 mg L^{-1}, which are suspected to have cases of crippling fluorosis and need detailed epidemiological study. Analytical findings of Balochistan and Punjab provinces give rise to the prediction that fluoride is more commonly found in the groundwater as depicted graphically in Fig. 7. However, detailed monitoring of the contaminated regions and surroundings is highly recommended to formulate a mitigation strategy.

Most of the fluoride found in groundwater is naturally occurring from the breakdown of rocks and soils or weathering and deposition of atmospheric volcanic particles. Fluoride can also come from runoff and infiltration of chemical fertilizers in agricultural areas, septic and sewage treatment system discharges and liquid waste from industrial sources. Pakistan is an agricultural country, and 96 % of its total water resources are being used by agriculture sector. Fertilizer consumption has increased threefold during the past 30 yr in the country. It reached one million nutrient tonnes in 1980–1981, two million tonnes in 1992–1993 and three million

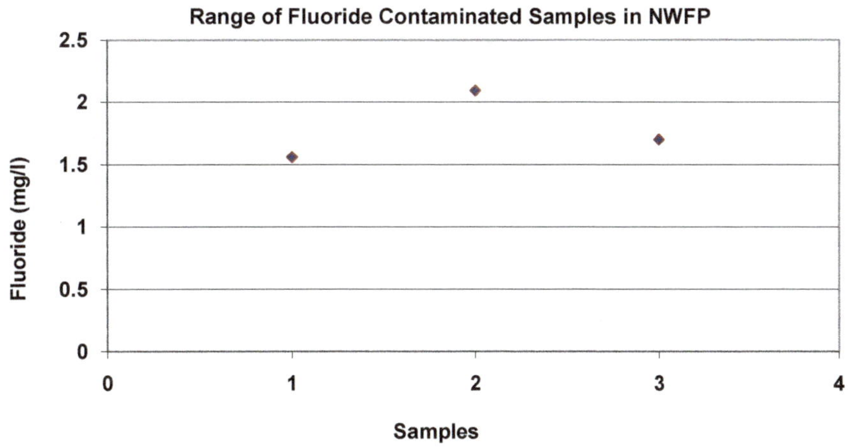

Figure 6. Range of fluoride-contaminated samples (> 1.5 mg L^{-1}) in NWFP.

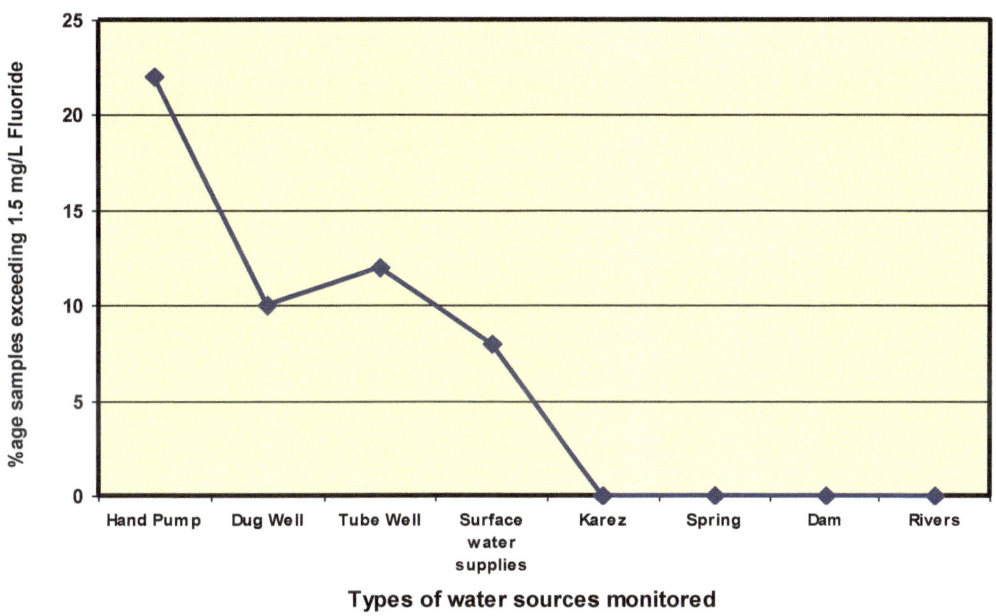

Figure 7. Distribution of fluoride in surface and groundwater sources.

tonnes in 2002–2003 and thus may be an important contributing factor of increased fluoride contamination (FAO, 2007). Moreover, fluorides are found at significant levels in a wide variety of minerals, including fluorspar, rock phosphate, cryolite, apatite, mica, hornblende and others (Murray, 1986). Fluorite (CaF_2) is a common fluoride mineral of low solubility occurring in both igneous and sedimentary rocks. Rock phosphates are converted into phosphate fertilizers by the removal of up to 4.2 % fluoride (Murray, 1986). In groundwaters, however, low or high concentrations of fluoride can occur, depending on the nature of the rocks and the occurrence of fluoride-bearing minerals. Concentrations in water are limited by fluorite solubility, so that in the presence of 40 mg L^{-1} calcium it should be limited to 3.1 mg L^{-1} (Hem, 1989). Balochistan province of Pakistan is rich in mineral deposits and diverse mineralogy, and in this region higher fluoride concentrations may therefore be expected in groundwater from calcium-poor aquifers and in areas where fluoride-bearing minerals are common. Fluoride concentrations may also increase in groundwater in which cation exchange of sodium for calcium occurs. High groundwater fluoride concentrations associated with igneous and metamorphic rocks such as granites and gneisses have been reported from India, Pakistan, West Africa, Thailand, China, Sri Lanka, and southern Africa. In China, endemic fluorosis has been reported in all 28 provinces, autonomous regions and municipalities except Shanghai. Both shallow and deeper groundwater is affected; in general the deeper groundwaters have the higher concentrations. To cover up the gaps, institutional arrangements, capacity building efforts, legislations and policy development with reference to water quality monitoring and surveillance activities as well as diagnosis and management of water-related diseases are highly required to be managed in order to resolve the water quality problems like excessive fluoride and possible health implications.

Acknowledgements. We would like to gratefully acknowledge the contributions of technical and non-technical staff of National Water Quality Laboratory, PCRWR involved in sampling, laboratory analysis and coordination. The authors are grateful to Muhammad El-Feteh and Tameez Ahmad from WES Programme UNICEF Pakistan for financial assistance for the study.

References

APHA-AWWA-WEF: Standard methods for the examination of water and wastewater, American public health association, American water works association and water environment federation, 18th Edn., Washington, DC, 1–33:4–82, 1992.

Edmunds, W. M. and Smedley, P. L: Groundwater geochemistry and health: an overview, in: Environmental geochemistry and health, edited by: Appleton, J. D., Fuge, R., and McCall, G. J. H., Geological society special publication No. 113, 91–105, 1996.

FAO: The Fertilizer sector, in: FAO corporate document repository, http://www.fao.org/docrep/007/y5460e/y5460e07.htm#TopOfPage (last access: 15 December 2004), 2007.

Hem, J. D.: Study and interpretation of the chemical characteristics of natural water, Water Supply Paper 2254, 3rd Edn., US Geological Survey, Washington, D.C., 263 pp., 1989.

Kahlown, M. A., Tahir, M. A., and Hifza, R.: Water Quality Status of Pakistan, Pakistan Council of Research in Water Resources, Islamabad, Pakistan, Technical report Series 121-2008, 79–80, 2008.

Khan, A. A., Whelton H., and O'Mullane, D.: Map of natural fluoride in drinking water of Pakistan, Int. Dent. J., 52, 7–10, 2002.

Kurttio, P. N., Gustavsson, N., Vartiainen, T., and Pekkanen, J.: Exposure to natural fluoride in well water and hip fracture: A cohort analysis in Finland, Am. J. Epidemiol., 150, 817–824, 1999.

Mangla, B.: India's dentists squeeze fluoride warnings off tubes, New Sci., 131, p. 16, 1991.

Murray, J. J.: Appropriate use of fluorides for human health, World Health Organization, Geneva, 1986.

Pak-SCEA: Pakistan; Strategic country environmental assessment report: Rising to the challenges, South Asian Development and Social Development Unit, the World Bank, 50 pp., 2006.

Pakistan Water Partnership: Framework for action for achieving the Pakistan water vision 2025, Islamabad, 2000.

Susheela, A. K., Das, T. K., Gupta, I. P., Tandon, R. K., Kacker, S. K., Ghosh, P., and Deka, R. C.: Fluoride ingestion and its correlation with gastro – intestinal discomfort, Journal Fluoride, 25, 7–12, 1992.

USNRC: US National Research Council, Health effects of Ingested fluoride, National Academy Press, Washington, D.C., 1993.

WHO: World Health Organization, Guideline for Drinking-Water Quality, 2nd Edn., Vol. 2, Health criteria and other supporting information, World Health Organization, Geneva, 2004.

Yelena, V., Marina, T., Viktor, A., and Rein, M.: Fluoride in drinking water, Tallinn Technical University, Estonia, 2001.

9

Non-residential water demand model validated with extensive measurements and surveys

E. J. Pieterse-Quirijns[1], E. J. M. Blokker[1], E. van der Blom[2], and J. H. G. Vreeburg[1,3]

[1]KWR Watercycle Research Institute, P.O. Box 1072, 3430 BB Nieuwegein, the Netherlands
[2]Uneto-VNI, P.O. Box 188, 2700 AD Zoetermeer, the Netherlands
[3]Delft University of Technology, Faculty of Civil Engineering, P.O. Box 5048, 2600 GA Delft, the Netherlands

Correspondence to: E. J. Pieterse-Quirijns (ilse.pieterse@kwrwater.nl)

Abstract. Existing Dutch guidelines for the design of the drinking water and hot water system of non-residential buildings are based on outdated assumptions on peak water demand or on unfounded assumptions on hot water demand. They generally overestimate peak demand values required for the design of an efficient and reliable water system. Recently, a procedure was developed based on the end-use model SIMDEUM to derive design-demand-equations for peak demand values of both cold and hot water during various time steps for several types and sizes of non-residential buildings, viz. offices, hotels and nursing homes. In this paper, the design-demand-equations are validated with measurements of cold and hot water patterns on a per second base and with surveys. The good correlation between the simulated water demand patterns and the measured patterns indicates that the basis of the design-demand-equations, the SIMDEUM simulated standardised buildings, is solid. Surveys were held to investigate whether the construction of the standardised buildings based on the dominant variable corresponds with practice. Surveys show that it is difficult to find relationships to equip the standardised buildings with users and appliances. However, the validation proves that with a proper estimation of the number of users and appliances in only the dominant functional room of the standardised buildings, SIMDEUM renders a realistic cold and hot water diurnal demand pattern. Furthermore, the new design-demand-equations based on these standardised buildings give a better prediction of the measured peak values for cold water flow than the existing guidelines. Moreover, the new design-demand-equations can predict hot water use well. In this paper it is illustrated that the new design-demand-equations lead to reliable and improved designs of building installations and water heater capacity, resulting in more hygienic and economical installations.

1 Introduction

Non-residential users have a significant impact on drinking water demand and exhibit a diurnal pattern which can be completely different from residential users (Loureiro et al., 2010; Blokker et al., 2011). In the Netherlands, non-residential water demand amounts to approximately 28 % of the total distributed water. Small-scale business users, defined as buildings using less than 10 000 $m^3\,yr^{-1}$, use 12 % of the total distributed water. Small-scale users include shops, hotels, restaurants, schools, offices, health care institutions, and campsites. Large-scale business users are defined as customers that use more than 10 000 $m^3\,yr^{-1}$ and are mainly concerned with industrial activities (Baggelaar and Geudens, 2008). Despite the significant impact of non-residential users, limited information relating to their expected water demand exists. It is known that the behaviour of the type of users varies: an office employee, a hotel guest, a cleaner, kitchen personnel and a patient in a nursing home exhibit variability in demand, in time of demand, in frequency of demand, etc. Some users have similar behaviour to residential users, whereas other have completely different behaviour (Blokker et al., 2011; Pieterse-Quirijns et al., 2009). Studies on water consumption in commercial and industrial sectors,

Table 1. Old guidelines for maximum momentary flow of cold water in various non-residential buildings (Scheffer, 1994).

Type of non-residential building	Relation for peak demand MMF_{cold} ($L\,s^{-1}$)
Office – toilets with cistern	1.464 + 0.0019* (number of employees)
Office – toilets with flushing valve	2.603 + 0.0031* (number of employees)
Hotel business	$q\sqrt{n} + 40\,\%^{1}$
Hotel tourist	$q\sqrt{n} + 70\,\%$
Retirement home	1.177 + 0.0092* (number of beds)
Nursing home	2.257 + 0.0130* (number of beds)

ad [1]: $q\sqrt{n}$-method: $q\sqrt{n} = 0.083 \cdot \sqrt{\Sigma}$ LU, LU is number of loading-units. Every type of tapping point is credited with loading-units equal to 0.083 $L\,s^{-1}$ (300 $L\,h^{-1}$). A toilet cistern is, for example, typically 0.25 LU.

i.e. hotels and mining residences, show that water consumption is influenced by cultural and social norms (Rankin and Rousseau, 2006). Water consumption not only varies between countries but also between comparable buildings in the same country. It is influenced by many factors, such as season and occupancy (Rankin and Rousseau, 2006; Barberán et al., 2013; Cobacho et al., 2005). In hotels, a significant part of water consumption is due to the needs of guests in their rooms. Therefore, it depends on occupancy rates, kind of facilities, habits of guests and personnel (Cobacho et al., 2005).

This lack of information creates problems during the design of water infrastructure and in the management of water demand (Ilemobade et al., 2010). To guarantee a high expected water demand, pipe diameters in distribution networks are generally oversized. However, it is understood that this overcapacity causes water quality problems, especially discolouration (Vreeburg, 2007). This lack of information also complicates the design of drinking water systems inside buildings as well as the optimal selection of the type and capacity of water heating systems. Badly designed systems can cause stagnant water with health risks, and are less energy efficient and therefore more expensive to operate. The occurrence of *Legionella* in non-residential buildings is an internationally recognised problem (Van der Kooij et al., 2005) that is related to too high temperatures in the cold water system or too low temperatures in the hot water system, that can be caused by badly designed hot water circulation systems.

For the design of the drinking water distribution system outside as well as inside buildings, understanding the peak value of the total water demand, the instantaneous peak demand or maximum momentary flow (MMF_{cold}), is essential (Loureiro et al., 2010; Blokker and Van der Schee, 2006). In the Netherlands, existing guidelines provide the MMF_{cold} for non-residential buildings of various types and sizes (Table 1). These guidelines are based on measurements carried out between 1976 and 1980. Total water demand was measured with small time step during several weeks in buildings of different sizes (Werkgroep momentane waterverbruiken, 1985). Presumably, the guidelines are no longer suitable for modern non-residential buildings, that have altered considerably in the past decades with respect to appliances and number and behaviour of users. For example, rest rooms are now equipped with water saving plumbing fixtures and urinals; in offices the coffee lady has been replaced with machines. Moreover, the working habits have changed, with more flexibility in working hours, with open-plan offices and a difference in ratio of men and women and an aim at maximum occupancy. Hotels are more luxurious and in some cases are extended with facilities for a conference or theatre. Nursing homes have changed in the recent years, aiming at small-scale services and residents' independency. The residents have expectations concerning privacy and continuing their personal way of life, resulting in qualitative enhancement of facilities and equipment to improve quality of life. The outlined problems, caused by overcapacity, and new developments in *Legionella* control, suggest that the existing guidelines do not give an accurate insight into peak water demand.

Hot water consumption profiles have a significant influence on the hot water system design (Rankin and Rousseau, 2006). Understanding hot water demand is essential to select the correct type of water heater as well as the design capacity of the hot water device. In addition to the peak demand of hot water, i.e. maximum momentary flow (MMF_{hot}), the hot water use (HWU) in different time periods is required for the choice and design of hot water devices. The desired time period depends on the type of selected water heater system (Scheffer, 1994). For the characteristic values of hot water demand (MMF_{hot} and HWU) there are no national and international guidelines, similar to those for cold demand in Table 1. Suppliers of heating systems use their own company specific guidelines.

Because of the described problems with the existing guidelines and because no guidelines exist for the hot water demand, new models for characteristic values of cold and hot water for non-residential buildings are required. In 2010, a procedure was developed to derive design-demand-equations for the peak demand values of both cold and hot water for various types of non-residential buildings, i.e. offices, hotels and nursing homes (Pieterse-Quirijns et al., 2010). The design-demand-equations are based on water demand

Table 2. Functional rooms and the type of user who take water in the functional rooms in three categories of non-residential buildings, office, hotel and nursing home.

Functional room	Type of users (and reason for variable occupancy)		
	office	hotel	nursing home
Meeting area	Employees (fulltime/part-time working), visitors	Guests (meeting, conference, theatre), employees	Visitors, employees
Lodging	–	Hotel guests (tourists, business people)	Residents
Restaurant	Kitchen personnel	Kitchen personnel	Kitchen personnel
Fitness room	Employees using the fitness room	Hotel guests using the fitness room	–
Technical/other	Number of buckets filled for cleaning, washing machines, ...		

patterns simulated by SIMDEUM, an end-use model which simulates residential and non-residential water demand patterns (Blokker et al., 2011). In the procedure, SIMDEUM simulates diurnal water demand patterns, for a specific non-residential building with a certain value of its dominant variable. This dominant variable characterises the size of a building, such as the number of employees in an office or the number of beds in a nursing home. A crucial part of the procedure is the standardisation of each type of building, meaning that for a specific value of the dominant variable, a building is constructed with the corresponding number and characteristics of appliances (like toilets, showers, etc.) and of users (like kitchen personnel, visitors, etc.). The relevant information is not easy to find, since there is much freedom in the design of sanitary installations. Moreover, the number and behaviour of present non-residential water users is barely known. In Sect. 2 of this paper, the background is described more extensively.

Before the SIMDEUM based design-demand-equations can be applied in practice and replace existing guidelines, they need to be validated. For the validation, measurements of cold and hot water diurnal demand patterns are required on a per second base for various buildings, to ensure an accurate determination of peak demands. This is the first time that water demand patterns (especially hot water demand) will be measured on such a small timescale in the Netherlands. The validation procedure consists of two steps.

In the first step, the assumptions of how to standardise buildings, on which the design-demand-equations are based, are validated with measurements and surveys. The validation step consists of two phases. Firstly, average simulated diurnal demand patterns of the standardised buildings are compared with measurements of cold and hot water patterns on a per second base for 6 different buildings. Secondly, surveys will show whether the construction of the standardised buildings based on the dominant variable corresponds with practice. When correspondence is established between standardised buildings and reality, in both predicted water demand patterns and construction of the building, the solid base of SIMDEUM based design-demand-equations is shown in a proof of concept, i.e. when the construction is reliable, the prediction of the expected cold and hot water demand in the building will be reliable.

In the second validation step, the outcome of the design-demand-equations is compared to the measured peak demand values of cold and hot water. When the basis is solid, this step shows that the derivation of the design-demand-equations is reliable.

This paper describes the validation of the SIMDEUM based design-demand-equations. Moreover, the consequences for design of distribution systems and heating systems are illustrated, by comparing resulting designs for pipe diameter, water meter and heaters from developed design-demand-equations, old guidelines and measured practical values.

2 Background

2.1 Background of SIMDEUM for non-residential buildings

SIMDEUM stands for "**SIM**ulation of water **D**emand, an **E**nd-**U**se **M**odel". It is a stochastic model based on statistical information of end uses, including statistical data on water appliances and users (Blokker et al., 2010). SIMDEUM's philosophy is that people's behaviour regarding water use is modelled, taking into account the differences in installation and water-using appliances. This means that in each building, whether it is residential or non-residential, the characteristics of the present water-using appliances and taps (i.e. flow rate, duration of use, frequency of use and the desired temperature) are considered as well as the water-using behaviour of the users who are present (i.e. presence, time of use, frequency of use).

SIMDEUM for non-residential water demand follows a modular approach. Each building is composed of functional rooms, characterised by its typical users and water using appliances (Table 2). The characteristics of the users and the appliances are different for each type of building and are extensively described by Blokker et al. (2011). With this approach, water demand patterns over the day for cold and hot water demand can be simulated for a specific non-residential

Table 3. Different typologies of non-residential buildings within the categories offices, hotels and nursing homes and the dominant functional room and dominant variable in each category.

category	typology			dominant	
	typology name	users	Discerning appliances	functional room	variable
office	office A	male:female = 65 % : 35 %	toilet with 6 L cistern and urinal	meeting area	number of employees
	office B	male : female = 65 %:35 %	toilet with flushing valve and urinal		
	office C	male : female = 65 % : 35 %	toilet with 6 L cistern and no urinal		
	office D	male : female = 90 % : 10 %	toilet with 6 L cistern and urinal		
hotel	business (incl. conference)	business guests	7 types of shower	hotel rooms	number of hotel rooms
	tourist (incl. theatre)	tourist guests	7 types of shower		
nursing home	A: individual room	care needed residents	3 types of intensity of use bedpan washer	lodging and meeting area	number of beds
	B: four persons per room	care needed residents	3 types of intensity of use bedpan washer		
	C: combination of single rooms and four persons per room	care needed residents	3 types of intensity of use bedpan washer		
	D: self-contained apartments/assisted living	independent residents			

building. From these daily water demand patterns, the characteristic peak demand values of cold and hot water during various time steps can be derived. These characteristics form the basis for design guidelines.

2.2 Background of SIMDEUM based design-demand-equations for non-residential buildings

For three categories of small-scale non-residential buildings SIMDEUM based design-demand-equations were developed, viz. offices, hotels and nursing homes. These categories are selected based on the cooperation with the Dutch installation sector, where they were mostly interested in these buildings (Blokker et al., 2011). These categories are most often applied in practice. Moreover, they are characterised by a generalised construction or standardisation. This in contrast with, for example, a hospital or a prison, which are more specific and cannot be captured in a general approach. Within each category, different typologies of buildings were defined. The typologies can vary in types of appliances, like types of toilets, flow of showers, and in the type of users, like business or tourist hotel guests. In Table 3, the defined typologies within each category are presented. The definition of the typologies is arbitrary. In the presented various typologies, existing building types were used as reference, based on practical experiences given by Uneto-VNI, the Dutch association of installers.

The aim of the design-demand-equations is to predict the peak demand values (MMF_{cold}, MMF_{hot} and HWU in different time periods) for various types of offices, hotels and nursing homes of arbitrary size. The new design-demand-equations predict the peak demand values as a function of a (dominant) variable. Blokker et al. (2011) presented a sensitivity analysis for SIMDEUM for non-residential water demand in each of the considered categories. The dominant variable was established from this sensitivity analysis. The dominant variable is the functional room or user in a building that contributes most to the pattern and total amount of water use in a building. In offices, the main contributor (80–90 %) to water demand is the meeting area with the toilets. The dominant variable for offices therefore is the number of employees that use the toilets. In hotels, the main contributor is the hotel rooms (app. 80 %) for hotels without conference and theatre facilities and without laundry. The dominant

Table 4. Number of users and their water use probability in each functional room in a standardised office as function of the dominant variable, x the number of employees.

Functional room	Type of user	Number of users	Times of presence and water use probability in office*			
			t_1	t_2	t_3	t_4
Meeting area	employees and	x	08:00	12:15	13:00	17:00
	visitors	$10\,\% \cdot x$				
Restaurant	kitchen personnel	$0.0052 \cdot x + 2.24$	08:30	12:30	13:30	15:00
Fitness room	employees using the fitness room	$1.5\,\% \cdot x$	07:30	–	–	18:30
Technical/other	cleaners	$0.0265 \cdot x + 0.027$	17:00	–	–	18:30

* t_1 is starting time of water use probability, t_4 is the end. t_2 and t_3 characterise a peak in water use probability (Blokker et al., 2011).

Table 5. Number of appliances in each functional room of a standardised office as function of the dominant variable, x the number of employees.

Functional room	Appliance		Number of appliances as function of x (number of employees)	remarks
Meeting area	toilet ladies	$x \leq 300$	$x/12$	frequency = 4, 60 % of flushes
		$x > 300$	$x/18$	use the water saving option
	toilet gentleman	$x \leq 300$	$x/12$	frequency = 1, no use of water
		$x > 300$	$x/18$	saving option
	urinal	$x \leq 300$	$x/12$	frequency = 3
		$x > 300$	$x/18$	
	tap wash basin	$x \leq 300$	$2 \cdot x/12$	frequency = 4.5
		$x > 300$	$x/18$	
	coffee machine		$0.0232 \cdot x + 4.5$	frequency = 8
Restaurant	kitchen tap		$0.0052 \cdot x + 2.24$	frequency of usage depends on number of kitchen personnel and number of guests in the restaurant
	dish washer			
	x: 0–500		1	
	x: 501–1000		2	
	x: 1001–1500		3	
Fitness room	shower		$0.0132 \cdot x - 0.49$	with boundary condition > 0
Technical/other	cold and hot water		$0.0265 \cdot x + 0.027$	

variable for hotels is the number of rooms, which can be occupied by 1 or 2 guests, depending on the type of hotel. In nursing homes, the water use is not determined by a single dominant functional room. The washing of the residents and the toilet use of personnel and visitors all largely determine the water use. Since the number of personnel and visitors is determined by the number of beds (with all beds occupied), the latter is defined as the dominant variable. For each category the dominant functional room and the dominant variable are given in Table 3.

The procedure to derive the design-demand-equations for the peak demand values of cold and hot water for non-residential buildings is extensively described in Pieterse-Quirijns et al. (2010). For a specific value of the dominant variable, a standard building was constructed, i.e. each functional room is equipped with appliances and users. For this purpose, the number of appliances and users is established as a function of the dominant variable for each type of non-residential building. An example for a standardised office is given in Table 4 for the number of users and in Table 5 for the number of appliances. From 100 stochastic demand patterns simulated with SIMDEUM at different values of the dominant variable, the peak demand values were derived, i.e. maximum momentary flow (MMF) for total and hot water (MMF_{cold} and MMF_{hot}) and the maximum hot water use (HWU) during time periods of 10 min (HWU_{10}), of 1 h (HWU_{60}), of 2 h (HWU_{120}) and of 1 day (HWU_{day}). The resulting design demand parameters were defined as the

Table 6. Selected buildings for measurement of cold and hot water demand and the number of measured days and weekdays, to validate the developed design-demand-equations for offices, hotels and nursing homes.

category offices			category hotels			category nursing homes		
typology office A	number of employees (x)	total days/ weekdays	typology: business hotel	number of hotel rooms (x)	total days/ weekdays	typology: care needed residents	number of beds (x)	total days/ weekdays
office I	255	47/33	business hotel I	80	32/22	nursing home I	124	36/26
office II	2000	34/24	business hotel II	192	28/18	nursing home II	260 212 (hot)*	29/20

* One wing of nursing home II has its own boiler. The measured hot water use concerns 212 beds.

99-percentile of the 100 values of the different peak demand values. These peak demand values for several buildings could be described by simple linear relations as a function of the dominant variable. These linear relations form the design-demand-equations.

3 Methodology

3.1 Methodology to measure cold and hot water demand of non-residential buildings

For the measurements of cold and hot water demand an ultrasonic flow meter was used. The Proline Prosonic Flow meter is a clamp-on meter that can be installed without disturbing the water supply. Requirements for a reliable measurement are closely met, such as sensor distance dependent on pipe material characteristics and distances from bends and other appendages. To ensure an accurate measurement of peak demand, the logging frequency was set to 1 s. The logging precision was 0.5 %. The water demand was measured during minimal 20 weekdays for hotels and nursing homes and 30 working days for offices.

In each category of non-residential users, two buildings were selected with different values of the dominant variable. It appeared to be very difficult to find suitable buildings for the measurements. Narrow installation environments, the presence of buffers in the hot water circulation system, not enough straight pipes to install the flow sensor with desired distance from fittings, such as valves, T-pieces, elbows, etc. were the most common issues. Moreover, it appeared that Dutch offices are seldom equipped with a collective hot water system. Therefore, the measurement of hot water was not possible, and the developed design-demand-equations for offices can only be validated on cold water use. Since the hot water use in offices is very small, this is not a problem. The selected buildings are given in Table 6. During the measurements, full occupation was aimed for. However, this was difficult to achieve for the two hotels. The occupation varied between 4–100 % during weekdays.

3.2 Methodology validation step 1: validation of how to standardise buildings on which the design-demand-equations are based

3.2.1 Phase 1: methodology to compare cold and hot water demand patterns of non-residential buildings

The simulated patterns of the standardised buildings are compared with the measured daily patterns of cold and hot water. The comparison of the median demand patterns, allows to test the assumptions in the standardised buildings, concerning the presence of the users and their probability of water use during the day (see Table 4). Moreover, the height of the demand patterns gives information on the assumptions of average water use per person, defined in for example the frequency and use of water saving options of the toilets (see Table 5).

For each building, the medians of the measured diurnal patterns of cold and hot water demand are compared with the corresponding medians of simulated patterns of the standardised building. For an objective evaluation, the coefficient of determination R^2 is applied as statistical parameter. R^2 is a measure for the similarity between the patterns and calculated as:

$$R^2 = 1 - \sum_{i=1}^{N} (y_i - x_i)^2 \Big/ \sum_{i=1}^{N} (x_i - \overline{x})^2 \tag{1}$$

where x_i is the measured data, y_i is the simulated data and $\overline{x}$ is the mean of x_i. Good agreement between measurements and simulations is achieved when $R^2 > 0.70$ (Blokker et al., 2011).

3.2.2 Phase 2: methodology to survey the construction of standardised buildings

In each category of non-residential buildings, extensive surveys were held in at least four buildings of different typology. In the surveys the number and characteristics of users and appliances were investigated and compared with the construction of the standardised buildings, as is given for offices in

Table 7. Number of interviewed buildings in each category of non-residential buildings and the requested information in the surveys.

category	typology	Number of buildings	users	appliances	additional
office		8	Number Time of presence	Number Flow Type	Behaviour employees (2 offices): Time and frequency of use (toilets, urinals, coffee machine) Type of use (saving options) Use of appliances in kitchen (1 office): duration and frequency
hotel	business	5*	Number Time of presence	Number Flow Type	Behaviour business hotel guest: frequency and duration of shower and bath use
	tourist				Behaviour tourist hotel guest: frequency and duration of shower and bath use
nursing home	care needed residents	5	Number Time of presence	Number Flow Type	Intensity of use bedpan washer Frequency of use toilet and shower of residents
	self-contained apartments	3			

* Hotels were not fully business or tourist. Some hotels were characterised as business on week days and tourist in weekend or 70 % business and 30 % tourist hotel. In the construction of standardised buildings, this does not cause problems, since the difference in water use is mainly caused by the different behaviour of a tourist guest and business guest.

Tables 4 and 5. In Table 7, the number of investigated buildings, the requested information of users and appliances in each functional room (see Table 2) and the additional information in different categories is summarised.

It was very difficult to collect these data. In practice, no survey send out to selected building owners were returned. It was required to collect the requested information by interviews. This was a highly labor and time intensive process: finding cooperation was a lengthy process, the information could not always be given by one person, the information was of a too high level of detail.

Agreement between the outcome of the surveys and the construction of the standardised buildings, was established in graphs in a visual way, a statistical test was not possible due to the limited number of buildings.

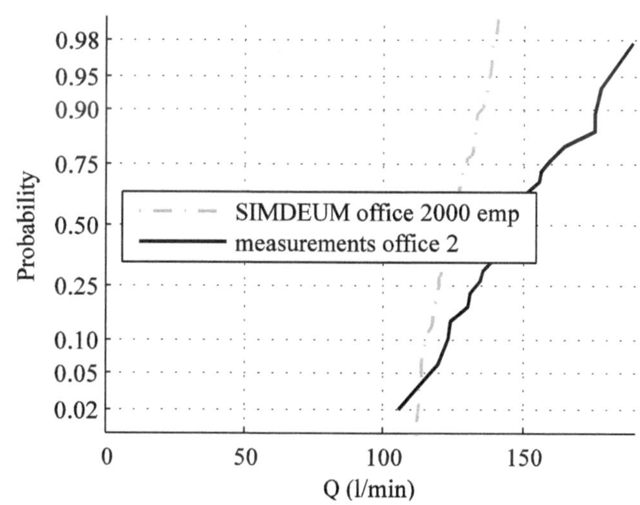

Figure 1. Frequency distribution of measured (24 weekdays at 1 s scale) and simulated (100 days at 1 s scale) peak demand values for office II: $Q = \mathrm{MMF}_{\mathrm{cold}}$ [$\mathrm{L\,min^{-1}}$].

3.3 Methodology validation step 2: validation of design-demand-equations for non-residential buildings based on SIMDEUM

For the validation of the design-demand-equations, MMF and HWU from the design-demand-equations are compared with measured peak demand values with the following methodology:

a. MMF and HWU are taken from SIMDEUM based design-demand-equations for the selected buildings (Table 6).

b. measured MMF and HWU were determined as $\overline{x} + 3 \cdot \sigma$, where $\overline{x}$ is the average value of the peak demand values from approx. 20–30 measured diurnal patterns and σ is the corresponding standard deviation. For offices and nursing homes, the frequency distributions of the measured peak demand values follow a standard normal distribution as is illustrated in Fig. 1 for office II and in Fig. 2 for nursing home I. Therefore, the resulting

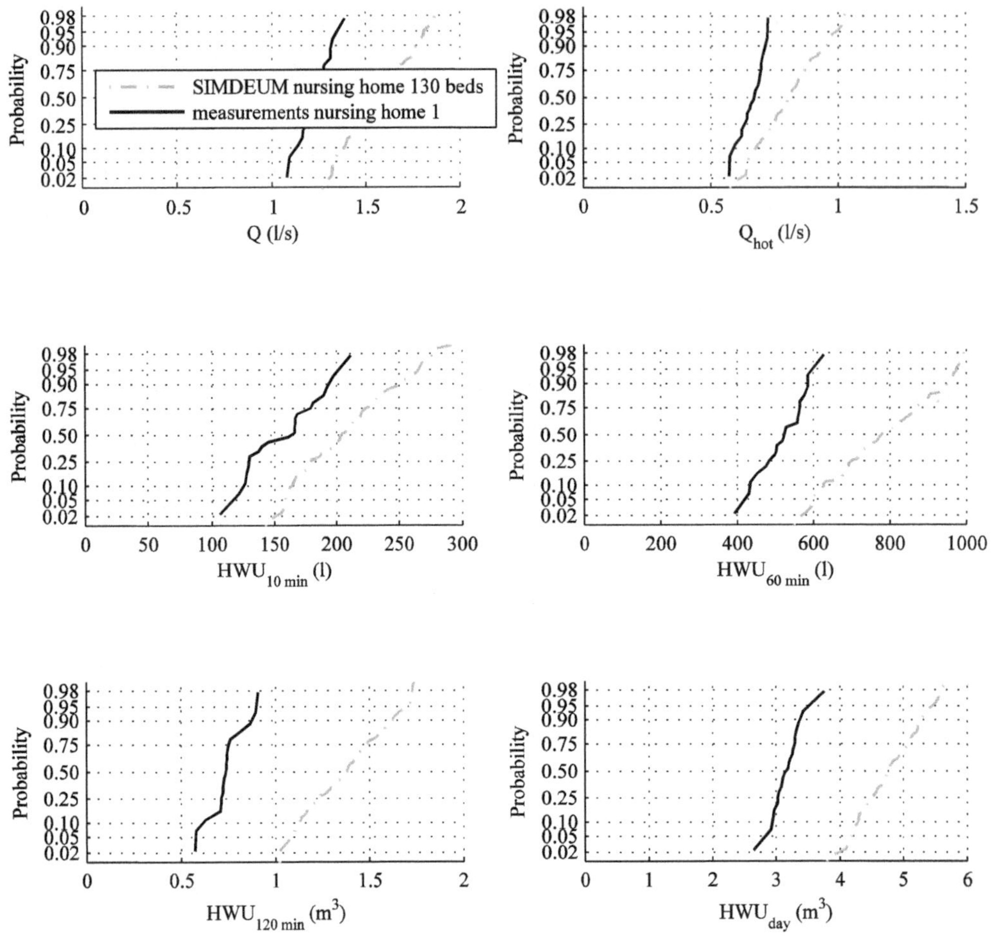

Figure 2. Frequency distribution of measured (26 weekdays at 1 s scale) and simulated (100 days at 1 s scale) peak demand values for nursing home I: **(a)** MMF_{cold} $[L\,s^{-1}]$, **(b)** MMF_{hot} $[L\,s^{-1}]$, **(c)** HWU_{10} [L], **(d)** HWU_{60} [L], **(e)** HWU_{120} $[m^3]$, **(f)** HWU_{day} $[m^3]$.

MMF and HWU represent the 99.7-percentile. For hotels, the measured peak demand values do not follow a normal distribution due to the highly variable occupation (Fig. 3). The figure shows that the highest values of the peak demand fall in the last 10 % of the measured peak demand values. In this region of the frequency distribution plot fall the water demand values of days with full occupation. The maximum value of the parameters (MMF and HWU) do not always coincide with the days with full occupation nor do they occur on the same day, e.g. MMF_{cold} and HWU_{10} take place at different days and even on days with low occupation. Therefore, the outcomes of the design-demand-equations are compared with the maximum value of each peak demand characteristic and with the average of the measured peak demand on the days with full occupation.

3.4 Methodology to investigate the consequences for design of distribution systems and heating system

The consequences of the new SIMDEUM based design-demand-equations are investigated by comparing the designs of the distribution systems and the heating system with designs in practice:

1. *compare design distribution system based on MMF with designs in practice*

   The design of distribution systems is based on MMF, obtained with a time step of 1 s. The maximum velocity (v_{max}) of flowing water in pipes, determines the selected diameter. The internal diameter of pipes is related to MMF (in $[L\,s^{-1}]$) and v_{max} by:

$$d_{in} = \sqrt{\frac{MMF/1000}{0.25 \cdot \pi \cdot v_{max}}} \cdot 1000\,[mm] \tag{2}$$

   where v_{max} is the maximal allowed design velocity set at $2\,m\,s^{-1}$, d_{in} is the internal pipe diameter in mm. The calculated internal diameter is compared with a limited

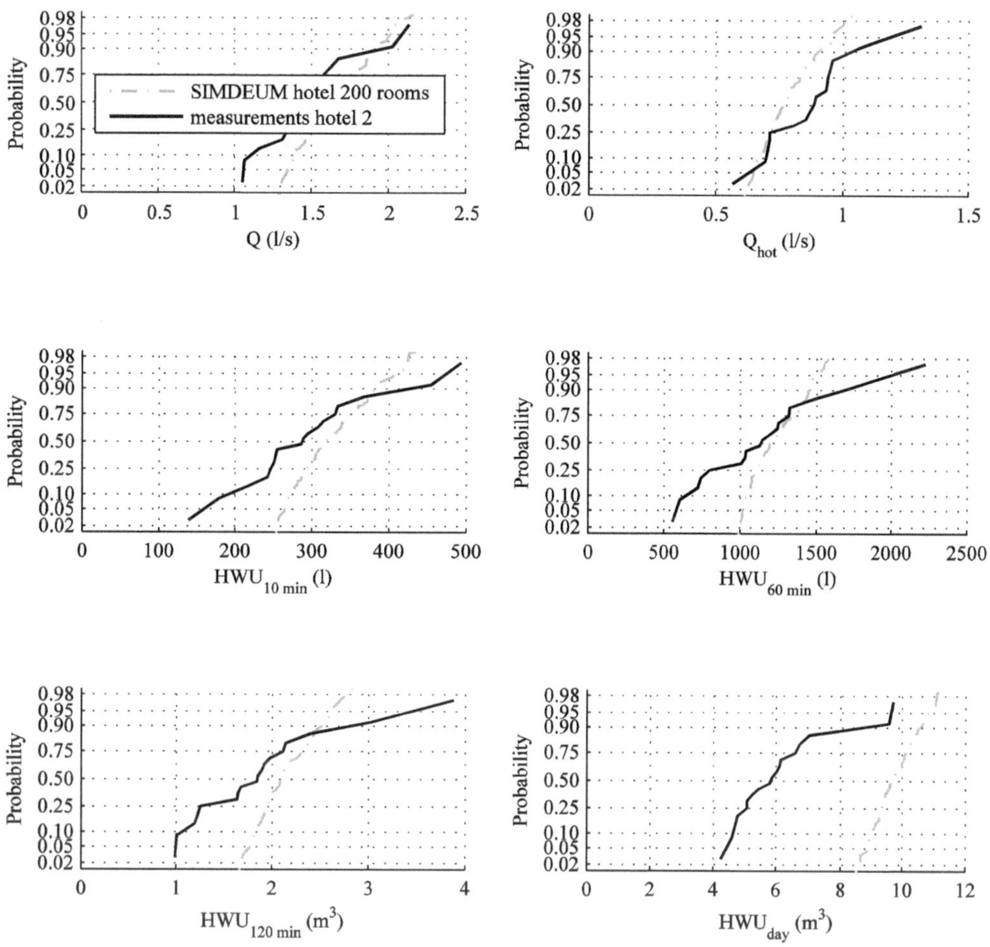

Figure 3. Frequency distribution of measured (18 weekdays at 1 s scale) and simulated (100 days at 1 s scale) peak demand values for hotel II: **(a)** MMF_{cold} [L s^{-1}], **(b)** MMF_{hot} [L s^{-1}], **(c)** HWU_{10} [L], **(d)** HWU_{60} [L], **(e)** HWU_{120} [m^3], **(f)** HWU_{day} [m^3].

number of practical copper pipe diameters, applied in drinking water systems. The smallest copper pipe fulfilling the calculated internal diameter is selected in design.

2. *compare design heating system based on MMF_{hot} and HWU with designs in practice*

The outcome of the SIMDEUM based design-demand-equations for the characteristics of hot water demand, MMF_{hot}, HWU_{10}, HWU_{60}, HWU_{120} and HWU_{day} are applied in general design tools to determine the desired volume (V in [L]) and power (P in [kW]) of a hot water charging system (ISSO-55, 2001). The resulting dimensions of the heating systems based on SIMDEUM are compared with dimensions proposed by different suppliers of heating systems for various buildings.

4 Results and discussion

4.1 Validation step 1: validation of how to standardise buildings on which the design-demand-equations are based

4.1.1 Phase 1: validation of cold and hot demand patterns of standardised non-residential buildings

The simulated patterns of the standardised buildings are compared with the measured daily patterns of cold and hot water. In Fig. 4 an example is given for one building in each category. The visually good agreement between the daily patterns is confirmed by the values of R^2 which is larger than 0.70 for offices, for cold water in nursing home I and for both cold and hot water in the nursing home II (Table 8). The lower value for hot water in nursing home I, is caused by the cleaning of the wash bowls on one location and at one specific moment in the afternoon, which was not applied in standardised nursing home. The lower value for hotels is most probably due to the

Table 8. Statistics of the comparison between measured and simulated water demands for offices, business hotels and nursing homes and their corresponding standardised buildings.

category offices	R^2	category hotels	R^2		category nursing homes	R^2	
	cold		cold	hot		cold	hot
office I	0.90	business hotel I	0.54	0.49	nursing home I	0.71	0.38
office II	0.96	business hotel II	0.45	0.50	nursing home II	0.82	0.71
			0.61	0.59			

varying occupation of the hotel. The morning peak appears later in reality and there is a difference in evening peak.

This phase shows that the assumptions made for the presence of users, their probability of water use during the day and for average water use per user are reliable for offices and nursing homes. For hotels, the presence of hotel guests and their probability of water use during the day is not estimated very well, following from the differences in morning and evening peak demand. On the other hand, the average water use per user is estimated in a reliable way, following from the height of the demand patterns.

This phase also shows the advantage of using the physically based SIMDEUM for the development of design-demand-equations, since differences can be interpreted and explained.

4.1.2 Phase 2: validation of construction of standardised non-residential buildings

The construction of the standardised buildings is validated by comparing the number and characteristics of appliances and users in all functional rooms (Table 2). The surveys mainly give information on the ratio between users and appliances. As an illustration, the results for offices are presented in Fig. 5, where the outcome of the surveys for number of users are shown together with the relations of the standardised building (Table 4). Figure 5a shows that the number of users in the dominant functional room, i.e. the meeting area (Table 3) corresponds well with practice. Also the behaviour of the users, like the frequency of toilet use, the use of coffee machine and the time of presence correspond well. In practice, the water saving options of the toilet flush are used less frequently than assumed in the standardised building. On the other hand, the number of users in the other functional rooms, such as the restaurant, the fitness room and the technical room (Fig. 5b) show higher variations and are less accurate. Moreover, the data from the surveys show that it is not possible to extract a more appropriate standard from the available data. The same is true for the appliances: the number and types of appliances in the dominant functional room, like the number of toilets for men and women, the number of wash basins and the number of coffee machines correspond well with practice. Only the number of urinals is lower in practice than assumed in the standardised office. This means that in the standardised offices a higher simultaneous use of urinals is possible, which might result in a higher peak demand predicted by the design-demand-equations. The assumptions for the number of appliances in other functional rooms are reasonable. Sometimes, not enough information can be extracted from the surveys to draw any conclusions.

The surveys indicate that the equipment of the standardised office is reliable for the dominant functional room, the meeting area. The functional room, that contributes most to the pattern and total amount of water use in a building is equipped corresponding with practice: the number of users and their water using behaviour as well as the number of appliances correspond with the surveyed offices. The agreement between the simulated and measured demand patterns of $R^2 > 0.90$ (Fig. 4a) means that a reliable equipment of the functional room in the standardised office is sufficient for a realistic prediction of the water demand of a building. Moreover, it shows that as long as the ratio between users and appliances is not extreme, for example when there are not enough toilets for the users causing queues, the exact ratio is not crucial. This gives SIMDEUM the possibility to be less exact, while still predicting the water demand in a reliable way.

The same tendency is found for hotels and nursing homes. The dominant functional rooms in the standardised buildings, respectively the hotel rooms for hotels and the lodging and meeting area for nursing homes have a reliable equipment regarding the number of users and appliances. The other functional rooms are less accurate and it is not possible to find better relations to equip the other functional rooms in the standardised buildings with users and appliances from the surveys. The reliable equipment of the dominant functional rooms is sufficient for a reliable prediction of the expected cold and hot water demand in the building (Fig. 4b and c). In short, the proof of concept is established for all considered categories of non-residential buildings (Pieterse-Quirijns et al., 2013a, b, c).

4.2 Validation step 2: validation of design-demand-equations for peak demand values for cold and hot water

The measured peak demand values for all non-residential buildings are presented in Table 9. This table also includes

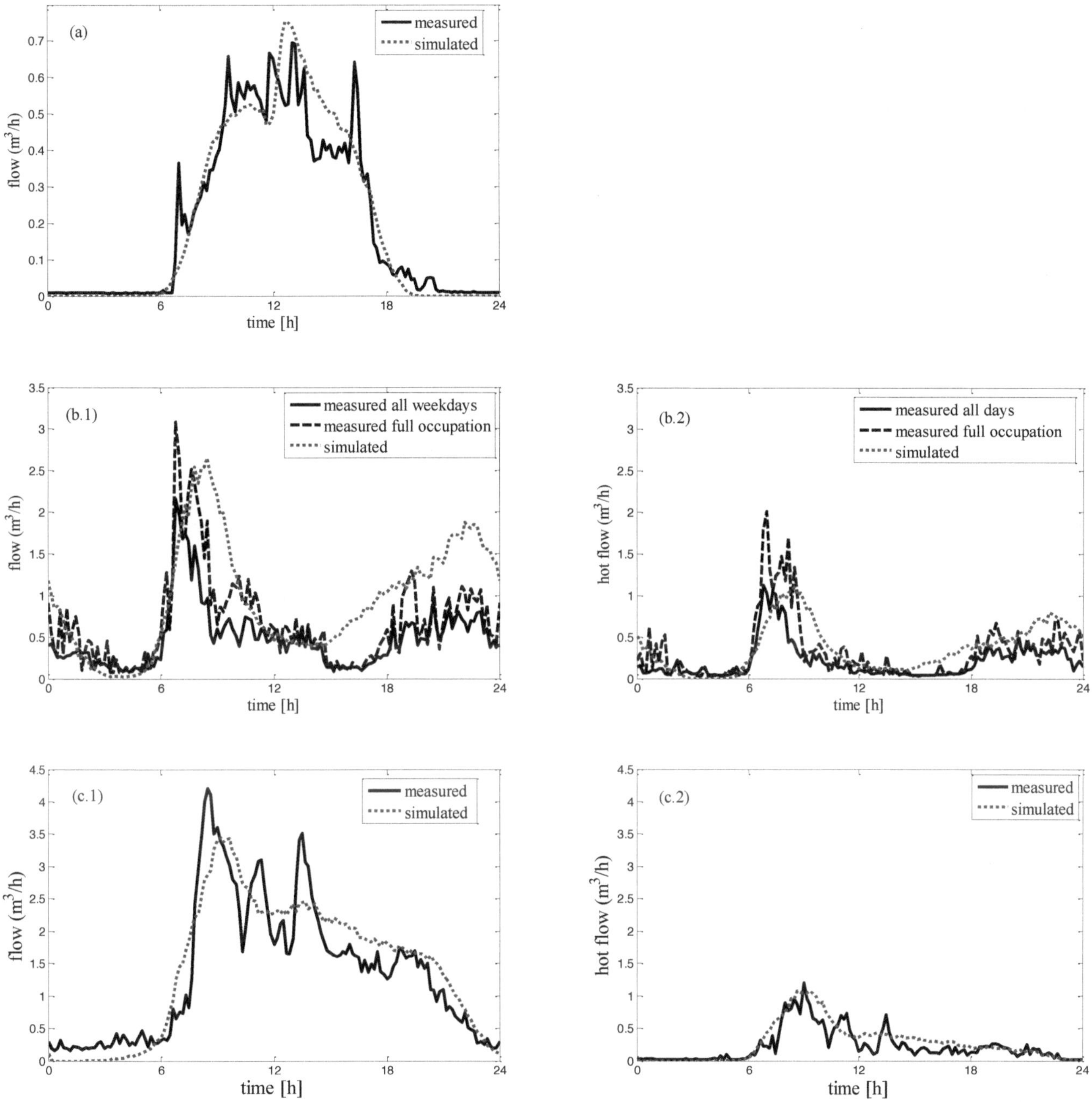

Figure 4. Comparing average measured and simulated demand of cold water of office I **(a)**, cold water **(b.1.)** and hot water **(b.2.)** of a business hotel (II) during weekdays and cold water **(c.1.)** and hot water **(c.2.)** of a nursing home (II) with care needed residents during weekdays.

the predicted peak demand values by the newly developed design-demand-equations, based on SIMDEUM. For offices of different sizes, the SIMDEUM based design-demand-equations predict the MMF_{cold} well. They slightly underestimate the MMF_{cold} with 10 %.

For hotels, the derivation of peak demand values from the measured water demand patterns is difficult, due to the varying occupation. 30 measuring days appears to be too short to achieve a statistically sound procedure because of the additional variation due to variable occupation (Fig. 3). For statistically sound conclusions for hotels with varying occupations a longer measuring period is recommended. Careful comparison shows that the deviation between measured peak demand and the outcome of the design-demand-equations varies from an underestimation of maximal 30 % to an overestimation of maximal 15 %. The MMF_{cold} can be predicted fairly well, while the hot water demand appears to be (slightly) underestimated.

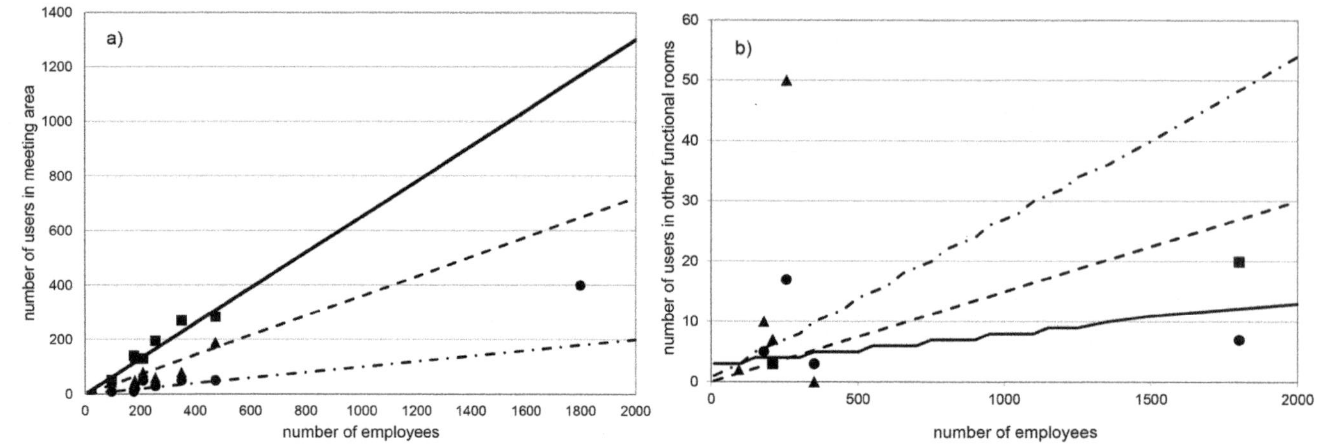

Figure 5. Number of users in the standardised office of design-demand-equations (lines) and in surveyed offices (data points) in the dominant functional room of offices, meeting area **(a)** with male employees (- and ■), female employees (- - and ▲) and visitors (-.- and •) and in other functional rooms **(b)** with kitchen personnel (- and ■), employees using the fitness room (- - and ▲) and cleaners (-.- and •).

The new design-demand-equations predict a higher cold and hot water demand of the nursing homes (2–72 %), except for MMF_{hot} in nursing home II, that is underestimated with 16 %. The higher predicted water demand can be caused by the assumed higher number of personnel in the standardised nursing homes (following from phase 2 in validation step 1). Moreover, in nursing home I the different practice in cleaning the wash bowls plays a role.

This paper shows that the solid basis of SIMDEUM and the standardisation of buildings for offices, hotels and nursing homes result in a reliable prediction of peak demand values by the design-demand-equations derived with the procedure based on SIMDEUM. With this paper it is shown that the procedure leads to good results for three types of non-residential buildings. The way we demonstrate this, is by not only showing that the procedure leads to good end results (maximum flows and hot water uses), but also by showing that the underlying principle (the assumptions of the appliances and people in the buildings and the resulting diurnal patterns of cold and hot water) is sound. These results suggest that the procedure for the design-demand-equations will result in reliable prediction for other small-scale buildings, when the standardisation of the buildings is solid. Due to the modular approach of SIMDEUM and its physical basis, it is possible to construct a specific building and simulate its water demand following a few steps: define the functional rooms where water is used, and equip each functional room with its typical users and appliances. From each user, the time of presence, the time and frequency of use needs to be estimated, which can be done with logical thinking, experience or interviews. This paper illustrates that mainly the dominant user or functional room needs to be equipped well. For many small-scale buildings these can be determined quite easy: for example toilet rooms in schools, used by children and used by teachers, toilet in a shop, etc. In this way, SIMDEUM was used to develop diurnal profiles for several different buildings, like schools, shops, restaurants, sporting facilities etc. (Pieterse-Quirijns and Van de Roer, 2013). One example is described in Blokker et al. (2011), where a theatre hotel is shown.

4.3 Consequences for design of distribution systems and heating system

The design of the outside and inside distribution system is based on the maximum momentary flows (MMF_{cold} and MMF_{hot}) for pipes distributing the total amount of water and the hot water pipes respectively. In Table 9, the resulting pipe diameters are shown, based on the measured and the predicted peak demand values. The table shows that the newly developed design-demand-equations result in the same design as the measurements would. This indicates that the deviation between the design-demand-equations and the measurements falls within the design accuracy, given by the limited number of existing pipe diameters.

In the same table, the diameters resulting from water demands predicted with old guidelines (Table 1) and applied in practice are included. The old guidelines always result in a larger diameter than the new design-demand-equations, caused by a large overestimation of the expected water demand. The old guidelines overestimate the MMF_{cold} with 30–80 % for offices, with 70–170 % for hotels, and with 75–166 % for nursing homes, which is much larger than the deviations of the new design-demand-equations.

Moreover, the design-demand-equations also result in smaller diameters than applied in practice, while still meeting the desired comfort wish (resulting from the measurements). The table shows that the practical diameters are even larger than the diameters resulting from the old guidelines. Apparently, extra safety factors are applied in practice.

Table 9. Comparison of the measured peak demand values for cold and hot water with the outcome of the SIMDEUM based design-demand-equations ("new") and existing guidelines for cold water ("old") for non-residential buildings (offices, hotels and nursing homes) and the resulting pipe diameters compared with diameters used in practice ($d_{pract.}$). Deviation from measured peak demand values is shown between brackets.

Building	Peak demand parameter	Peak demand value measured	design-demand-equations		old guidelines		Consequences for of pipes [mm] selection diameter $d_{pract.}$	$d_{meas.}$	d_{new}	d_{old}
offices										
office I	MMF_{cold} [$L s^{-1}$]	1.1	1.0	(−11 %)	1.9	(+79 %)	54	35	35	42
office II	MMF_{cold} [$L s^{-1}$]	4.0	3.7	(−8 %)	5.3	(+32 %)	54	54	54	64
hotels										
based on maximal measured values (not on the same day)										
business hotel I	MMF_{cold} [$L s^{-1}$]	1.9	1.3	(−27 %)	3.1	(+69 %)	76	42	35	54
	MMF_{hot} [$L s^{-1}$]	0.8	0.8	(−10 %)	–		54	28	28	–
	HWU_{10} [L]	339	293	(−14 %)	–					
	HWU_{60} [L]	987	851	(−14 %)	–					
	HWU_{120} [L]	1500	1360	(−9 %)	–					
	HWU_{day} [L]	5560	4998	(−10 %)	–					
based on average measured values during full occupation										
business hotel I	MMF_{cold} [$L s^{-1}$]	1.5	1.3	(−11 %)	3.1	(+107 %)	76	35	35	54
	MMF_{hot} [$L s^{-1}$]	0.7	0.8	(+11 %)	–		54	28	28	–
	HWU_{10} [L]	255	293	(+15 %)	–					
	HWU_{60} [L]	844	851	(+1 %)	–					
	HWU_{120} [L]	1288	1360	(+6 %)	–					
	HWU_{day} [L]	5143	4998	(−3 %)	–					
based on maximal measured values (not on the same day)										
business hotel II	MMF_{cold} [$L s^{-1}$]	2.0	2.1	(+2 %)	4.8	(+139 %)	67	42	42	64
	MMF_{hot} [$L s^{-1}$]	1.3	1.1	(−18 %)	–		54	35	35	–
	HWU_{10} [L]	494	455	(−8 %)	–					
	HWU_{60} [L]	2225	1594	(−28 %)	–					
	HWU_{120} [L]	3882	2667	(−31 %)	–					
	HWU_{day} [L]	9732	10 574	(+9 %)	–					
based on average measured values during full occupation										
business hotel II	MMF_{cold} [$L s^{-1}$]	1.8	2.1	(+15 %)	4.8	(+170 %)	67	42	42	64
	MMF_{hot} [$L s^{-1}$]	1.2	1.1	(−10 %)	–		54	35	35	–
	HWU_{10} [L]	474	455	(−4 %)	–					
	HWU_{60} [L]	1994	1594	(−20 %)	–					
	HWU_{120} [L]	3457	2667	(−23 %)	–					
	HWU_{day} [L]	9662	10 574	(+9 %)	–					
nursing homes										
nursing home I	MMF_{cold} [$L s^{-1}$]	1.5	1.9	(+31 %)	3.9	(+166 %)	42	35	42	54
	MMF_{hot} [$L s^{-1}$]	0.8	1.0	(+27 %)	–		2 × 28	28	28	–
	HWU_{10} [L]	243	294	(+21 %)	–					
	HWU_{60} [L]	698	1041	(+49 %)	–					
	HWU_{120} [L]	983	1693	(+72 %)	–					
	HWU_{day} [L]	3830	5366	(+40 %)	–					
nursing home II	MMF_{cold} [$L s^{-1}$]	3.2	3.3	(+2 %)	5.6	(+75 %)	54+35	54	54	76
(260 beds cold)	MMF_{hot} [$L s^{-1}$]	1.9	1.6	(−16 %)	–		54	42	42	–
(212 beds hot)	HWU_{10} [L]	381	454	(+19 %)	–					
	HWU_{60} [L]	1432	1853	(+29 %)	–					
	HWU_{120} [L]	2354	3122	(+33 %)	–					
	HWU_{day} [L]	7198	11 861	(+65 %)	–					

Table 10. Dimensions of heating systems for different type of buildings, based on measured hot water demand, based on SIMDEUM based design-demand-equations and proposed by companies.

type of building	design based on measurements		design based on SIMDEUM		proposal company	
	V [L]	P [kW]	V [L]	P [kW]	V [L]	P [kW]
hotel I (small business)	500	30	500	35	1000	200
hotel II (large business)	1000	85	1000	60	4000	200
hotel III (tourist)			250	50	740	100
nursing home I	500	25	500	45		
nursing home II	1000	55	1000	75		
nursing home III: care needed residents			250	25	500	45
nursing home IV: self-contained apartments with independent resident			500	37.5	1000	100

With the supply of the same amount of water, the smaller diameters will result in better water quality with respect to hygiene, due to higher velocities and less stagnancy. Moreover, reducing pipe diameters, also reduces the actual surface areas of pipework material to support biofilm growth, whilst also reducing the sizes of bracketing, ancillaries and insulation material. This contributes to a more sustainable use of materials and reduced installation costs (Angus et al., 2010).

The resulting dimensions of the water heating systems based on SIMDEUM are compared with dimensions proposed by different suppliers of heating systems, presented in Table 10. To know which dimensions are required to meet the comfort in a building, the measured hot water demands, when available, are also applied in the same design tool for a heating system. The dimensions based on measured hot water demands are also given in Table 10. For a small business hotel, the actual measured hot water demand, during full occupation requires a heating system with a volume of 500 L and a power of 30 kW. The dimensions resulting from the by SIMDEUM predicted hot water demand are 500 L and 35 kW. The supplier, on the other hand, proposes for this hotel a heating system of 1000 L and 200 kW. This comparison shows that SIMDEUM does not underestimate the dimensions of the heating system, while the supplier's overestimation is large. This tendency is generally found for different types of buildings (Table 10). The suppliers of heating systems propose heaters with too large capacities, both in volume and power that do not match with the actual hot water demand. The proposed capacities are 2 and sometimes 4 times larger than needed. This shows that SIMDEUM based design-demand-equations result in heating systems that still fulfil the desired comfort wish, while being more energy efficient. The smaller heaters require less energy to heat the system and will have less energy losses. Moreover, the smaller design of the heating system reduces the stagnancy of water, which might lead to less hygienic problems.

The paper shows that SIMDEUM forms a reliable basis to predict non-residential cold and hot water demand. The new SIMDEUM based design-demand-equations lead to smaller diameters and reduced capacity of heating systems. These designs diverge from the common practice in the Netherlands. In this paper it is shown that the common practice leads to oversized systems. In other countries the same tendency is expected, although international guidelines do not exist in the public domain. Therefore, it is recommended to apply the SIMDEUM based procedure to develop design-demand-equations for peak demand values for various building types in specific countries to improve the designs. Due to the physical basis of SIMDEUM, the presented procedure is easily transferable to other countries when specific information on users and appliances is available. Also in other countries, we succesfully used SIMDEUM for simulating the water demand for residential water use.

The US National Bureau of Standards approach to determine the MMF (the so called Hunter's curves) has been incorporated into design codes for plumbing systems. Present day experience indicates that Hunter's curves often overestimate actual water demands. There is recognition of a need to update methods for estimating peak water demands in plumbing systems to better reflect today's high efficiency water fixtures and realistically account for non-congested conditions in a variety of end use applications (e.g. residential, commercial, institutional, etc.). Buchberger et al. (2012) suggest an update of Hunter's method, following an analytical approach. It would be worthwhile to compare these analytical estimates for the peak demand with the ones obtained from the measurements and from the SIMDEUM based design-demand-equations.

5 Conclusions

Recently, new SIMDEUM based design-demand-equations were developed for cold and hot water demand of non-residential buildings. In this paper, these design-demand-equations are validated with measurements of both cold and hot water on a per second basis for different buildings and with surveys.

Firstly, the assumptions of building standardisation, on which the design-demand-equations are based, were validated with surveys and measurements of both cold and hot water on a per second basis for six different non-residential buildings, offices, hotels and nursing homes. The validation shows that the model predicts the cold and hot water daily demand patterns reasonably well to good. The correlation of the simulated patterns with the measured patterns indicates that the basis of the design-demand-equations is solid.

Moreover, the construction of standardised buildings with users and appliances is validated with at least four surveys per category of non-residential buildings. The validation proves that with a proper estimation of the number of users and appliances in only the dominant functional room of the standardised buildings, SIMDEUM renders a realistic cold and hot water diurnal demand pattern.

With the correspondence between standardised buildings and reality, in both diurnal water demand patterns and construction of the building, the solid base of SIMDEUM based design-demand-equations is proven by the proof of concept, i.e. when the construction is reliable, the prediction of the expected cold and hot water demand in the building will be reliable.

Secondly, the outcome of the design-demand-equations is validated. The validation shows that the design-demand-equations predict the cold and hot water demand peak values reasonable to good. The deviation of the predicted demand values from the measured ones is mostly less than 15 %. Sometimes a higher deviation is found, but less than 30 %. Only the deviation of the predicted hot water demand of nursing homes is higher, caused by a higher assumed number of personnel or another cleaning practice. These deviations are much smaller than the deviations of existing guidelines for MMF_{cold} which overestimate by 30–170 %. Moreover, the new design-demand-equations give insight into hot water use, where no existing public guidelines are available.

The new design-demand-equations improve the design of the building installation, expressed as the selected pipe diameters, and of the heating system. First of all, the design-demand-equations lead to the same pipe diameter and the same heating system as based on the measured peak values. This means that the design based on the new rules is reliable. Secondly, the selected pipe diameters are smaller than the ones used in practice and the ones predicted by the existing guidelines. This indicates that the common practice leads to oversized systems, with corresponding potential quality problems. Also the heater capacity can be reduced based on the outcome of the design-demand-equations. SIMDEUM based design-demand-equations reduce the design of a heater capacity with a factor 2 to 4 compared to suppliers proposals, while still meeting the desired need and comfort. Thus, the improved insight of the new design-demand-equations will lead to an energy efficient choice of the hot water systems, and thus save energy. Moreover, the smaller design of the heating system reduces the possibility of stagnant water, that might reduce the chance on hygienic problems.

References

Angus, P., Ingle, S., King, D., and Turner, J.: The effects of using water velocity as a technique to control biofilm development in water supply systems, Water Supply and Drainage for Buildings, CIB W062, Sydney, Australia, 8–10 November, 106–124, 2010.

Baggelaar, P. K. and Geudens, P. J. J. G.: Prognose landelijke drinkwatervraag t/m 2005, Vewin, Rijswijk, The Netherlands, 2008.

Barberán, R., Egea, P., Gracia-de-Rentería, P., and Salvador, M.: Evaluation of water saving measures in hotels: a Spanish case study, Int. J. Hosp. Manag., 34, 181–191, 2013.

Blokker, E. J. M. and Van der Schee, W.: Simulation of water demands provides insight, Water Supply and Drainage for Buildings, CIB W062, Taipee, Taiwan, 18–20 September, 2006.

Blokker, E. J. M., Vreeburg, J. H. G., and Van Dijk, J. C.: Simulating residential water demand with a stochastic end-use model, J. Water Res. Pl.-ASCE, 136, 19–26, 2010.

Blokker, E. J. M., Pieterse-Quirijns, E. J., Vreeburg, J. H. G., and van Dijk, J. C.: Simulating non-residential water demand with a stochastic end-use model, J. Water Res. Pl.-ASCE, 137, 511–520, 2011.

Buchberger, S. G., Blokker, E. J. M., and Cole, D.: Estimating peak water demands in hydraulic systems I – current practice, Water Distribution System Analysis, WDSA 2012, Adelaide, Australia, 24–27 September 2012.

Cobacho, R., Arregui, F., Parra, J. C., and Cabrera Jr., E.: Improving efficiency in water use and conservation in Spanish hotels, Water Sci. Technol., 5, 273–279, 2005.

Ilemobade, A. A., Van Zyl, J. E., and Van Zyl, H. J.: New guidelines for non-domestic water estimation in South Africa, in: Integrating Water Systems CCWI 2009, CRC Press/Balkema Leiden, Sheffield, UK, 543–549, 2010.

ISSO-55: ISSO – publicatie 55; Tapwaterinstallaties voor woon – en utiliteitsgebouwen, Stichting ISSO, Rotterdam, The Netherlands, 2001.

Loureiro, D., Coelho, S. T., Rebelo, M., Ramalho, P., Alegre, H., Covas, D. I. C., and Pina, A.: Correlating water consumption behaviour with billing, infrastructure and socio-demographic factors, in: Integrating Water Systems CCWI 2009, CRC Press/Balkema Leiden, Sheffield, UK, 105–111, 2010.

Pieterse-Quirijns, E. J. and Van de Roer, M.: Verbruikspatronenbibliotheek, KWR, Nieuwegein, The Netherlands, 2013.

Pieterse-Quirijns, E. J., Blokker, E. J. M., and Vogelaar, A. J.: Modelleren van niet-huishoudelijk waterverbruik; waterverbruik van kantoren, hotels, zorginstellingen en veehouderij, KWR, Nieuwegein, The Netherlands, 2009.

Pieterse-Quirijns, E. J., Blokker, E. J. M., Van der Blom, E., and Vreeburg, J. H. G.: Modelling characteristic values for non-residential water use, Water Distribution System Analysis, WDSA 2010, Tuscon, AZ, USA, 12–15 September, 2010.

Pieterse-Quirijns, E. J., Beverloo, H., and Van Loon, A.: Validatie rekenregels voor waterverbruik hotels, KWR, Nieuwegein, The Netherlands, 2013a.

Pieterse-Quirijns, E. J., Beverloo, H., and Van Loon, A.: Validatie rekenregels voor waterverbruik kantoren, KWR, Nieuwegein, The Netherlands, 2013b.

Pieterse-Quirijns, E. J., Beverloo, H., Van Loon, A., and Kwakkel, M.: Validatie rekenregels voor waterverbruik zorginstellingen, KWR, Nieuwegein, The Netherlands, 2013c.

Rankin, R. and Rousseau, P. G.: Sanitary hot water consumption patterns in commercial and industrial sectors in South Africa: impact on heating system design, Energ. Convers. Manage., 47, 687–701, 2006.

Scheffer, W. J. H.: Het ontwerpen van sanitaire installaties, Misset uitgeverij bv, Arnhem, The Netherlands, 1994.

Van der Kooij, D., Veenendaal, H. R., and Scheffer, W. J. H.: Biofilm formation and multiplication of *Legionella* in a model warm water system with pipes of copper, stainless steel and cross-linked polyethylene, Water Res., 39, 2789–2798, 2005.

Vreeburg, J. H. G.: Discolouration in drinking water systems: A practical approach, Ph.D. thesis, Delft University of Technology, Delft, The Netherlands, 2007.

Werkgroep momentane waterverbruiken: maximum momentane waterverbruiken, Kiwa N.V., Nieuwegein. mededeling nr. 93, 1985.

10

Low-cost multi-stage filtration enhanced by coagulation-flocculation in upflow gravel filtration

L. D. Sánchez[1], L. M. Marin[1], J. T. Visscher[2], and L. C. Rietveld[3]

[1]Universidad del Valle, Facultad de Ingeniería, Instituto Cinara, Cali, Colombia
[2]JTF-Rotterdam BV, Rotterdam, The Netherlands
[3]Delft University of Technology, P.O. Box 5084, 2600 GA Delft, The Netherlands

Correspondence to: L. D. Sánchez (luis.sanchez@correounivalle.edu.co)

Abstract. This paper assesses the operational and design aspects of coagulation and flocculation in upflow gravel filters (CF-UGF) in a multi-stage filtration (MSF) plant. This study shows that CF-UGF units improve the performance of MSF considerably, when the system operates with turbidity above 30 NTU. It strongly reduces the load of particulate material before the water enters in the slow sand filters (SSF) and therewith avoids short filter runs and prevents early interruption in SSF operations. The removal efficiency of turbidity in the CF-UGF with coagulant was between 85 and 96 %, whereas the average efficiency without coagulant dosing was 46 % (range: 21–76 %). Operating with coagulant also improves the removal efficiency for total coliforms, *E-coli* and HPC. No reduction was observed in the microbial activity of the SSF, no obstruction of the SSF bed was demonstrated and SSF runs were maintained between 50 and 70 days for a maximum head loss of 0.70 m. The most important advantage is the flexibility of the system to operate with and without coagulant according to the influent turbidity. It was only necessary for 20 % of the time to operate with the coagulant. The CF-UGF unit represented 7 % of total construction costs and the O&M cost for the use of coagulant represented only 0.3 %.

1 Introduction

Water quality and quantity from surface sources are changing due to the deterioration of watersheds caused by deforestation, erosion, and the discharge of untreated wastewater (e.g., in Colombia only 3.1 % of the total volume of wastewater produced is treated, CGR, 2009). These changes are intensified by the global climate change causing longer dry periods on the one hand and more intense rainfall on the other (Bates et al., 2008). The two most serious problems concern the peaks in turbidity level and high *E-coli* concentrations of long duration. These changes are affecting the existing water treatment plants, causing higher operation and maintenance (O&M) requirements and even interruptions in their operation (Bates et al., 2008). These growing water quality problems are not unique for Colombia but imply a significant challenge in the Andean region, because surface water is the main source in the water supply systems. In Colombia about 80 % of the water supply systems are based on surface water supply (Ministerio de Desarrollo, 1998). The problems particularly affect water supply systems in rural areas and small towns, many of which even lack adequate water treatment.

Multi-Stage Filtration (MSF) is one of the more promising and reliable water treatment options for small communities. This technology uses a combination of gravel filtration (GF) in combination with slow sand filtration (SSF). Upflow gravel filtration (UGF) is the most common pre-treatment system used for MSF in Colombia (Sánchez et al., 2006a). In UGF the water passes through the gravel bed from bottom to top. During this passage impurities are retained in the filter. Upflow filtration has the advantage that the heavier particles are removed first at the bottom of the filter. Burganos et al. (1994) have reported that upflow units have an increased collection efficiency at small and medium pore inclination angles. This concept is interesting for the theoretical analysis of particle motion and deposition, but complex to manage in practice. When the time comes to clean the filters, the accumulated particles can be removed by opening the drainage valve, allowing gravity flow to drain and clean the filter.

Research carried out by Cinara over more than 15 yr showed that different MSF alternatives that were tested, including UGF (filtration rate, $V_f = 0.6\,\mathrm{m\,h^{-1}}$) and SSF (operating at $0.15\,\mathrm{m\,h^{-1}}$), were able to produce effluents with a low microbial risk (Galvis, 1999). Research also explored the use of coagulation and flocculation with UGF, called CF-UGF, focusing, in pilot plants, on the combination with rapid filters (RF). The results showed a reduction in the consumption of coagulants by up to 30 % compared to the conventional system of coagulation, flocculation and high rate sedimentation (Cinara and IRC, 1996).

Other research into CF-UGF has emphasized the laboratory variables and the removal efficiencies. Richter and Moreira (1981) reported that a flocculation time of 3–5 min in a UGF is equivalent to a time of 15 min in the jar test under laboratory conditions and 25 min in non-compartmentalized flocculation units in full-scale plants. Santamaría (1999) showed that, using UGF, the flocculation time can be reduced by up to 60 % compared to mechanical flocculators; Salazar and Ocampo (1999) found that in CF-UGF, producing the same water quality would require between 10 and 20 times less retention time compared to a sludge blanket clarifier; Kawamura (1985), working on pilot units with UGF and RF, reported a turbidity removal of 50 %; Ahsan (1995) found that horizontal gravel filtration (HGF) with coagulation removes more particles compared to the HGF without chemicals. In addition, studies with different packed gravel beds found that a stratified bed is more efficient than a uniform bed (Attakoya et al., 1991). Di Bernardo and Sabogal (2009) have further refined some of the parameters for design and operation and maintenance, but have only applied these at pilot scale. The use of CF-UGF with MSF has had few full-scale applications. Full-scale experiences using UGF based on conventional technology with rapid filtration as a final stage were reported by Kardile (1981), working with a $V_f = 4\text{–}10\,\mathrm{m\,h^{-1}}$ and turbidity levels between 300–500 NTU, achieving construction costs between 30 to 50 % less than conventional systems of equal capacity. Bhole (1981) reported a velocity gradient in a truncated pyramid filter of $G = 1230\,\mathrm{s^{-1}}$ at the bottom layer and $35\,\mathrm{s^{-1}}$ at the upper layer with a $V_f = 11.3\,\mathrm{m\,h^{-1}}$. More recent studies indicated that aluminium residual has not shown any effects on biological activity in SSF, when coagulation with aluminium sulphate has been used, Dorea and Clarke (2006). This indicates that the addition of coagulant with UGF has potential to improve the performance of MSF during variations in influent water quality, conserving biological processes in the following stages. Consequently full-scale evaluations are necessary to better understand the design variables and operation and maintenance conditions.

The community of Colinas de Arroyo Hondo, located in a rural area of Yumbo municipality, Colombia, had a treatment plant functioning with the processes of coagulation, flocculation, sedimentation, rapid filtration and disinfection with chlorine and ultraviolet light. However, in field studies on water quality in the distribution network, biofilms were found in the pipes, which generated problems of re-growth of micro-organisms in the water supplied to the users (average values of 2183 CFU $(100\,\mathrm{ml})^{-1}$ for heterotrophic bacteria, 7 CFU $(100\,\mathrm{ml})^{-1}$ for *E-coli*, and 39 CFU $(100\,\mathrm{ml})^{-1}$ for total coliforms were also found at four points in the distribution network), (Sánchez et al., 2006b). In the treatment plant, which had been operating for 4 yr, failures were identified in its functioning, allowing solids and micro-organisms to pass into the distribution system. After an investigation, the treatment system was redesigned and rebuilt to CF-UGF with MSF. This new system has been in operation for 7 yr at the time that this evaluation study was carried out.

This paper assesses the performance and the design aspects of the CF-UGF units followed by UGF and SSF, defining influent turbidity levels to allow operation with and without coagulant, taking advantage of the SSF as the final stage of filtration for the removal of microorganisms.

2 Materials and methods

2.1 Set-up of the treatment system

In Fig. 1 the set-up of the CF-UGF MSF plant is shown. The system operation is performed by pumping raw water to the plant and then, after treatment, pumping it into the distribution network. The system uses two pumping flow rates: $6\,\mathrm{l\,s^{-1}}$, which is the maximum capacity for 12 h during the day and $3\,\mathrm{l\,s^{-1}}$ for another 12 h at night. The treatment plant consists of 5 components. Raw water passes through the flow control unit in the dynamic roughing filter (DRF), whose main function is to protect the next steps from excessive loads of suspended solids and turbidity. Filtered water flows into a rapid mixing chamber (RMC), where the coagulants are dosed. Afterwards, the water enters the CF-UGF stage which consists of 2 units in parallel, where the processes of coagulation, flocculation, sedimentation and filtration of destabilized particles occur. Water is collected in a front weir to later enter the UGF where remaining flocculated particles are removed in the different layers of gravel. From there, the water enters the SSF stage for final removal of suspended particles and microorganisms.

The RMC unit allows a proper rapid mixing time and a velocity gradient, while the CF-UGF unit facilitates flocculation and deposition of particles. The hydraulic RMC operates through a rectangular weir that allows free flow of water to form a hydraulic jump for mixing the coagulant with the raw water. Above the RMC a channel provided with holes is installed to distribute the coagulant and initiate the destabilization of the particles, which will be removed in the CF-UGF

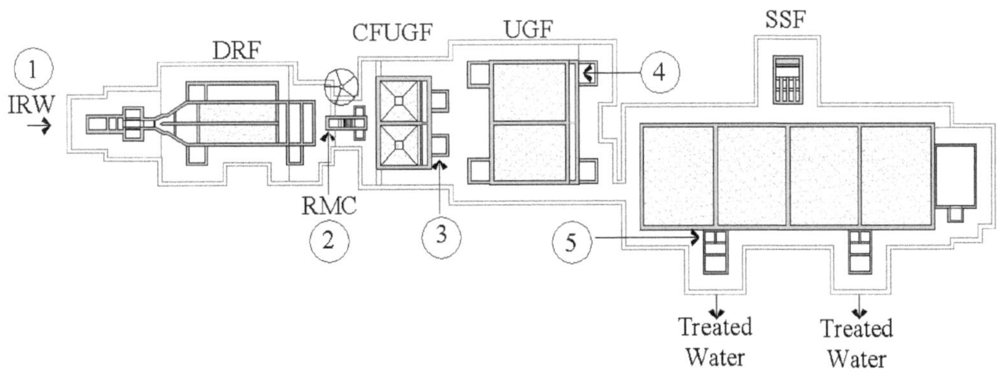

Figure 1. Layout of the treatment system (1–5: monitoring points of water quality).

Table 1. Parameters and formulae.

Parameter	Formulae	Value
RMC		
Length to the mixing point	$L_m = 1 + 0.5h_v(P_v/h_v)^{0.54}$ (1)	0.17 m
Velocity gradient for rapid mixing	$G = (\gamma E_p/\mu T_m)^{0.5}$ (2)	1282 s^{-1}
Average mixing time (T_m)	$T_m = L_r/V_{ap}$ (3)	0.31 s
CF-UGF		
Average velocity gradient CF-UGF (Di Bernardo and Sabogal, 2009)	$G = \sqrt{\frac{(\gamma)(V_{ap})(J_{mg})}{(\mu)(\varepsilon_o)}}$ (4)	28.5–3.1 s^{-1}
Unit head loss in the porous medium (Ergun, 1952)	$J = \frac{150\nu(1-\varepsilon_o)^2(V_{ap})}{g(\varepsilon_o)^3(C_s)(D_{mg})^2} + \frac{1.75(1-\varepsilon_o)(V_{ap})^2}{g(\varepsilon_o)^3(C_s)(D_{mg})}$ (5)	0.0016–0.0009 m m^{-1}
Reynolds number in the porous medium (Dinoy, 1971)	$R = \frac{\rho x V_f x k^{1/2}}{\mu}$ (6)	732–41

Where: L_r = hydraulic jump length; L_m = length between the base of the weir and the mixing point (m); h_v = table of water over the weir (m); P_v = height of water from the base of the weir and fill up sheet of water (m); E_p = loss of energy in the channel unit coagulation (m); T_m = average time of mixing (s); γ = specific weight of water (N m^{-1}); μ = absolute viscosity (N m s^{-2}); V_{ap} = approach velocity (m s^{-1}); ε_o = porosity of clean filter; J = loss of unit load (m m^{-1}), ν = kinematic viscosity (m^2 s^{-1}). g = gravity constant (m s^{-2}), C_s = coefficient of sphericity; D_{mg} = average grain size, ρ = density of water (kg m^{-3}), Re = Reynolds number, k = permeability (cm^{-1}).

and the UGF. The RMC is divided into 3 compartments, one for arrival and energy dissipation, one for mixing and one for the outlet of the coagulated water to the CF-UGF. Table 1 lists the parameters considered for the design of the unit, the formulae and the values obtained during the operation, while Fig. 2 shows the layout of the CF-UGF unit.

The CF-UGF step consists of 2 units in parallel and forms the third component in the treatment scheme. The system has 4 layers of gravel bed and was designed as a truncated pyramid to facilitate variation in the velocity gradient, producing a variable gradient from the highest to the lowest value from the bottom to the surface of the unit. In Table 2 the values of the velocity gradient for two flow operations are listed, depending on the properties of the filter bed and the average fluid velocity in each gravel layer, the cross-sectional area of the filter and the head loss in the bed. The calculation of the velocity gradient was done by Eq. (4) (see Table 1) and the head loss by Eq. (5) proposed by Ergun (1952), which is valid for any flow regime as long as the bed is not fluidized (Di Bernardo and Sabogal, 2009).

The flocculation gradient is greater at the bottom of the bed, decreasing towards the top of the filter, basically to promote the formation of flocs (Fair et al., 1984). It should be noted that the velocity gradient values were lower than those reported by Ahsan (1995) (200–300 s^{-1}) because the filtration rate in CF-UGF was lower. The Reynolds numbers

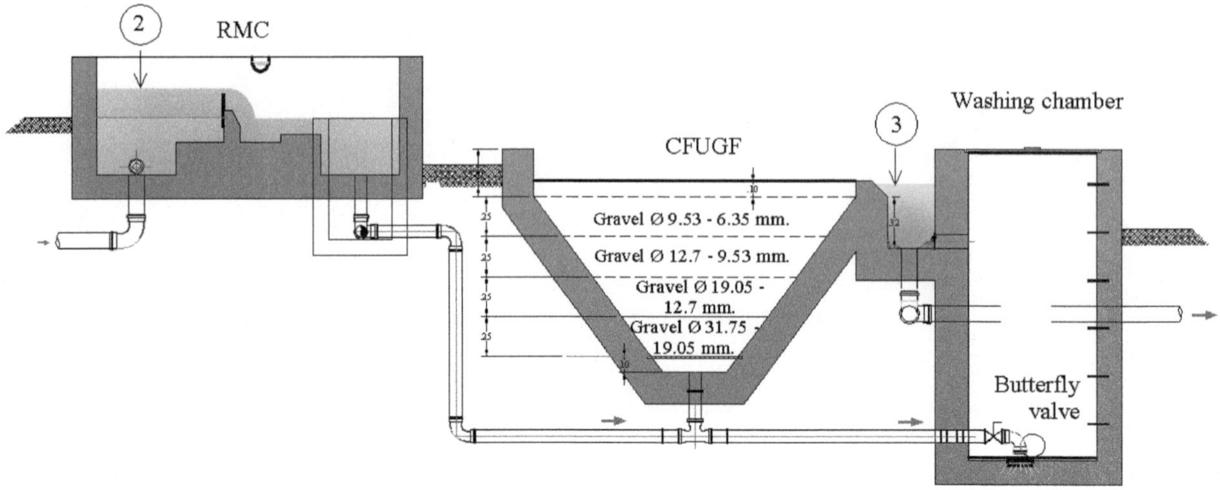

Figure 2. CF-UGF Unit Scheme (2–3: monitoring points of water quality).

Table 2. Velocity gradients in the CF-UGF for the 2 flows of operation.

Filter media size mm	ε	D_{mg} (m)	Flow $3\,l\,s^{-1}$				Flow $6\,l\,s^{-1}$			
			V_{ap} ($m\,s^{-1}$)	J ($m\,m^{-1}$)	G s^{-1}	Re	V_{ap} ($m\,s^{-1}$)	J ($m\,m^{-1}$)	G s^{-1}	Re
31.7–19.0	0.31	0.0317	0.00375	0.0016	14.2	366	0.0075	0.0032	28.5	732
19.0–12.7	0.38	0.0191	0.0014	0.0007	5.3	124	0.0029	0.0015	11	257
12.7–9.5	0.40	0.0111	0.00075	0.0009	4.2	43	0.0015	0.0019	8.5	86
9.5–6.3	0.42	0.0079	0.00045	0.0009	3.1	20	0.0009	0.0018	6.3	41

$T = 22.8\,°C, \gamma = 9737\,N\,m^{-1}, \mu = 9.44\times10^{-4}\,N\,m\,s^{-2}; \nu = 9.47\times10^{-7}\,m^2\,s^{-1}, C_s = 0.81$.

indicate that the units work in the hydraulic transition regime; values obtained above 10 indicate a stable inertial regime for flocculation in the porous medium, as reported by Wright (1968). For the calculation of the head loss in the filter the Ergun equation was used.

The calculations for each barrier of the treatment system were done on the basis of the design parameters presented in Table 3 and the dimensions of each barrier in Table 4.

2.2 Monitoring water quality and system operation

During the study, the raw water was monitored for the following parameters: turbidity, true colour, *E-coli*, total coliforms, heterotrophic bacteria plate count (HPC) and pH. The water quality parameters and related methods are listed in Table 5. Measurements of the head loss (h_f) over the filter bed were done for short periods of operation (4–6 h), when coagulants were applied. However, when operating without coagulant, daily measurements were done during periods of 8 days, according to the schedule defined by the plant operators for cleaning the CF-UGF and UGF. This is because with time large quantities of solids will accumulate within the gravel bed and then filter resistance will increase gradually and the water level within the inlet chamber will rise to a maximum h_f. To facilitate the measurements of h_f, steel rules were installed in the inlet chamber, taking care that the zero of the rule coincides with the water level in the chamber for the no-flow condition.

The dose of coagulant for operation of the CF-UGF system was previously defined according to studies developed by Cinara (2004). In these studies two types of jar tests were carried out: (a) to define basic parameter such as rapid mixing time, gradient of rapid mixing, slow gradient mixing, slow mixing time, and sedimentation time; the rapid mixing intensity and the slow mixing during the jar test were expressed in the velocity gradient G (s^{-1}) following the method described in CEPIS (2004) and were used for a first approximation of the velocity gradient in the CF-UGF and (b) one second set of jar tests were done with raw water of the source to define the optimal dose of coagulant (turbidity up to 100 NTU), following the method presented by Di Bernardo and Sabogal (2009), where rapid mixing is done using jar test equipment with 2-L glass jars. The coagulant used was 50 % liquid aluminium sulphate diluted with water to 2 %, and was added for a period of 60 s at a velocity gradient greater than 300 rpm. After rapid mixing, water was extracted and filtered

Table 3. Treatment barriers and design parameters.

Criteria	Treatment Stage				
	DRF	RMC	CF-UGF	UGF	SSF
Design period (years)	15	15	15	15	15
Operation time (h)	24	24	24	24	24
Number of units in parallel	2	1	2	2	4
Flow per unit ($l\,s^{-1}$)	3	6	3	3	1.5
Filtration rate ($m\,h^{-1}$)	3		3.2–27	1.0	0.30
Backwashing rate ($m\,h^{-1}$)	20		20	20	
Area by unit (m^2)	3.6		4	10.8	18
Gravel					
Length (m)	0.6		1	1.1	0.2
Size (mm)	25.4–3.2		19–6.3	25.4–3.2	12.5
Sand					
Length (m)	–	–	–	–	0.85
d_{10} (mm)					0.15–0.35
C_u	–	–	–	–	2–3.5

Table 4. Dimensions of each barrier.

Treatment barrier	Number of units	Dimensions L (m)	A (m)	H (m)	Flow ($l\,s^{-1}$)	Material structure
Input chamber	1	5.15	0.6	0.7	6.0	Concrete
DRF	2	4.0	0.9	0.8	3.0	Concrete
RMC	1	2.2	0.4	0.5–0.7	6.0	Concrete
CF-UGF*	2	2.0	2.0	1.4	3.0	Concrete
UGF	2	3.8	2.84	1.3	3.0	Concrete
SSF	4	3.7	4.7	1.75	1.5	Concrete

L: length; A: width; H: depth; * Bottom Area: 0.16 m^2 Surface Area: 4.0 m^2

Table 5. Water quality parameters and analysis method.

Parameters	Method	Limit of detection
Total coliform	9222B, filtration × membrane	0
E-coli	9222B, filtration × membrane	0
HPC	9215A, discharge in plate	0
Turbidity	2130B	0.1
Aluminium	3500-Al B	0.03
pH	4500 H +	2
True colour	2120C	1
TOC	5310 B	0.18
(APHA, AWWA, WPCF, 2005)		

in a funnel containing filter paper Whatman 40 (pore size 8 μm), to obtain a volume sufficient to realize analyses.

Tracer tests were performed according to the methodology described by Pérez and Galvis (1990), in order to understand the hydraulic performance of the CF-UGF and UGF units. Trials were conducted following the experimental stimulus-response method, in which a tracer, easily detectable and not involved in any of the physical and chemical processes that may alter the actual fluid hydrodynamics and with a known concentration, is injected into the influent (Rocha et al., 2000). The concentration curve of the tracer was analysed to determine the portion of plug flow, dead zones, and the fraction that works as a completely mixed flow. These tests were conducted using sodium chloride. The substance was dosed continuously through a constant hydraulic head dispenser with a sodium chloride concentration between 50–100 $mg\,L^{-1}$. The dosing period was three times the theoretical retention time of each unit and the response was measured at the output of each unit by means of electrical conductivity. Measurements were taken after the CF-UGF unit every 2 min. Results were analysed with the mathematical model Wolf-Resnick, the Morril Index (relationship between the time between the 90 % and the 10 % passage of the tracer), and the model of completely mixed reactors in series (CMRS), Pérez and Galvis (1990). Wolf-Resnick model

Eq. (7) indicates that by plotting the fraction of tracer remaining in the filter $(1 - F(t))$ versus tt_o^{-1} (relationship between measured time and the theoretical retention time), it is possible to estimate the values θ and $\tan\alpha$ (the slope of the straight line), and to identify the characteristics of the reactor using Eqs. (8) to (11). CMRS model was analyzed using Eq. (12), where n is the number of reactors in series and CC_o^{-1}, is the relationship between the concentration of tracer which remains in the reactor at a time and the concentration of tracer applied.

$$\text{Log}1 - F(t) = -\tan\alpha[(tt_o^{-1}) - p(1-m)] \tag{7}$$

$$\text{Tan}\alpha = (0.434 \cdot p)/(\theta \cdot (1-p)) \tag{8}$$

$$\text{Plug flow } (\theta): \theta = p(1-m) \tag{9}$$

$$\text{Dead zones } (m): m = 1 - (\theta/p) \tag{10}$$

$$\text{Mixed flow } (M_f): M_f = (1-p)(1-m) \tag{11}$$

$$\text{CMRS model}: \frac{C}{C_o} = n\left[\frac{(n \cdot \frac{t}{t_o})^{n-1}}{(n-1)!}\right]\left(e^{-n \cdot \frac{t}{t_o}}\right) \tag{12}$$

The porosity (ε_0) of the filter material was determined following the procedure defined by Ives (1990): first, the mass (M) occupied by the sample of gravel in a container of known volume and the apparent volume (V) occupied by the gravel in the container were both measured; then the density (ρ_s) of the gravel was determined by the ratio between the mass of the sample and the volume occupied by the sample. The porosity was calculated by Eq. (13).

$$\varepsilon_0 = 1 - M/\rho_s V \tag{13}$$

The filter cleaning procedure was the following: the water inlet to the unit was interrupted; without removing the supernatant water the surface of the gravel bed was cleaned manually with a shovel, stirring the surface layer of the filter to remove solid material adhering to the gravel; the supernatant water with the removed deposits from the top of the filter bed was discharged through a front weir; then the filter was prepared for removal of the deposits in the filter by adding water to the unit by opening the flow control valve in the inlet chamber to restore the supernatant water layer and increasing its height to a level of water between 20–25 cm above the gravel in the main compartment; this increased height adds some pressure for the cleaning procedure, in which fast drainage was carried out by quickly opening and closing (some 10 times) the butterfly valve on the underdrains; the unit was thereafter refilled with water and the cleaning and drainage procedure was repeated for adequate cleaning of gravel; then the filter was put back into operation.

The flow rate (Q) of the draining procedure in the UGF units was determined by the following procedure: (a) the unit was filled to the maximum level of the top; (b) the water flow into the filter was interrupted; (c) the surface area was measured (A); (d) the butterfly valve was opened until the water level lowered by 5 cm; (e) the declining water level (Δh) was measured over time (t) and (f) the initial flow rate was set by the expression $Q = \Delta h \cdot A/t$ ($m^3\,s^{-1}$).

3 Results and discussion

3.1 Water quality of the source water

The water source is a small mountain river (the Arroyo Hondo River) which drains an area that has problems of deforestation and erosion, strong activity in the basin of rocky material exploitation for construction and discharge of untreated wastewater. The behaviour of microbiological parameters over a 3-month period indicated that the source has, according to Lloyd and Helmer (1991) and WHO (2011), a high microbiological risk, because faecal coliform values at all times exceeded 1000 CFU $(100\,\text{ml})^{-1}$. *E-coli* bacteria were below 4.2 logs CFU $(100\,\text{ml})^{-1}$ for 95 % of the time, but were never less than 3.2 logs CFU $(100\,\text{ml})^{-1}$. The average values of HPC and total coliforms were 5.3 logs CFU $(100\,\text{ml})^{-1}$ and 5.1 logs CFU $(100\,\text{ml})^{-1}$, respectively.

The behaviour of turbidity in the source was measured for a period of one year. This parameter, which is easy to measure, is a good indicator to control the system and facilitates decision-making by the operator. The results indicated that the turbidity in the source did not exceed 100 NTU for 97 % of the time, while the turbidity was lower than 25 NTU for 75 % of the time. Minimum values of 3 NTU were recorded during the summer period and maximum values of 350 NTU during the rainy season. Turbidity peaks were of short duration (4–6 h) but sometimes lasted up to 24 h. For 95 % of the time the true colour level in the river was below 25 UPC and at no time the level dropped below 5 UPC.

3.2 Coagulant dosage

The basic parameters defined for the operation with coagulant were as follows (Cinara, 2004): rapid mixing time 60 s, gradient of rapid mixing 300 rpm ($G = 280\,s^{-1}$), slow gradient mixing of 60 rpm ($G = 28\,s^{-1}$), slow mixing time 25 min, and sedimentation time of 20 min. Figure 3 presents the coagulant dose of aluminium sulphate for different turbidity levels, applied to the operation with coagulant based on previous studies developed by Cinara (2004). The dosage behaves as a logarithmic function and a little variation in the dose of coagulant is presented for affluent turbidities between 60–100 NTU. The optimum pH was in the range of 6.6–7.6 for an alkalinity between 59–133 mg $(l\,CaCO_3)^{-1}$. The pH range of the water source has facilitated an efficient and low cost operation because only very small changes occurred during the coagulation-flocculation process, thus avoiding the need for pH adjustment. The dosing conditions as shown in Fig. 3

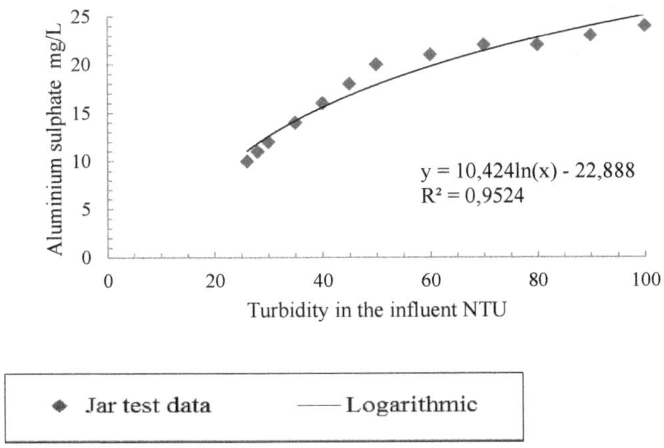

Figure 3. Aluminium sulphate dose as a function of influent turbidity in CF-UGF Unit.

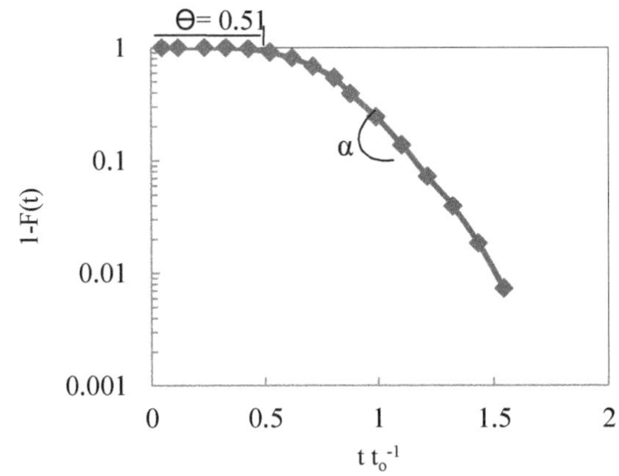

$\text{Log } 1\text{-} F(t) = -0.95\ (t\ t_o^{-1}) + 1.24\ (r^2 = 0.902)$

$\text{Tan}\alpha = 0.95;\ \theta = 0.51;\ p = 0.53;\ m = 0.03;\ \text{Mf} = 0.46$

Figure 4. Hydraulic Wolf-Resnick model results for the CF-UGF.

are clearly within the range suggested by CEPIS (2004), of 3–30 mg L^{-1} of aluminum sulfate and a pH close to 7, which suggest that prevailing coagulation mechanism is due to charge neutralization of the aluminium hydroxide. Dosing of aluminum sulphate is carried out by a dispenser at the point of greatest turbulence in the RMC; the concentration of solution of aluminum sulphate was 2 %, which is in line with the recommendation of CEPIS (2004) which suggests a coagulant concentration between 1–2 % for water treatment plants. This level of concentration in combination with sufficient turbulence, allows for a good coagulant dispersion which facilitates its coming into quick contact with a large number of particles (Di Bernardo and Sabogal, 2009).

3.3 Hydraulic behaviour of CF-UGF units

Figure 4 shows the results of the mathematical model Wolf-Resnick. The model results for CF-UGF indicate that the system worked with a plug flow fraction of 51 %, a mixed fraction of 46 % and a dead zone fraction of 3 % ($r^2 = 0.90$). Taking into account the dead zones in the CF-UGF, the velocity gradient in the unit varied between 3.2 and 29.4 s^{-1}, which is close to the value calculated in Table 2 and obtained in the jar test by Cinara (2004) to define the coagulant dose. The Morrill Index (MI) was 1.82, which suggests, according to experiments by Pérez and Galvis (1990), the presence of plug and mixed flow in the CF-UGF unit. Figure 5 presents the results of the CMRS model. The continuous lines show the hydraulic behaviour with n reactors in series, while the dotted line represents the measurements at the CF-UGF unit. When comparing the results of the theoretical model with experimental data, the hydraulic behaviour of the reactor CF-UGF tends to $n = 6$ reactors in series (see continuous black curve, Fig. 5), confirming the presence of a relative plug flow. The mean residence time (t_m) was estimated from the experimental curve, determining the centroid under the curve, the mean residence time for the CF-UGF was 19.7 min for the flow of 31 s^{-1} while theoretical retention time was 19 min. These differences may be explained by the presence of dead zones and implies that the curve of distribution residence time has a tail and therefore the time will be displaced in the time axis (Fig. 5), the fluid elements that were trapped in the dead zones is conducted very slowly and will have a much larger residence time.

3.4 Removal of turbidity in the operation without coagulant

The operation of the CF-UGF without coagulant dosing was used in dry periods. Figure 6 shows the frequency of turbidity in raw water and after different treatment barriers, including DRF, CF-UGF, UGF and SSF. The CF-UGF units produced an effluent between 2.5–7.5 NTU, with mean removal efficiency of 46 %, while for 78 % of the time the UGF showed turbidity levels lower than 1 NTU and another 22 % were between 1.5–2 NTU. SSF units processed water with 0.3 NTU for 98 % of the time. CF-UGF and UGF always produced water with turbidity levels below 10 NTU, which is the guideline value of inflow water to the SSF units, according to Di Bernardo and Sabogal (2009) and to Galvis et al. (1999) who add the requirement that filtration rates should be lower than 0.20 m h^{-1} in SSF units. In this case it was a little higher because the system operated at filtration velocities of between 0.15 and 0.30 m h^{-1} (3–61 s^{-1}).

3.5 Removal of turbidity in the operation with coagulant

Figure 7 shows the frequency of turbidity in raw water and different treatment barriers, when the CF-UGF is operated with coagulants. The application of coagulant in the RMC varied as indicated in Fig. 3. In the effluent CF-UGF units, for 97.2 % of the time turbidity levels lower than 10 NTU

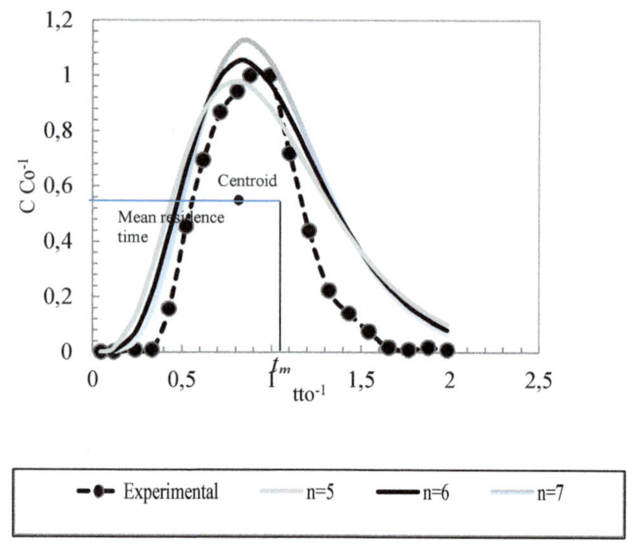

Figure 5. CMRS model for CF-UGF.

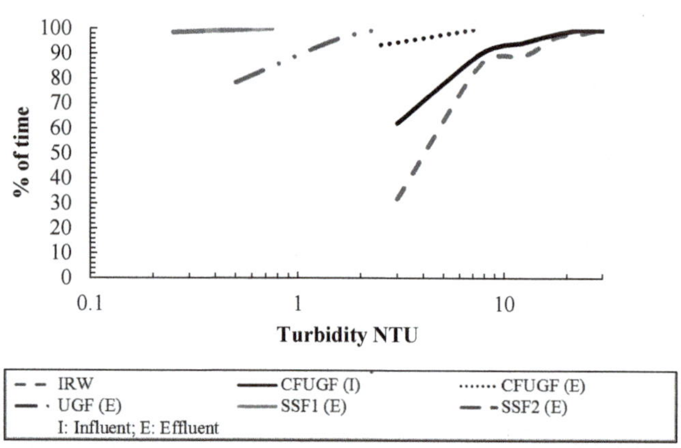

Figure 6. Frequency of turbidity levels (operation without coagulant).

were achieved, while the maximum value of turbidity in the effluent in this step ranged between 15–20 NTU for 1.4 % of the time. The UGF showed turbidity levels lower than 6 NTU for 100 % of the time, thereby facilitating the operation of the SSF, which produced water with turbidity below 1 NTU for 98 % of the time. The addition of coagulant in the CF-UGF enabled water with turbidity levels below 10 NTU after UGF to be obtained. Increments in turbidity levels, which occur basically in the rainy season, could be managed in the treatment plant by the CF-UGF unit, which contributed to an effective operating system, preventing reductions in the SSF filtration runs. Additionally, stops in the operation of the treatment plant were prevented.

The removal efficiency of turbidity in the CF-UGF with coagulant was between 85 and 96 %, which is higher compared to operating without coagulant, and average efficiency of turbidity in the CF-UGF was 46 %, ranging between 21–

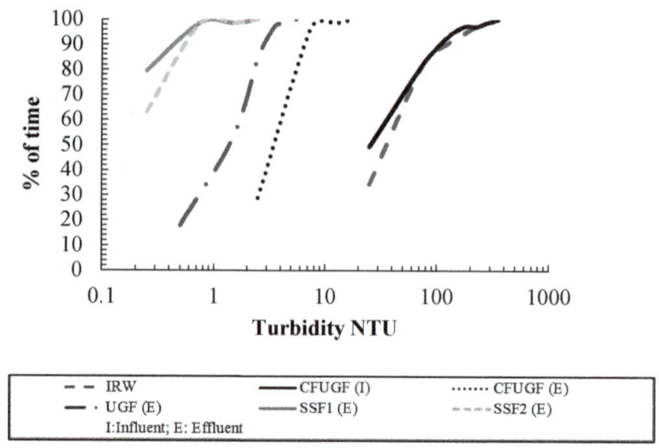

Figure 7. Frequency of turbidity levels (operation with coagulant).

76 %. The removal efficiencies in CF-UGF with velocity gradients between 28.5–3.1 s^{-1} (for 3 and 6 l s^{-1}) was consistent with the findings of Di Bernardo and Sabogal (2009), who established that the gradient must be less than 100 s^{-1}. The efficiency in the removal of turbidity can be explained by the good opportunities for contact with particles in the gravel bed, which is consistent with Richter (1987), Attakoya et al. (1991), and Di Bernardo and Cruz (1994).

3.6 Removal of microorganisms in the operation without coagulant

When operating without aluminium sulphate, the duration of the filter run was 8 days in the CF-UGF and UGF, regardless whether it achieved the maximum head loss or not (normally head loss is less than 5 cm in 8 days, to a maximum value of 15 cm). The typical removal of microorganisms for a filter run is presented in Table 6 for samples taken in the influent of CF-UGF unit and the effluent of the UGF. The removal efficiency in the CF-UGF and UGF steps was 0.16 logs for total coliforms, 0.16 logs for *E-coli* and 0.17 logs for HPC, lower than that reported by Galvis et al. (1999), which was probably due the effect of a higher filtration rate. SSF achieved a total reduction of *E-coli* of 3.2 log units, facilitating the work of chlorination as a security barrier. These results are consistent with the WHO (2011) which indicated that the range of log removal of bacteria for SSF must be between 2–6 under presence of schmutzdecke and appropriate: grain size, flow rate, operating conditions (mainly temperature, pH), cleaning and refilling and in the absence of short circuiting.

The reduction of the filtration velocity in the UGF from 1.0 to 0.5 m h^{-1} contributed to improved efficiency in the removal of microorganisms. The average removal for total coliforms and *E-coli* was 0.57 and 0.5 logs respectively in CF-UGF and UGF steps, while HPC reached a removal of 0.64 logs. These reductions are close to those reported by Galvis et al. (1999). The average removal in SSF was 3.0 logs. The

Table 6. Microbiological behaviour without the use of coagulant (UGF: $V_f = 1.0\,\mathrm{m\,h^{-1}}$, SSF: $V_f = 0.30\,\mathrm{m\,h^{-1}}$).

Descriptive statistics	Raw water			CF-UGF and UGF effluent			SSF effluent		
	TC	EC	HPC	TC	EC	HPC	TC	EC	HPC
No data	9	9	7	9	9	7	10	10	9
Average	15 131	2262	202 629	12 227	2161	139 625	12	0	744
Maximum	25 000	3600	403 500	19 400	6700	272 000	68	0	1900
Minimum	8000	1250	120 000	6100	800	73 000	2	0	100
STD deviation	6733	914	123 549	5893	1859	80 237	19.7	0	651.6
Average log CFU 100^{-1} $\mathrm{ml^{-1}}$ removal units				0.16	0.16	0.17	3.3	3.2	2.3

TC: Total coliforms (CFU $(100\,\mathrm{ml})^{-1}$) EC: *E-coli* (CFU $(100\,\mathrm{ml})^{-1}$) HPC: heterotrophic play count bacteria (CFU $(100\,\mathrm{ml})^{-1}$), operational flow $6\,\mathrm{l\,s^{-1}}$.

Table 7. Microbiological behaviour with the use of coagulant (UGF: V_f $1.0\,\mathrm{m\,h^{-1}}$, SSF: V_f $0.3\,\mathrm{m\,h^{-1}}$).

Descriptive statistics	Raw water			CF-UGF and UGF effluent			SSF effluent		
	TC	EC	HPC	TC	EC	HPC	TC	EC	HPC
No data	5	5	5	5	5	5	5	5	5
Average	13 240	2780	110 740	6070	1125	41 100	2.8	0	540
Maximum	16 050	5600	160 000	10 100	2150	47 100	4	0	800
Minimum	8900	1200	79 300	3900	1000	35 000	1	0	200
STD deviation	2746	1684	31 494	2402	465	5290	1.3	0	219
Average log CFU 100^{-1} $\mathrm{mL^{-1}}$ removal units				0.44	0.40	0.44	3.4	3.1	3.9

TC: Total coliforms (CFU $(100\,\mathrm{ml})^{-1}$) EC: *Eschirichia coli* (CFU $(100\,\mathrm{ml})^{-1}$); HPC: heterotrophic play count bacteria (CFU $(100\,\mathrm{ml})^{-1}$), operational flow $6\,\mathrm{l\,s^{-1}}$.

efficiency of the SSF was not influenced by changes in the filtration rate, probably due to the high level of maturity of the filters.

3.7 Removal of micro-organisms, operation with coagulant

Table 7 presents the results of operating with coagulant, the data correspond to samples taken in the influent of CF-UGF unit and the effluent of the UGF for a period of 6 h, because when turbidity was less than 30 NTU the coagulant dosing was stopped. The dose of aluminium sulphate corresponded to $10\,\mathrm{mg\,L^{-1}}$, with an operation flow of $6\,\mathrm{l\,s^{-1}}$. Overall, the pre-treatment with CF-UGF and UGF contributed to the reduction of microbiological load: average 0.44 log removal for total coliforms, 0.40 log removal of *E-coli*, and 0.44 log removal for HPC. Only the CF-UGF unit contributed with average efficiencies for total coliforms equivalent to 0.19 log, 0.12 logs for *E-coli* and 0.15 log for HPC. The last stage of treatment, SSF, allowed a total reduction of 3.4 log of total coliforms, 3.1 log for *E-coli*, and 3.9 log of HPC. This suggests that the dosage of aluminium sulphate did not affect the biological activity in pre-treatment and SSF, which is consistent with that reported by Dorea and Clarke (2006).

When comparing operation with and without coagulant an increase in the average efficiency of removal of microorganisms, between 0.16–0.17 log to 0.40 to 0.44 log was observed, i.e. in the operation with coagulant the removal efficiency for total coliforms, *E-coli* and HPC, was 2.75, 2.5 and 2.6 higher respectively compared to the operation without coagulant in the CF-UGF and UGF units.

3.8 Aluminium, pH, colour and organic matter

Residual aluminium in the system was low in the effluent of CF-UGF and 53 % was removed in the UGF, so that the concentration in the influent of the SSF varied between 0.07 and $0.09\,\mathrm{mg\,L^{-1}}$, and the average effluent concentration was $0.04\,\mathrm{mg\,L^{-1}}$ (STD $0.005\,\mathrm{mg\,L^{-1}}$). This value was lower than the WHO (2011) guidelines, which recommended less than $0.2\,\mathrm{mg\,L^{-1}}$ for aluminium in drinking water. In a pilot study that examined the impacts of chemical pre-treatment by gravel filters on SSF Dorea and Clarke (2006), reported an average aluminium concentration of $0.041\,\mathrm{mg\,L^{-1}}$ in the effluent of the SSF and indicated that the chemical improved the overall treatment efficiency. However as indicated by these authors careful control of the coagulation step is needed to avoid carry-over of aluminium to the SSF as this might contribute to possible filter clogging even though turbidities of less than 10 NTU are achieved. In our study however premature clogging did not occur at all and filter runs of SSF were maintained between 50 and 70 days with a maximum

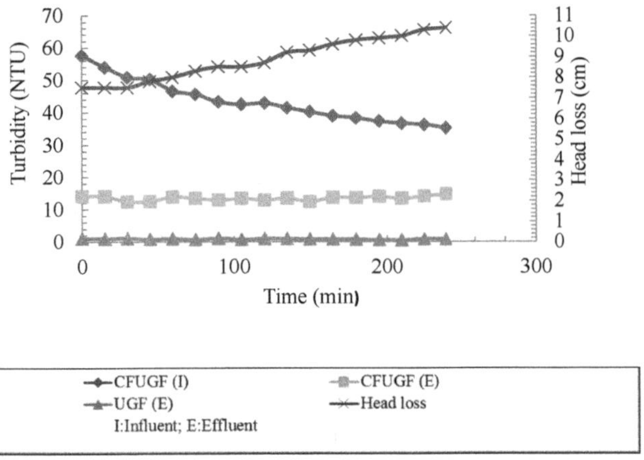

Figure 8. Turbidity in CF-UGF and UGF units and head loss on CF-UGF.

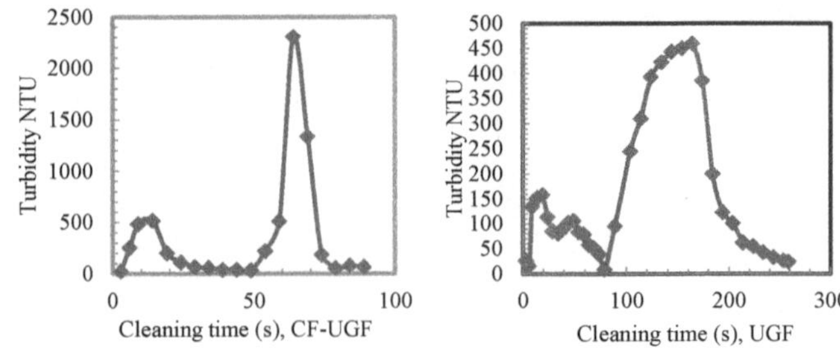

Figure 9. Behaviour of turbidity on a deep clean-up of CF-UGF and UGF.

head loss of 0.70 m, which is in line with the range of 20–60 days reported by Schulz and Okun (1984), the minimum of 45 days recommended by Cleasby (1991) and the range of 46–178 days recommended by Galvis et al. (1999). The pH in the influent varied between 8.2 and 8.5 and between 8 and 8.2 in the effluent of CF-UGF, which is expected not to affect the biological development of the Smutzdecke in the SSF, respect Galvis et al. (1999) reported pH in the range of 7.1–8.0 for operation of SSF and indicate that the adsorption of virus the sand improves with increasing ion concentration and valence of the cations in solution. The true colour reduction recorded an average efficiency of 54 % in the CF-UGF stage and 57 % in the UGF stage, and the net efficiency of the true colour reduction of the CF-UGF and the UGF stage together was 76 %. The organic matter content measured as total organic carbon (TOC) was low, the influent had an average value of 1.1 mg L^{-1} (±0.075), and the removal efficiency in the CF-UGF and UGF step together was 9 %, with an efficiency at the end of the treatment of 28 %.

3.9 Operation and maintenance

The treatment plant operated without coagulant for turbidity levels below 30 NTU. When influent turbidity was greater than 100 NTU, operators interrupted the operation to reduce the turbidity load on the plant and, depending on water needs, operated with coagulant, reducing the filtration rate by half. When rain events occurred and the influent turbidity was greater than 30 NTU, a dosage of coagulant was applied. Figure 8 shows the behaviour of turbidity in CF-UGF and UGF and head loss in the CF-UGF units for an event of short duration when the turbidity increased above 30 NTU up to a maximum value of 58 NTU, and operation with coagulant was necessary. The monitoring of the head loss was only done in the CF-UGF units. In the UGF unit, there was no change in head loss detected in the relatively short period of the event. The total loss over a period of 4 h of operation was 3.5 cm and did not achieve the maximum value of 20 cm. Table 8 summarizes the operating and monitoring conditions.

Based on the information registered by the operators of the water treatment plant, cleaning frequency for the CF-UGF and the UGF units was every 8 days (without a coagulant dosage). During the rainy period the CF-UGF units, operating with aluminium sulphate, registered cleaning frequencies in the range of 6–100 h. The cleaning frequency of 6 h was obtained when influent turbidity reached values of above 100 NTU, while the 100-h operation mode was observed when influent turbidity levels were between 30–60 NTU.

The behaviour of the cleaning of the CF-UGF in terms of turbidity is presented in Fig. 9. In the CF-UGF, cleaning was done by operating the butterfly valve in such a way that 10 shock waves were created. It can be observed that, for the CF-UGF, in the first 15 s of the discharge a turbidity peak occurred, for 35 s the water was clear, and then after 85 s a second peak was observed. For the UGF, the solids discharge started immediately to reach a second peak after 150 s, about 5 times higher than the first peak. The behaviour of the water quality in the discharge of the wash water was different from that reported by Wolters (1988) and Cinara and IDRC (1993). When the valve was opened suddenly, the particles which stayed on the filter media experience a change in velocity, whose effect may be to drag the particles to transport to the drainage system.

The flow rate of the drain water for cleaning the filter was variable over time; measurements were done for different heights of water level relative to the position of the butterfly valve. The maximum backwashing rate for the UGF was 9.4 m h^{-1} while for the CF-UGF it was 9.1 m h^{-1}.

3.10 Investment, operation and maintenance costs

Tables 9 and 10 list the construction costs (year 2011) and operation and maintenance costs (year 2010). The CF-UGF stage and the pre-treatment by UFG represent respectively 7 % and 28 % of the total construction costs of the water treatment plant (see Table 10). The cost per m^3 of produced water was US$ 0.05 (discount rate of 12 % for Colombia and project horizon of 15 yr). Per capita investment costs are US$ 18, for an average consumption of 150 l c^{-1} d^{-1}. The

Table 8. Summary of operating and monitoring conditions.

Variable	Value
Operational parameters	
Influent turbidity for coagulant dosing	> 30 NTU
Maximum turbidity of operation	Normally 100 NTU, sometimes can operate with peaks above 100 NTU for short periods of time (4–6 h).
Cleaning period CF-UGF and UGF, operation without coagulant.	Each 8 days
Time cleaning of CF-UGF	15 min
Time cleaning of UGF	45 min
Monitoring parameters	
Coagulant used	Liquid aluminium sulphate, type A, 50 % concentration
Filtration run of the CFUGF with application of coagulant	6–100 h
Percentage of time with coagulant dosage	20 % yr
Percentage of operating time without coagulant	80 % yr
Maximum period of time recorded dosing coagulant	100 h
Maximum duration of registered plant shutdown	24 h
Percentage of stops in the year	3 %
Maximum head loss in CF-UGF units	20 cm
Maximum head loss in UGF units	15 cm
Maximum backwashing rate CF-UGF	9.1 to 7.5 $m h^{-1}$
Maximum backwashing rate UGF	9.4 to 8.4 $m h^{-1}$

costs of O&M for the use of coagulant are low, representing only 0.3 % of total O&M costs, because the operator only doses in periods of high turbidity, making the technology attractive. The highest costs of O&M represent pumping energy and staff. The O&M costs of US$ 0.264 m^{-3} is low (depreciation was included), when compared to the costs of US$ 1.04 m^{-3} (includes the average investment costs) for the utility of the city of Cali, close to the community. The O&M costs in a gravity system would reduce to US$ 0.14 per m^3.

4 Conclusions

CF-UGF is a relatively new technology that has been applied in a few cases with rapid sand filtration. In this study CF-UGF has been used in combination with MSF technology comprising UGF and SSF. This study shows that combining CF-UGF with MSF greatly contributed to the removal efficiency of the system without negatively affecting the biological activity of the treatment system in terms of the efficiency of microorganism removal in the UGF and SSF when coagulant was dosed. This strongly contributes to the operational flexibility of the system as it allows to dose coagulant only when high influent turbidity peaks occur.

CF-UGF improved the operation of MSF compared to only UGF, when the system operated with turbidity levels above 30 NTU, facilitating the performance of the SSF by reducing the load of particulate material to avoid short filter runs and possible interruptions in treatment plant operation. The removal efficiency of turbidity in the CF-UGF with coagulant dosing was between 85 and 96 %, which is higher compared to operation without coagulant dosing; average efficiency of turbidity in the CF-UGF was 46 %,

Table 9. Initial investment costs.

Stage of treatment	*Cost (US $)	%
DRF	9821	16
CFUGF	4554	7
UGF	17 002	28
SSF	29 783	49
Total cost (US $)*	61 160	
Cost $l\,s^{-1}$ (US $)	10 193	
Per capita cost (US $)	18	

* Costs up to February 2011.

Table 10. Operation and maintenance costs.

Item	Monthly cost (US $)*	%
Cost of coagulant (aluminium sulphate)	13	0.3
Cost of the chlorine	365	9.5
Staff costs (including benefits and social security)	1399	36.4
Electrical energy costs	1661	43.3
Costs of materials and equipment	249	6.5
Costs for water quality analysis	155	4
Total monthly cost O&M (US $)	3842	
O&M costs per m^3 produced (US$ m^{-3})	0.25	

* Costs to December 2010.

ranging between 21–76 %. The addition of coagulant in the CF-UGF allowed for obtaining water with turbidity levels below 10 NTU after UGF, which contribute to the effective operation of the SSF. The overall system produced water with turbidity below 1 NTU for 98 % of the samples that were taken in the research period. In the operation with coagulant the removal efficiency for total coliforms, *E-coli* and HPC, was 2.75, 2.5 and 2.6 higher respectively compared to the operation without coagulant in the CF-UGF and UGF units. No reduction was observed in the microbial removal efficiency of the SSF, no obstruction of the SSF beds were demonstrated and SSF runs were maintained between 50 and 70 days for a maximum head loss of 0.70 m.

The hydraulic behaviour of CF-UGF indicated that the system worked with a plug flow fraction of 51 %, a mixed fraction of 46 % and a dead zone fraction of 3 %. The comparison between the theoretical model and experimental data indicated that hydraulic behaviour of the reactor CF-UGF tends to $n = 6$ reactors in series, confirming the presence of a relative plug flow. The mean residence time was 19.7 min for the operating flow of for the flow of $3\,l\,s^{-1}$; theoretical retention time was 19 min.

It was only necessary to operate the system with coagulant for 20 % of the time. The CF-UGF run time was 6–100 h depending on raw water turbidity. In the operation without coagulant, the run time for the CF-UGF and UGF was 8 days. Whereas the designed drainage flow was established at $20\,m\,h^{-1}$ in practice this level was not reached and the real flow according to the measurements in the units was about $9\,m\,h^{-1}$ indicating that more research is needed on the effect of the drainage system during the cleaning operation.

The CF-UGF unit represented only 7 % of the total construction costs, and the pre-treatment CF-UGF and UGF represented 35 % of total costs, while the cost of m^3 produced by the MSF with the CF-UGF system was US$ 0.05, for a per capita investment of US$ 18. The O&M costs for the use of coagulant represented only 0.3 % of the total O&M costs. The production cost was US$ $0.264\,m^{-3}$ for the operation with pumping. In a gravity system the cost is reduced to US$ 0.14 per m^3.

Acknowledgements. The authors acknowledge the support and trust received from the community and the governing board of Arroyo Hondo for the construction and evaluation of the system. They are also grateful for the fieldwork done by Carlos Mejia (FHIS, IRC-YZ program) and Diana Posso (Sanitary Engineer program, Univalle) and the plant operators Alfonso Escandón and Eulogio Mósquera. Financial and staff support for the study were received from the water supply research group of the Cinara Institute, Universidad del Valle, which included the use of the research station in Puerto Mallarino.

References

Ahsan, T.: Process Analysis and Optimization of Direct Horizontal – Flow Roughing Filtration, Ph.D. thesis, Delft University of Technology, A.A. Balkema De, Rotterdam, The Netherlands, 193 pp., 1995.

APHA, AWWA, and WPCF: Standard Methods for the examination of water and wastewater, USA, 21a Edn., 2005.

Attakoya, A., Prasad, V. S., and Nagaraja, S. V.: Performance study of granulated flocculators, Indian Jour. Evir. HLTH., 33, 498–504, 1991.

Bates, B. C., Kundzewicz, Z. W., Wu, S., and Palutikof, J. P. (Eds.): El Cambio Climático y el Agua. Documento técnico del Grupo Intergubernamental de Expertos sobre el Cambio Climático, Secretaría del IPCC, Geneva, 224 pp., 2008.

Bhole, A. G.: Design and fabrication of a low cost water treatment plant for rural areas in India, Aqua, 5, 315–320, 1981.

Burganos, V. N., Paraskeva, C. A., Christofides, P. D., and Payatakes, A. C.: Motion and deposition of non-Brownian particles in upflow collectors, Sep. Technol., 4, 47–54, 1994.

Centro Panamericano de Ingeniería Sanitaria y Ambiental (CEPIS): Tratamiento de agua para consumo humano, Filtración Rápida, Perú, 2004.

Cinara: Estudio de tratabilidad planta de potabilización de la parcelación Colinas de Arroyohondo; proyecto: ASOPROCOLAR , Cali, Colombia, 27 pp., 2004.

Cinara and International Development Research Centre (IDRC): Proyecto limpieza hidráulica de filtros gruesos, Final report, Cali, Colombia, 110 pp., 1993.

Cinara and IRC: Potencial de la filtración gruesa en múltiples etapas (FiGME) en la optimización de plantas de tratamiento con filtración rápida de agua químicamente coagulada, Cali, Colombia, 1996.

CGR – Contraloría General de la República: Estado de los Recursos Naturales y del Ambiente, República de Colombia, 180 pp., 2009.

Cleasby, J. L.: Source Water Quality and pretreatment option for slow sand filters, Chapter 3, in: Task committee on slow sand filters, New York, USA, 1991.

Di Bernardo, L. and Cruz, C. H.: Influencia de altas tasas de filtración en el desempeño de sistemas de filtración directa ascendente, in: proceeding of XXIV Interamerican Congress of Ingeniería Sanitaria y Ambiental, Buenos Aires, Argentina, 1994.

Di Bernardo, L. and Sabogal, L. P.: Seleção de tecnologias de tratamento de água, Editora LDIBE LTDA, Sao Carlos Brasil, I, 878 pp., 2009.

Dinoy, A. A.: Friction Factor and Reynolds Number Relationship in Flow through Porous Media, ME thesis, AIT, Bangkok, Thailand, 1971.

Dorea, C. C. and Clarke, B. A.: Impact of chemical pretreatment in Slow Sand Filtration, in: proceeding of Recent progress in slow sand and alternative biofiltration processes, IWA publishing, 215–223, 2006.

Ergun, S.: Fluid Flow Through Packed, Colum. Chem. Eng. Prog., 48, 89–94, 1952.

Fair, G. M., Geyer, J. C., and Okun, D. A.: Purificación de aguas y tratamiento y remoción de aguas residuales, Fifth reprint, Editorial Limusa Mexico, II, 764 pp., 1984.

Galvis, G.: Development and evaluation of multistage filtration plants; an innovative, robust and efficient water treatment technology. Ph.D. thesis, Guilford, UK, CEHE, University of Surrey, 228 pp., 1999.

Galvis, G., Latorre, J., and Visscher, J. T.: Filtración en múltiples etapas, tecnología innovative para el tratamiento de agua, Universidad del Valle, Instituto Cinara, Cali, Colombia, International water and sanitation centre, IRC, The Netherlands, UNESCO, united nations office for science and culture, 197 pp., 1999.

Ives, K. J.: Testing of filter media, J. Water SRT Aqua, 39, 144–151, 1990.

Kardile, J. L.: Development of simple and economic filtration methods for rural water supplies, Aqua-Fenicca, 1, 226–229, 1981.

Kawamura, S.: Two stage filtration, J. AWWA, 77, 42–47, 1985.

Lloyd, B. and Helmer, R.: Vigilancia y control de la calidad del agua para consumo humano en la zona rural, Published for the WHO and UNEP by Logman Scientific and Technical, 1991.

Ministerio de Desarrollo: Inventario nacional del sector de agua potable y saneamiento básico, Vol. I Infraestructura física de los sistemas. Ministerio de Desarrollo Económico, Viceministerio de Desarrollo Urbano, Dirección de servicios públicos domiciliarios, Colombia, 13–67, 1998.

Pérez, J. and Galvis, G.: Análisis de flujos y factores que determinan los periodos de retención, Centro de Ingeniería Sanitaria y Ciencias del Ambiente, CEPIS, Organización Panamericana de la Salud, OPS, II, 112 pp., 1990.

Richter, C.: Sistemas simplificados de floculación, in proceeding of Seminario Internacional sobre Tecnología Apropiada para Potabilización del Agua, ACODAL, seccional Valle del Cauca, Cali, Colombia, August, 1987.

Richter, C. A. and Moreira, B. R.: Floculadores de Pedras. Experiencia en Filtro Piloto, Reporte interno Sanepar, Parana, Brasil, 1981.

Rocha, M. A., Normando, H., and Onofre, C.: Aplição de tracadores em filtros anaeróbicos para avaliação das características hidrodinámicas, in: Proceeding of XXVII Congresso Interamericano de EngenhariaSanitária e Ambiental, Porto Alegre, Brasil. Associação Brasileira de Engenharia Sanitária e Ambiental – Região Rio Grande do Sul (ABES/RS), 2000.

Salazar, M. and Ocampo, J.: Estudio sobre la eficiencia hidráulica y remoción de material suspendido en clarificadores ascendentes de gravas estratificadas, Universidad del Valle, Cali, Colombia, 1999.

Sánchez, L. D., Burbano, L. M., and Sánchez, A.: Multi stage filtration to prevent biofilm growth in a distribution network, in: proceeding of recent progress in slow sand and alternative biofiltration processes, IWA publishing, 511–518, 2006a.

Sánchez, L. D., Sánchez, A., Galvis, G., and Latorre, J.: Multi-Stage Filtration, Thematic overview paper 15, IRC International Water and Sanitation Centre, 64 pp., 2006b.

Santamaría, J.: Comparación de la floculación mecánica y la floculación en clarificadores ascendentes en gravas estratificadas, Universidad del Valle, Cali Colombia, 97 pp., 1999.

Schulz, C. R. and Okun, D. A.: Surface water treatment for communities in developing countries, John Wiley & Sons, New York, 299 pp., 1984.

WHO: Guidelines for Drinking Water Quality. World Health Organization, Ed. WHO library, 4th Edn., 564 pp., 2011.

Wolters, H.: Roughing Filtration a Literature Study, MSc thesis Report, TU Delft, 173 pp., 1988.

Wright, D. E.: Non-linear flow through granular media, Proceeding American Society Civil Engineering, Journal Hydraulic, 94, 851–872, 1968.

11

CLIPS based decision support system for water distribution networks

K. Sandeep[1] and K. Rakesh[2]

[1]Research scholar, IIT Delhi, Delhi, India

[2]Faculty, Department of Civil Engineering, Indian Institute of Technology, Delhi, India

Abstract. The difficulty in knowledge representation of a water distribution network (WDN) problem has contributed to the limited use of artificial intelligence (AI) based expert systems (ES) in the management of these networks. This paper presents a design of a Decision Support System (DSS) that facilitates "on-demand" knowledge generation by utilizing results of simulation runs of a suitably calibrated and validated hydraulic model of an existing aged WDN corresponding to emergent or even hypothetical but likely scenarios. The DSS augments the capability of a conventional expert system by integrating together the hydraulic modelling features with heuristics based knowledge of experts under a common, rules based, expert shell named CLIPS (C Language Integrated Production System). In contrast to previous ES, the knowledge base of the DSS has been designed to be dynamic by superimposing CLIPS on Structured Query Language (SQL). The proposed ES has an inbuilt calibration module that enables calibration of an existing (aged) WDN for the unknown, and unobservable, Hazen-Williams C-values. In addition, the daily run and simulation modules of the proposed ES further enable the CLIPS inference engine to evaluate the network performance for any emergent or suggested test scenarios. An additional feature of the proposed design is that the DSS integrates computational platforms such as MATLAB, open source Geographical Information System (GIS), and a relational database management system (RDBMS) working under the umbrella of the Microsoft Visual Studio based common user interface. The paper also discusses implementation of the proposed framework on a case study and clearly demonstrates the utility of the application as an able aide for effective management of the study network.

1 Introduction

The day-to-day operation of a water distribution network is managed by skilled operators who use experience and other heuristics to adjust and control network elements such as pumps and valves in their desire to achieve satisfactory performance of the water supply system. However, following an unforeseen and an unusual emergency such as a burst water main, prompt and effective measures from the network manager is often taken for granted by the consumers. On their part, the network managers usually have to depend upon scant information that is reported by consumers impacted by these exigencies (Machell et al., 2010).

Managers of water utilities design most of their interventions and other decisions based on intuition, rule of thumb heuristics and trial and error. These operative decisions are often vague and inconsistent as they lack objectivity and, as a result, these often result in inefficient network management. Further, on account of distribution complexities and excessive reliance on a heuristics driven management paradigm that is often unconvincing and seemingly feeble in substance without an accompanying validation, there is no guarantee that resources are utilized effectively (Raghvendran et al., 2007).

Difficulties in managing a typical water supply system arise on account of (i) absence of a well defined policy framework; (ii) difficulty in gathering information that is coherent and objective because, as is often the case, information is based on individual perception and experience; (iii) complexity of a typical water supply system on account of its extensive and spatially distributed character and accompanied by a variety of control mechanisms; and (iv) frequent changes in the network topology (Leon et al., 2000). Collectively, these foregoing issues present great difficulties in efforts to develop mathematical models of such inherently complex systems (Walski, 1993). Recent developments in AI technology make it possible to encode knowledge and reasoning in a structured computer program that enables it to mimic the human thought process and its application, albeit within a narrow but well defined domain specific framework, during a problem-solving process. Such a tool, generally referred to as either a knowledge-based system (KBS) or an Expert System (ES) (Chau, 2004), is designed to assist in solving problems that would otherwise require the skills and expertise of a knowledgeable "flesh and blood" professional by applying embedded heuristics and other pertinent rules of thumb.

The paper presents descriptive details of a DSS that has been developed to assist managers of WDN, especially where water supply is intermittent and demand driven and not pressure driven. Water distribution networks operate for a limited duration, and because the amount of water each user is able to collect depends on the available pressure at their connection and the duration of the service, the user demand is often not fully met. To improve the service standards, water utilities are struggling hard to develop aids in the form of computer based analysis and management software. Expert systems, as an artificial intelligence (AI) based approach, is one such approach that managers of water utilities are increasingly gravitating towards. The suggested tool is but a computer program that seeks to provide solutions to the same problems which were traditionally solved only by a human expert by archiving knowledge in the form of rules and other heuristics and then utilizing this pre-programmed set of strategies and the knowledge base in a manner that is akin to human reasoning. The AI field has always fascinated researchers who have applied it in many diverse fields (Cheng et al., 2002; Wang et al., 2004; Muttil and Chau, 2007). In the field of water management, some important applications of expert systems include EXPLORE (Leon et al., 2000), OASIS (Goforth and Floris, 1991), CRITQUING Expert System (Shepherd and Ortolano, 1996), IITWSEXP (Khosa et al., 1995), Expert System for treated water distribution (Bunn and Helms, 1999), Network Management System For Water Distribution System (Raghvendran et al., 2007), and Intelligent Control System For a Municipal Water Distribution Network (Chan et al., 1999). Most of these initiatives included components to transfer knowledge from the heuristic domain to the knowledge base of the expert system, while some of the other developments have applied fuzzy logic to process information and suggest "best practice" guidelines for the network manager.

Review of literature further reveals that presently available expert systems have an extremely limited scope and are confined to only those specific tasks that come as preprogrammed knowledge and are therefore unable to reason broadly across fields of peripheral expertise. These ES employ static knowledge domains without the ability to store newer facts and, therefore, are unable to learn and evolve. Moreover, the difficulty in knowledge representation of a WDN problem has contributed to the limited use of expert systems (ES) in the management of WDN. WDN are seldom new and are associated with large amount of data for its elements namely pipes, demand nodes, supply nodes and pump etc. As a result of the natural process of ageing, and with frequent interventions necessitated by our perceived need to meet various performance goals, water supply networks undergo physical changes that significantly impinge on their hydraulic responses. To analyze such a continuously changing system, it is necessary to store information related to WDN elements in a logical and systematic way, thereby preserving and retrieving the cross relations between them. The DSS, therefore, has been designed to have SQL as RDBMS.

2 CLIPS expert shell and domain integration

Knowledge base constitutes the core of any DSS and for the latter to be an effective surrogate for any expert, it is essential that the heuristic knowledge is available in the form of easily understandable rules. The public domain expert shell, CLIPS, provides an ideal platform wherein expert knowledge may be codified as rules related to operation and management of WDN. To supplement its rules management capability, CLIPS also works as an inference engine that enables it to perform the inference procedure whereby rules are interpreted to trigger various actions as is deemed appropriate. This mechanism employs the embedded pre-existing rules based knowledge as "facts" to drive the firing mechanism of the inference engine and thereby produce a recommended conclusion to a particular problem.

A feature of CLIPS is that this expert shell does not have its own dedicated database and, as a result, facts are volatile and are purged from its memory as soon its execution is terminated. To overcome this intrinsic limitation, an external RDMS must be added for a seamless integration with CLIPS. This then creates the desired two-way communication capability in the DSS whereby all facts can be transferred into the database during closing and, conversely, these facts can also be re-asserted back into the memory bank of CLIPS during program execution. In comparison to the previous efforts as documented in the literature, the structure and design of the proposed DSS is a significant improvement as it combines network modelling capabilities with ES. Hydraulic modelling requires complex and extensive calculations to be carried out and this capability has been achieved in the proposed DSS by integrating the MATLAB computation platform with the latter system. The present study attempts to develop a DSS that incorporates a dynamic knowledge acquisition system driven by simulated runs of a hydraulic model, suitably calibrated and validated for the given water utility. In addition to CLIPS as an AI tool and the

MATLAB computational platforms, and in order to vest it with the capability to handle spatially distributed attributes of the study network, the proposed DSS uses open source GIS coupled together with a RDBMS. Together, these seemingly disjointed components work under the umbrella of a common user interface with cross platform integration being facilitated through a dynamic link library (DLL) in the Visual Studio environment.

3 Problems of water distribution system management

Municipal water distribution systems are: (i) spatially extensive; (ii) composed of multiple pipe loops to maintain satisfactory levels of redundancy; (iii) governed by non-linear hydraulic equations; (iv) designed complex hydraulic control devices such as valves and pumps; and (v) complicated by numerous layout, pipe sizing, and pumping alternatives. In addition to the palpable impairment in hydraulic performance that results as a network inevitably ages, it is natural that with a steady improvement in living standards accompanied by continuously increasing population, aspirations for better reliability in service and periodic inclusion of newer areas within the municipal limits especially in metropolitan cities together make these systems difficult to manage.

Often, these impacts result in performance levels that are well below expectations. Further, WDN are managed by experts, who, over the years of their association and responsibility, acquire an empirical knowledge of the system, but characteristically this knowledge remains largely confined to their respective personal domains. In the event of any new information and/or emergence of a new problem, these experts apply simple heuristics to design corrective measures and cognitively seek to predict network performance. Understandably, therefore, the assurance of a satisfactory response of the study network to suggested interventions is often based more on hope rather than on a validated belief.

With specific reference to issues of availability of domain knowledge, managers of water utilities face numerous difficulties on account of (i) expert scientific knowledge is not readily accessible; (ii) expert scientific knowledge is not available in a user friendly manner; and (iii) the tacit, undocumented expertise required for operation of a water supply system may be lost when experienced personnel leave service for whatever reason. These and similar concerns have seen resurgence in consumer demands that water engineers are suitably trained to apply current, technology based tools objectively and intelligently, and with the recent advances in AI technology, these personnel can be trained to fulfill this requirement. This indeed follows from the widely held view that there usually are very few specialists who have a thorough understanding of all the issues pertinent to network modelling for flows, pressures, and/or water quality. The proposed KBS has the potential to fill the gaps in knowledge that exist between researchers and practitioners (Chau, 2004).

Knowledge of basic network modelling concepts is indeed central to the decision-making process within water utilities not just as an aid to pipe network analysis but also for its efficient planning, design, and operation. This allows simulation of the design network and its examination under a variety of current and future operating conditions including possible design interventions. A number of off-the-shelf packages are available that allow tailor-made simulation models to be constructed for a water utility's specific requirements. Popular packages include EPANET (US Environmental Protection Agency), AQUIS (7T), Infoworks (Wallingford software), SynerGEE (Advantica), and WaterGEMS (Bentley).

These software packages enable development of mathematical models for a WDN that combine the physical configuration of these networks with physics-based equations that relate pressure and flow for each operational element (Machell et al., 2010). For the model to be acceptable as a reasonable representation of the actual system, the results from the model must bear close resemblance to the actual performance of the hydraulic system (ECAC, 1999). This, therefore, necessitates a prior model calibration.

4 The DSS

The aim of development of a DSS is to codify the heuristic expert knowledge in a form that is easily understood by the computer and also enables a novice to effectively manage a given WDN under unforeseen exigencies. Towards this end, possible real world scenarios corresponding to various design interventions are generated and corresponding results are incorporated in the knowledge base in order to enhance the scope and reach of the DSS. The development process involves (i) understanding the physical component elements of a typical WDN and their functionality, (ii) study of already existing DSS, (iii) understanding common terminology used by operators and experts, (iv) knowledge of current strategies for management of a water utility.

Since the DSS is to be applied on a WDN, it is imperative to have a modal case study of a WDN. Anytown Network (Walski et al., 1988; Kapelan, 2002), as shown in Fig. 1, is taken as the model case study. The network consists of 35 pipes assembled in a fashion that they form 19 loops. As the present study deals only with steady state analysis, the tanks at nodes 501 and 502 in the original network have been changed to reservoirs.

The proposed DSS is developed as a personal computer (PC) based application in the Visual Studio.Net environment. Visual Studio.NET is a complete set of development tools for building ASP Web applications – XML Web services, desktop applications, and mobile applications, etc., and all use the same integrated development environment (IDE), which allows them to share tools and facilitates in the creation of

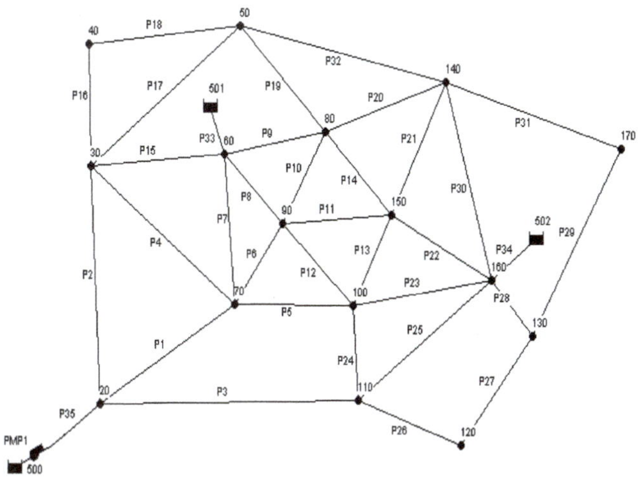

Figure 1. The anytown network.

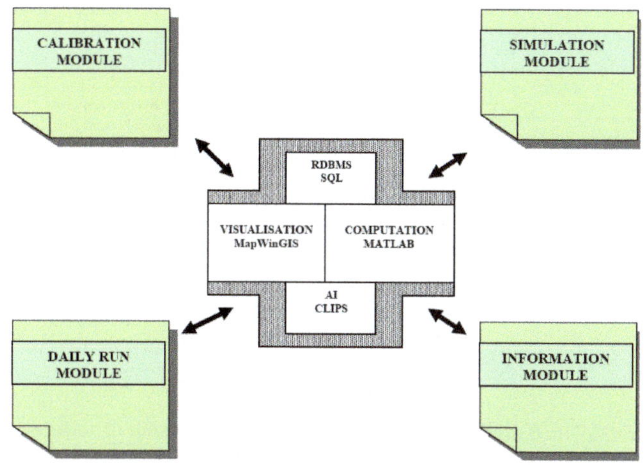

Figure 2. Basic framework of Decision Support System.

mixed-language solutions. This framework provides a clear, object-oriented, extensible set of classes that enables one to develop rich Windows applications.

A DSS framework for a system as spatially extensive as a WDN has to reflect the underlying complexity of the system for realism and yet be simple to use. Accordingly, the complex functionality of these and similar systems is best disaggregated into a limited number of manageable facets in the form of interconnected modules and linked together in a structured manner to yield the overall DSS. Accordingly, within this modular framework, the proposed DSS has been designed with object modules for Network Calibration, Network Simulation, Daily Run Operation, and Network Information. The complex task of computations, data handling, visualization, and decision/inference of each module is handled through MATLAB, SQL server, MapWinGIS, and CLIPS. Figure 2 shows that platforms such as SQL, MATLAB, CLIPS, and MapWinGIS enable exchange/transfer of data between different modules of the DSS under the overall umbrella of Visual Studio.Net platform. In addition to the modules, there are additional supporting tools that have been integrated within the expert DSS. These supporting tools are listed in the following text along with a brief accompanying description.

4.1 Relational database management system

A typical WDN may be characterized in terms of a diverse set of attributes and, importantly, these attributes are complimentary in nature with no attribute capable of acting as a surrogate for one or more of the other identified attributes. While some attributes are static in nature (for example link diameters and elevations of the identified critical elements of the distribution system), many of the other attributes are dynamic in nature and may assume different values through time. The skill with which these data are stored, retrieved, and/or used largely determines the utility and acceptability of any proposed DSS.

Typically, data pertaining to the components of WDN must be collected from different sources and need to be processed and stored within a well-structured framework in order to ensure availability to modellers, planners, and decision makers on demand and, at the same time, amenable to easy discrimination by the processing system. Relational databases allow such interrelationships to be preserved within the overall database management framework.

In the present study, a relational database management system (RDBMS) for WDN has been developed using Structured Query Language (SQL). All the necessary tables have been created in SQL and these are linked to each other through a primary key.

4.2 CLIPS knowledge base and inference engine

CLIPS (acronym for C Language Integrated Production System) is a computer language designed and developed by the US Space Agency, NASA, specifically for writing expert system (ES) applications. CLIPS is a forward-chaining rule-based language that resembles OPS5 and ART, other widely known rule-based development environments. Figure 3 shows the basic components of CLIPS and the basic functionality of each of these components is given as below:

1. *User Interface*: the mechanism by which the user and the ES communicate.
2. *Fact-list*: a global memory for data inserted as pipe-facts, node-facts, valve-fact, etc. into the memory of CLIPS.
3. *Knowledge-base*: contains all the rules used by the ES.

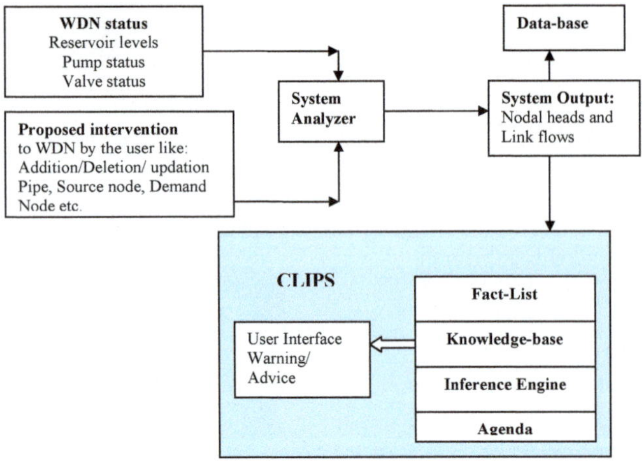

Figure 3. Components of CLIPS and its implementation in ES.

4. *Inference engine*: makes inferences by deciding which rules are satisfied by facts, prioritizes the satisfied rules, and executes the rule with the highest priority.

5. *Agenda:* a prioritized list created by the inference engine of instances of rules whose patterns are satisfied by facts in the fact-list.

CLIPS is a fact driven environment where rules are fired depending upon the pattern matching. CLIPS does not have its own database and facts stay in the memory of the CLIPS during its active state only. As soon as the CLIPS environment is closed, all facts are removed from its memory. Therefore, a mechanism was required through which facts are inserted into the memory from the database and vice-versa. This was achieved through creation of classes and sub-classes in Visual Studio environment. CLIPS has been integrated with Visual Studio through DLL. CLIPS is fact driven language that works on pattern matching mechanism of the facts. It is therefore necessary to insert facts based on a certain template only. In CLIPS, the "deftemplate" construct is used to create a template which can then be used to access fields by name. Main class of all facts is "Fact". To facilitate transfer of data from CLIPS to Visual Studio and vice-versa in particular template, only two child classes have been created for each of the objects as pipe, source node, demand node, etc. Figure 4 shows two child classes for pipe object of WDN. First child class ("DRAFPipe") captures the value of each of the slots ("pipe_id", "snode", etc.), as shown in Fig. 4, values from database or user through user interface. Second child class ("DRPipe") saves the captured values in the format defined by template.

The chaining mechanism of CLIPS is programmed to fire rules if the inserted fact matches any of the already existing facts. Further, for CLIPS to become operational as an ES shell, the following two kinds of integration are required. The interface required for such an integration to be possible

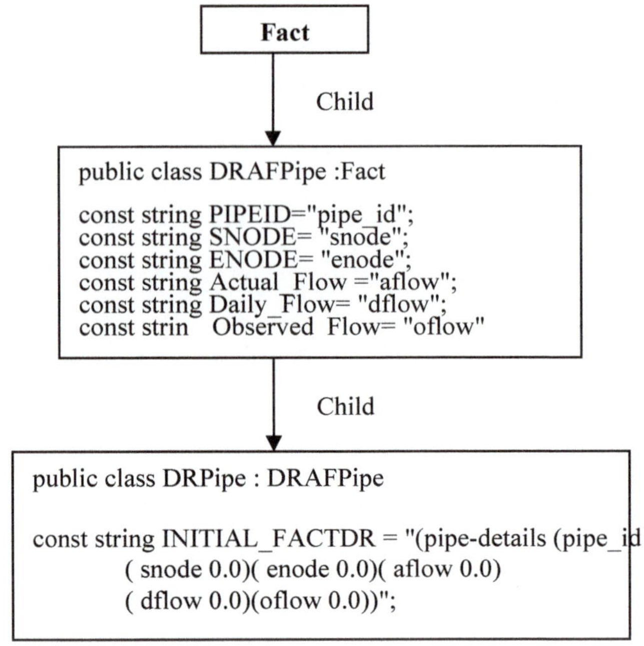

Figure 4. Class and child Class for fact of Pipe object.

enable (i) CLIPS to be embedded in other software platforms, and (ii) calling external functions from CLIPS. In the present study, integration of CLIPS within the Visual Studio.Net platform has been accomplished through ClipsNet.dll.

4.3 MATLAB platform

CLIPS and SQL server are not vested with the capability to perform complex mathematical computations. Therefore, in order to empower the DSS with computational capabilities, the MATLAB computing engine (acronym for MATrix LABoratory) has been integrated with the DSS in this study. The MATLAB engine incorporates the LAPACK and BLAS libraries, which provide access to state-of-the-art in software standards for matrix computation. The proposed DSS uses standalone COM components created using the MATLAB compiler. Each of the COM components is used in the Visual Studio.Net platform as a class that can perform mathematical operations on data imported from SQL and restore derived results back to the ".Net" environment.

4.4 Hydraulic network solver

The possibility of linking public domain network solver, EPANET (Rossman, 1993), through DLL was examined and it was felt that an inbuilt solver would be economical in terms of computer time. Therefore, based on the Todini and Pilati (1988) improved gradient method, a hydraulic network solver was developed in MATLAB and incorporated in Visual Studio.Net through a COM component.

4.5 Open source GIS platform

MapWinGIS is an open source GIS and an application programming interface (API) distributed under the Mozilla Public License (MPL) that was built upon the Microsoft.Net Framework 2.0. The MapWinGIS components and end user application support manipulation, analysis, and viewing of geo-spatial data in many standard GIS data formats. Hence, MapwinGIS is a mapping tool, a GIS modelling system, and a GIS API in a re-distributable open source.

Further, and importantly, the creation, visualization, and manipulation of the shape files of various constituent elements of the study network have been achieved through integration of "ax-Map object" of MapWinGIS with Visual studio.Net.

5 DSS modules

The DSS being developed has modules on calibration, simulation, planning, and daily run.

5.1 Network calibration module

Various automatic network calibration algorithms have been proposed in the literature and include: (i) methods based on analytical equations (Walski, 1983); (ii) simulation models (Boulos and Ormsbee, 1991; Gofman and Rodeh, 1982; Ormsbee and Wood, 1986; Rahal et al., 1980); and (iii) optimization methods (Coulbeck and Orr, 1984; Ormsbee, 1989). Techniques based on analytical equations require significant simplification of the network through skeletonizations. As a result, such techniques may only get the user close to the correct results. Conversely, both simulation and optimization approaches take advantage of using a complete model. In the recent past, the focus of researchers has shifted from traditional analytical techniques to heuristic techniques such as Genetic Algorithm (GA), Simulated Annealing (SA), Shuffled Frog-Leaping Algorithm (SFLA), and Ant Colony Optimization (ACO). Performance of ACOA, as a class of optimization techniques, has been quite encouraging and this technique has often been seen to outperform other heuristics-based approaches when applied to some benchmark problems (Maier et al., 2003; Zecchin et al., 2005).

Considering that the calibration process involves adjustment of some key system parameters like nodal demands and/or pipe-roughness values, prior knowledge of their rational values can act as a heuristic in stochastic search procedures such as GA and ACOA. In this study an automated calibration algorithm was developed for WDN and has been incorporated in the proposed DSS. The calibration algorithm has been designed using ACO approach and searches for optimal hydraulic performance by adjusting internal pipe roughness as a calibration parameter. As an aid, the algorithm has been designed to use age of the network pipes as prior information (PI) for guiding the search (Kulshrestha et al., 2011).

Accordingly, and in line with the foregoing discussion, a Calibration Module has been developed as an intrinsic component of the DSS. The calibration process has been developed based on ACO algorithm and systematically adjusts the Hazen Williams C-values in order to tune the hydraulic performance of the model, as expressed in terms of the derived distribution of nodal heads and link flows, with the network observations on corresponding state variables. The use of prior information (PI) on age and diameter of the various network components has been studied extensively in the study and is shown to aid the calibration process significantly. A detailed discussion on the development of the ACO-based calibration algorithm along with its evaluation and detailed sensitivity analysis is described separately in (Kulshrestha et al., 2011). The paper also presents a comparison of its performance with that of a MATLAB-based implementation of GA-based algorithm as well as the industry standard DARWIN CALIBRATOR of Benteley Systems, USA.

The significance of the role of some of the key parameters of the designed calibration algorithm in the calibration process has been studied extensively and their respective significance established on the basis of a detailed parametric study as reported in (Kulshrestha et al., 2011). The calibration algorithm parameters that have been the subject matter of scrutiny in the latter study include (along with their respective derived best values): (i) decision control parameters ($\alpha = 1.0$ and $\beta = 0.3$); (ii) initial pheromone value, $\tau_0 = 200$; (iii) pheromone persistence factor, $\rho = 0.95$; (iv) pheromone Reward factor, $R = 200$; (v) number of ants, $m = 10$; and (vi) maximum number of iterations, $n = 200$.

Basic working of the calibration module is shown in Fig. 5. Observations on the state variables of link flows (Q) and nodal heads (H) at different loading conditions are taken as input parameters. The study network is calibrated using ACOA with one set of observed boundary conditions and then validated with another, but independent, set of boundary conditions. The calibrated network is saved in the RDBMS for scenario generation in response to various unforeseen, but hypothetical, design interventions that may be necessitated in the foreseeable future.

5.2 Simulation module

A typical network simulation model is comprised of static asset information such as pipe lengths, diameters, connectivity, and network topography. In addition to the static information, the model is also armed with information about dynamic attributes that are expected to change with time. These include distribution of water demand across the network and, importantly, direct or indirect measures of growth in pipe roughness with age, amongst others. Most contemporary modelling packages utilize models that include every pipe down to the level of, but not including, customer

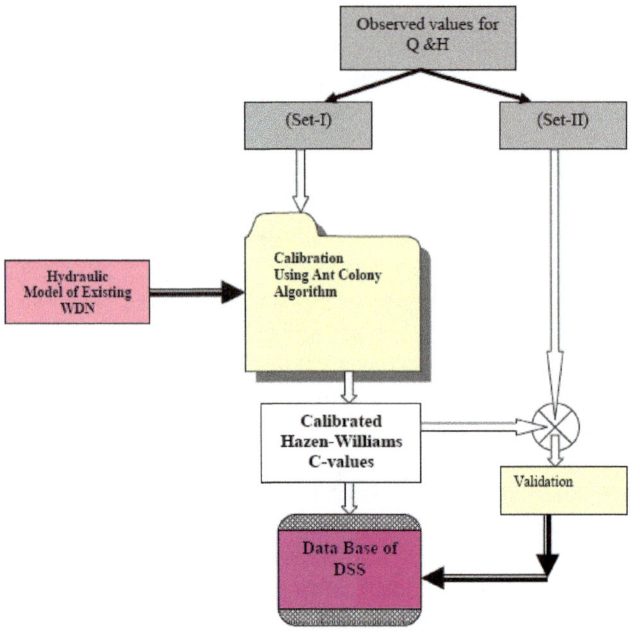

Figure 5. Schematic of calibration module.

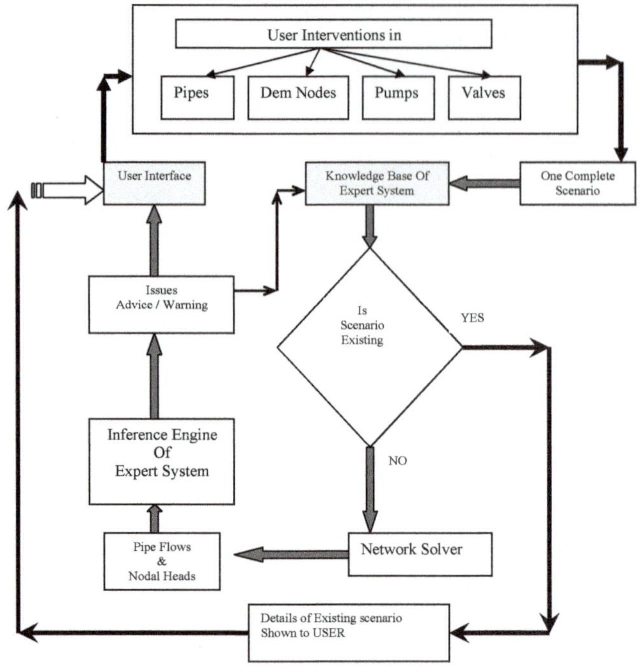

Figure 6. Schematic of simulation module of the DSS.

service pipes. Standard daily time-varying demand relationships are estimated for different customer types, and these are summed and allocated at pipe junctions (nodes) according to the distribution of customers. Simulation models are normally used to simulate flows and pressures over a fixed 24 h period (Machell et al., 2010) and are usually vested with the capability to generate a great number of reports. The variety of data generated sometimes makes analysis and decision making tedious.

The simulation module of the proposed DSS is designed to enable simulation studies for planning enhancements/upgrades for the WDN as well as solving network problems. To meet service standards for a growing and increasingly fastidious population, it becomes imperative to add newer areas under the coverage of the distribution network or, in some situations, either add a new parallel link or even replace an existing smaller diameter pipe with a larger diameter pipe. A simulation study on WDN can examine the impact of any proposed or even a hypothetical intervention and, upon recall from the memory bank of the DSS, can guide the network managers whenever similar interventions are deemed to be desirable in future.

Figure 6 shows the schematic of the simulation module. It merits mention that a suggested change in the network configuration constitutes a valid scenario and may already exist in the knowledge base after having been generated following a prior query. In such a case, the user has the option to either evaluate the scenario details or, in the event the information is not already present in the knowledge base in the absence of any past interest in such a scenario, the network solver is triggered for flows and nodal heads. The values of pipe flows and nodal heads are asserted as facts in inference engine of CLIPS, which fires certain rules to issue Advice/Warning to the user. The Advice/Warning along with its corresponding scenario is added to the Knowledge Base of the DSS as an upgrade for future reference, thus rendering the DSS as dynamic and amenable to upgradation/augmentation. In order to facilitate delivery of existing (previous memory) and current (user interventions) information, two templates, namely "memory" and "user", are created in CLIPS.

5.3 Planning module

Planners carefully research all aspects of a water distribution system and try to determine which major capital improvement projects are necessary to ensure the quality of service for the future. This process, called master planning (also referred to as capital improvement planning or comprehensive planning), may be used to project system growth and water usage for the next 5, 10, or 20 years. System growth may occur because of population growth, annexation, acquisition, or wholesale agreements between water supply utilities (Walski et al., 2001).

The ageing of pipes, and a consequent increase in internal pipe roughness, is always accompanied by loss of conveyance. Such a scenario is indeed realistic and confronts all water supply networks. In water utility practices, Hazen-Williams C-factor is the most commonly used parameter to represent internal roughness of water mains and by extension their carrying capacity and internal roughness of water mains (Walski et al., 1988), and the present study has also adopted this approach. The pipe roughness grows roughly linearly with time and the rate of roughness growth, a, is strongly influenced by the pH of water (For CI pipes).blackboxbf sentence break created here The corresponding model for growth of roughness may be written as (Colebrook and White, 1937):

$$e = e_o + at \tag{1}$$

In Eq. (1), e is the absolute roughness height, (L); e_0 is the initial roughness height, (L); a is the growth rate in roughness height, (L/T); and t is the time variable, (T). Using Colebrook and White (1937) and Swamee and Jain (1976) approximations, Hazen-Williams C_{HW} may respectively be written as (Walski et al., 1988).

$$C_{\mathrm{HW}} = 18.0 - 37.2 \log(X) \tag{2}$$

$$C_{\mathrm{HW}} = 33.3|\log(0.27\,X)|^{1.08} \tag{3}$$

where $X = (e_0 + at)/D$ and D is the diameter of pipe. The relationship between the annual roughness growth rate (a) and the Langelier saturation index (LI), with the latter controlled by factors such as the pH of water, its alkalinity, and calcium content, can be represented as Eq. (4) (Lamont, 1981):

$$a = 10^{-(4.08+0.38LI)} \text{ for LI} < 0 \tag{4}$$

Given the age of the pipe and its diameter of old pipe, it is possible to obtain an estimate of the Hazen-Williams C-factor. Based upon the above theoretical account, as the system undergoes the ageing process, the following things are expected to happen for any WDN:

1. With age of the WDN, the Hazen-Williams C-value will reduce.
2. The consequent head loss in the WDN will increase.
3. The total energy requirement in the WDN for the same input parameters will as well increase.

Therefore, from the management perspective, it is important to be forewarned on these changes and, at the same time, know in advance the manner in which its consequences are likely to manifest. As an example, the proposed planning module estimates Hazen-Williams C-values expected over various time horizons of 5, 10, 15, and 20 years and, for each such horizon, additionally computes (i) total energy requirement, (ii) total head loss across the network, (iii) head loss in each individual pipe, and (iv) temporal variation of Hazen-Williams C-values.

The module further has the capability to present a graphical depiction of the above information besides populating the databases of the proposed expert DSS. The schematic of planning module is shown in Fig. 7.

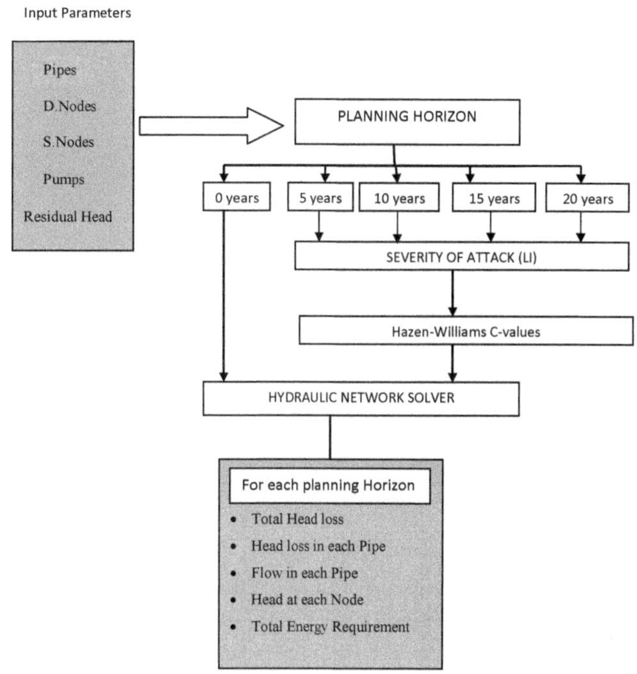

Figure 7. Schematic of planning module.

5.4 Daily run module

Many water utilities use human operators whose primary function is to monitor the pulse of the water distribution system and provide system control when needed. Automated control is increasingly becoming the preferred option and uses instrumentation and control equipment to control the distribution system automatically. Such control can be implemented either locally at the facility or throughout the system. Typically, simple operating rules are used to determine which component is operated and how. Systems that rely on advanced control use optimization algorithms, DSS, AI, or control logic to control the distribution system. It becomes clear that for this paradigm to be realistic, the actual system operation, in terms of its numerous diagnostic attributes, needs to be monitored continuously and logged for prospective reference. The proposed design of the DSS uses a Daily Run Module (DRM) that logs these operational attributes in order to be able to offer diagnostic inferences from the logged daily runs. Figure 9 depicts a selected screen shot of DRM. As shown in Fig. 8, this module has subordinate component sub-modules for the following tasks: (i) retrieve status of the study network on some previous date and time (Sub-Module-I); (ii) generate daily run scenario for a given set of input conditions (Sub-Module-II); and (iii) diagnose specific consumer complaints (Sub-Module-III). A brief note on each of these sub-modules follows:

- Sub-Module-I: refer to Fig. 9 with dotted circle "I". Each daily run history, for a given set of input conditions, is stored in the knowledge base of the DSS. The

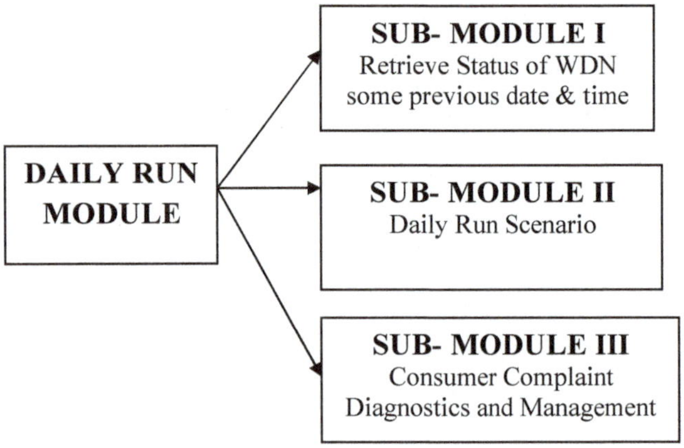

Figure 8. Different modules of Daily Run Module.

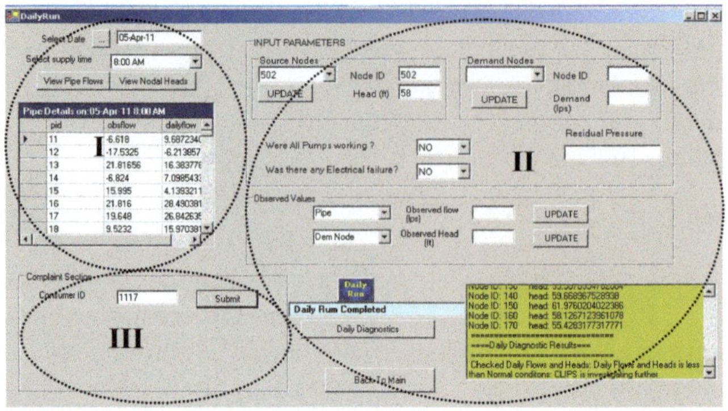

Figure 9. Screen shot of Daily Run Module of DSS showing sub-modules.

status of the WDN corresponding to any given date and time can be retrieved with the help of the graphical user interface (GUI) and runs independent of CLIPS inference engine.

- Sub-Module-II: refer to Fig. 9 with dotted circle "II". This sub-module assists a network manager in ascertaining status of the network corresponding to operator specified inputs on source node levels, pump, and electricity status. There is also an option available that allows comparison between observations on link flows and nodal heads, provided to the sub-module as prior specified data, and corresponding computed values as output from the network solver. The comparison is made by the inference engine of CLIPS, shown in Fig. 10, and is followed by generation of appropriate caution and/or advice. The inference engine fires rules at each decision stage, identifies cause, and also proposes action using the backward chaining mechanism. Figure 10 presents "if else" rules firing under different scenarios with message details. For example, if comparison of flows and heads with normal values is found OK, then message "No Action Required" is popped up; otherwise, "Rule 2" is fired and reservoir levels are compared.

- Sub-Module-III: This is the Daily Diagnostic Module and is designed to manage consumer complaints. The module uses backward chaining process for inferencing and starts by first identifying the consumer in terms of the network segment that he/she belongs to. Rules for this module are similar to those shown in Fig. 10. Using backward chaining mechanism, CLIPS is designed to identify possible cause of the complaint and, as a logical follow up, also suggests appropriate measures to be adopted to manage the situation.

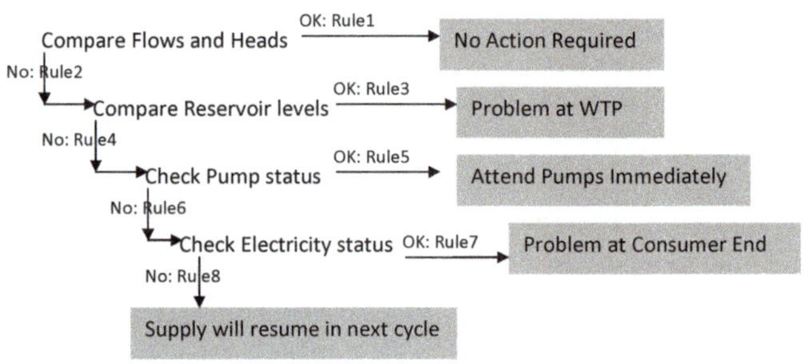

Figure 10. Working of CLIPS for Sub-Module II.

6 Case study

The proposed application has been implemented on an existing water distribution network that services a segment of Vasant Vihar (identified as Blocks "D", "E", and "F") in New Delhi. The study area is a predominantly residential area having a total population of 21 000. The network consists of 111 pipes and 86 demand nodes. Processing any new project on the proposed ES platform begins with data formatting and is done invoking the "New Network" stream of DSS. Figure 11 shows "New Project" form of DSS for formatting data pertaining to WDN.

By clicking "Add Base Map" button, the JPEG file of the desired network can be loaded on the form which would facilitate creation of shape files. Source Node, Demand Node, Pumps and Valves are added as nodes on the shape file "node.shp". A pipe is added as line in the shape file "pipe.shp". Upon conclusion of this task, the project network is then processed as "Existing Network" in DSS.

6.1 Calibration

The DSS adopts an optimization based approach for calibration and is enabled by the state-of-the-art ant colony optimization algorithm (ACO). As an added novelty, the calibration algorithm is also designed to accommodate prior

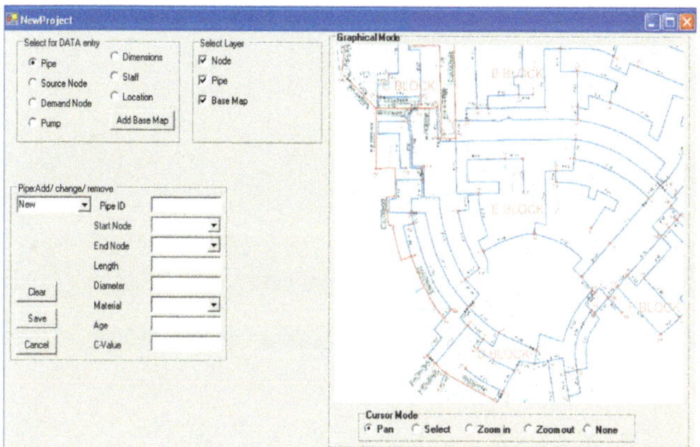

Figure 11. New Project form of ES.

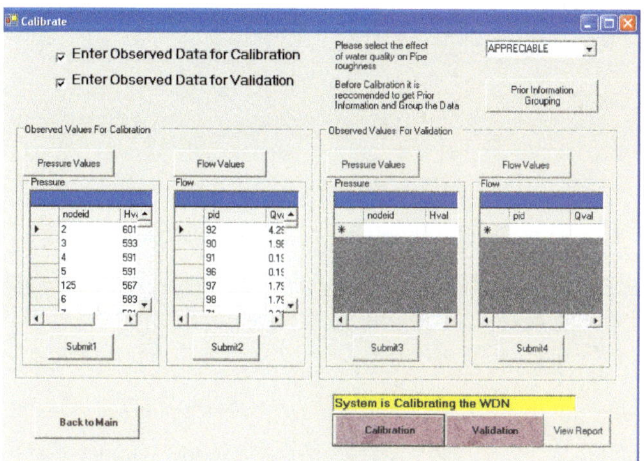

Figure 12. Screen Shot of the Calibration Module.

information on age of the network components as a guide leading the search process towards the desired optimality. As shown in Fig. 12, calibration is initiated by prompting the user to provide requisite information such as observed link flows and junction heads (these will form the basis for comparison between these values and model derived outputs).

Before proceeding for calibration, the module groups pipes based on diameter and age and computes prior information on Hazen-William C-values for each such group based on type of attack on the pipes on account of quality of water. The calibration module derives and assigns calibrated C-values to each pipe, and these results are archived as shown in Fig. 13

6.2 Simulation

The CLIPS engine has been designed to be an ideal foil for the simulation module and allows the DSS to exploit the full potential of the latter module. Each simulation outcome is checked for precedence, and only those scenarios that are

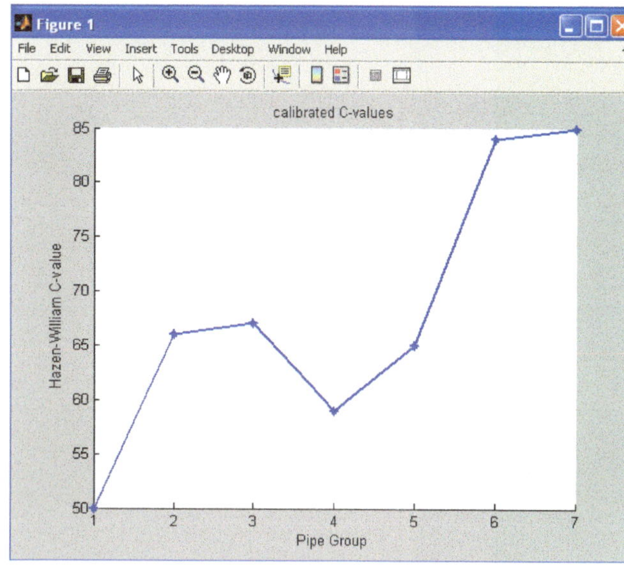

Figure 13. Calibration result for Vasant Vihar WDN.

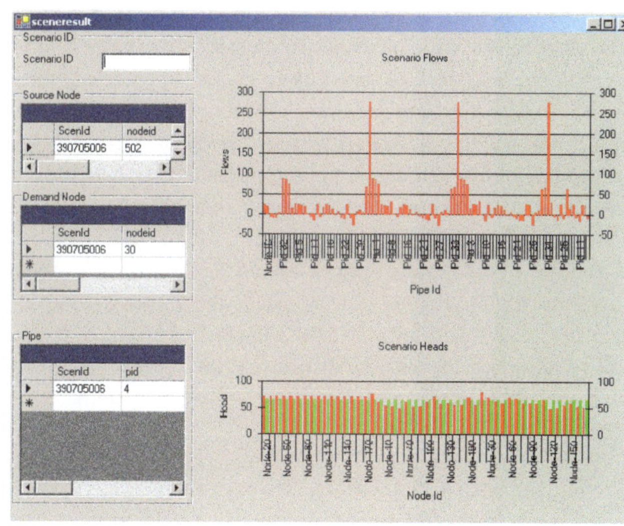

Figure 14. Existing scenario results of Simulation module.

not already available in the knowledge base are imbibed into it with a new, unique ID for possible future use. It is understood that the design of the proposed DSS has distinctive features such as: (i) the knowledge base (KB) is continuously upgraded; and (ii) the knowledge base is dynamic. Figure 14 presents a screen shot of the results propagated by simulation mModule in terms of link flows and nodal heads with the latter being displayed in comparison with residual head. Also, the left part of the figure displays changes in pipes, demand nodes, and source nodes that make up a particular scenario.

Simulation studies are undertaken to diagnose problems related to WDN. As an example, consider a scenario 1 in Fig. 15. Initially, source node 501 is maintained at reservoir level of 2.5 m. At this condition head at node 20 was 8.90 m.

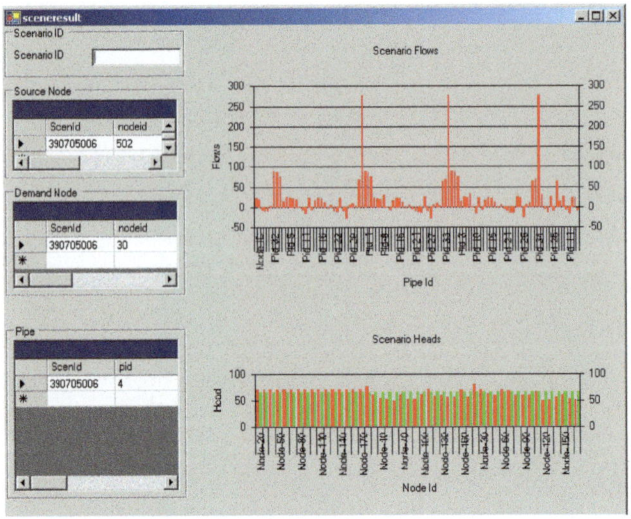

Figure 15. Scenario for problem identification and remedial measures.

Considering desirable residual head of 12.0 m at node 20, network administrator proposes to change pipe diameters of pipe number 135 and 34 (feeding to node 20) from existing 100 mm to 200 mm and 150 mm, respectively. Since infrastructure improvement work involves huge cost, the simulation module can be used to study the impact of such proposed changes. It can be seen from Fig. 15 that head at node 20 of Vasant Vihar WDN has improved to 13.61 (above residual head). The scenario can be saved for future reference. Whenever such scenario will be tried, the module shall issue a message "Similar scenario was earlier tried. Do you want to view the scenario?". If user chooses "Yes", then existing scenario details are populated; otherwise, scenario is processed as new scenario.

6.3 Planning

Planning module of the DSS has been designed to give an insight into the WDN's performance over planning horizon. Planning horizon has been taken as 20 years with a time step of 5 years (i.e. 5, 10, 15, and 20 years). The module is designed to help network managers take guidance on decisions for replacement of pipes and upgradation of pumping stations. In this regard, the module derives strength based on its ability to report the following:

- Variation of pipe flows and junction heads over the different planning horizon (Fig. 16).
- Variation of Hazen-William C-values for different groups of Pipes over the different planning horizon (Fig. 17).
- Variation of total energy requirement in kW and total head loss in the system over the different planning horizon (Fig. 18).

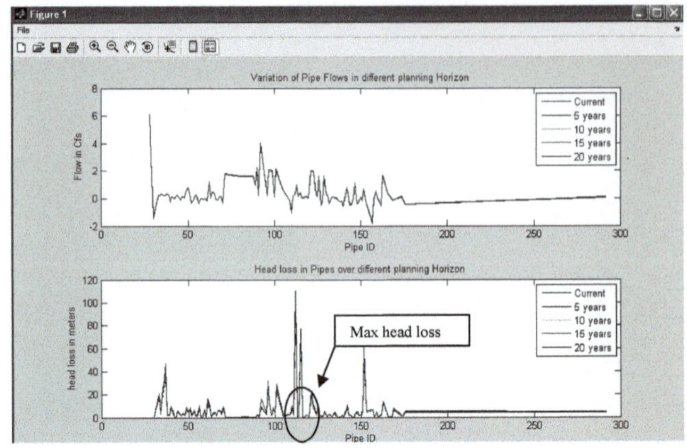

Figure 16. Variation of flows and heads over PH.

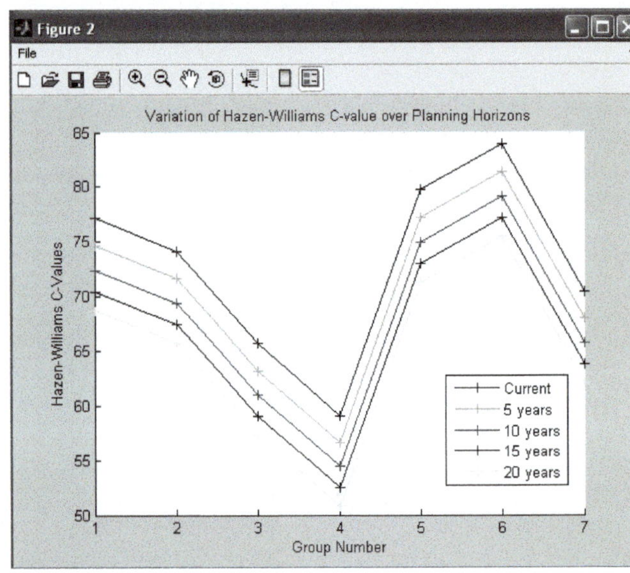

Figure 17. Variation of Hazen-William C-values for different groups.

6.4 Daily run

The DSS is built with a daily run module (DRM) and as already explained, this module consists of three sub-modules I, II, and III. Sub-Module-I prompts user to select date and supply time (for intermittent supply) for information on link flows and junction heads. For this already existing scenario, the module presents information as depicted in Fig. 8. The sub-modules II and III are tailored to invoke CLIPS inference engine to assert facts derived as a solution from the hydraulic network solver in its memory, and importantly, follow up with Advice/Suggestion/Warning as deemed appropriate as illustrated in Fig. 19.

The daily tun module is also endowed with an important functionality in terms of its ability to manage consumer

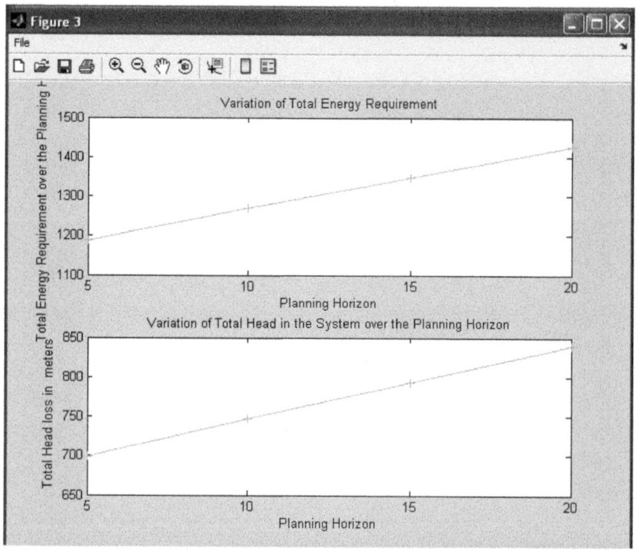

Figure 18. Variation of Total Energy and total head loss.

Condition	Message/ Warning/ Advice
CLIPS working for Sub-Module II	
If all Link Flows and Junction Heads normal	No Action Required by Network Administrator
If Link Flows and Junction Heads not normal then Check Reservoir Levels. If Reservoir Levels not OK	Contact Water Treatment Plant Managers for supply status
If Reservoir Levels Ok, check Pump Status. If there is any Pump failure	WDN Parameters were OK but there was pump failure. Please attend immediately
If Pump status Ok, check Electricity status. If Electricity status not OK	WDN Parameters were OK but there was Electricity failure. Please wait for next supply.
If Electricity status OK	WDN Parameters were OK. There might be problem at consumer end.

Figure 19. Advice/Suggestion/Warning issued by CLIPS.

complaints. Consider, for example, the water reservoir (ID 501) of the case study. The following information is lodged:

1. Reported head at the reservoir (ID 501) is 2.5 m instead of 3.0 m.
2. Consumer with ID 3423 lodges a complaint that "water supply to the user has been less than normal".
3. Reports of a pump failure have also been independently lodged.
4. There was no reported electricity failure.

From the network configuration, it is seen that the complainant is connected to pipe ID 71 (u/s node 3 and d/s node 30). The CLIPS inference engine uses backward chaining mechanism and reports appropriately reasoned diagnostic message as depicted in Fig. 20 which shows successive messages as would be received from the consumer complaint management module.

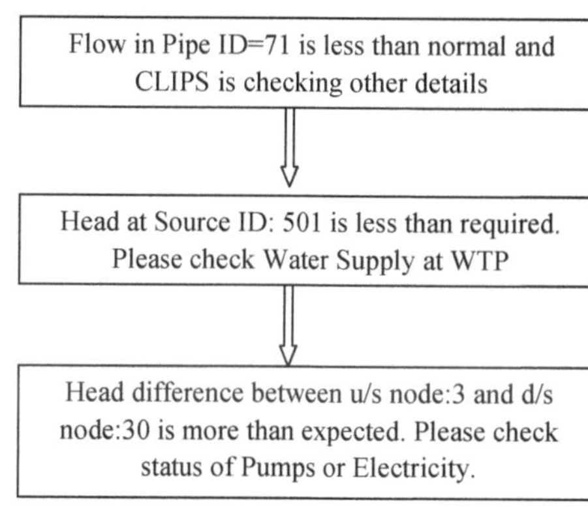

Figure 20. Messages received for addressing consumer complaint.

7 Conclusions

The aim of the development of DSS was to design a comprehensive operation and management aid for a water utility. The study has demonstrated the potential application of CLIPS as a core component around which such a management aid could be built. Present work suggests that CLIPS is indeed a powerful platform for development of an expert system.

Supporting platforms like MATLAB, SQL, and MapWinGIS have also been integrated together under a common umbrella of Visual Studio.Net environment. In this work, the CLIPS inference engine has been linked directly with Visual Studio.Net environment and indirectly with SQL and MATLAB. Such interlinking of these seemingly diverse set of computing platforms has resulted in a multiplier effect and, together, the study has been able to unlock the combined potential of such an application. In particular, the proposed framework acquires a totally dynamic flavor where a water supply system can be modelled, calibrated, and validated and its hydraulic performance simulated for a diverse set of operating requirements. The modules of DSS, as designed, can then acquire facts that emerge from such an exercise for further analysis and post processing by the inference engine and its resident and equally dynamic and flexible knowledge base.

Furthermore, the proposed DSS has been designed to perform other routine tasks such as directed simulation runs in response to any real or hypothetical intervention and to assess the given study network for consequent impacts. This helps in a priori generation of knowledge regarding network performance following such interventions.

CLIPS platform does not have permanent memory and an important contribution of the study has been the development of an external dynamic knowledge base with the mechanism for transferring newer facts into the knowledge base for subsequent use by the inference engine of CLIPS. The dynamic knowledge domain is designed to be self learning as it autogenerates newer knowledge for the DSS for future guidance.

The DSS, with its simulation and planning modules, is able to estimate, in advance, the expected Hazen Williams C- values for planning horizons of 5, 10, 15, and 20 years. Importantly, the proposed DSS avoids excessive dependence on interventions by highly trained expert personnel and with the help of a highly user friendly interface, even relatively novice operators are now vested with necessary tools for effective management.

Scope for further developments

In order to enhance the appeal of the proposed DSS for water supply systems that are in continuous operation and the system is essentially driven by demand, Extended Period Simulation (EPS) would be a desirable addition. Further, in present day scenario, it would be advantageous to harness the potential of Information Technology in water sector. Most water utilities now use some form of Supervisory Control and Data Acquisition (SCADA) system in their daily operations. The DSS would be an ideal partner tool for a SCADA based water distribution management system and, undoubtedly, would significantly widen the utility of the proposed Expert System.

References

Boulos, P. and Ormsbee, L.: Explicit network calibration for multiple loading conditions, Civ. Eng. Environ. Syst., 8, 153–160, 1991.

Bunn, S. and Helms, S.: Application of an expert system to control treated water distribution, ASCE, Annual Water Resource Planning and Management Conference, Phoenix, AZ, 1999.

Chan, C., Kritpiphat, W., and Tontiwachwuthikul, P.: Development of an intelligent control system for a municipal water distribution network, Proceedings of the 1999 IEEE Canadian Conference on Electrical and Computer Engineering, Edmonton, Alberta, Canada, 9–12 May 1999.

Chau, K.: Knowledge-based system on water-resource management in coastal waters, Water Environ. J., 18, 25–28, 2004.

Cheng, C. T., Ou, C. P., and Chau, K. W.: Combining a fuzzy optimal model with a genetic algorithm to solve multi-objective rainfall-runoff model calibration, J. Hydrol., 268, 72–86, 1999.

Coulbeck, B. and Orr, C.: A network analysis and simulation program for water distribution systems, Civ. Eng. Environ. Syst., 1, 139–144, 1984.

Colebrook, C. and White, C.: The reduction of carrying capacity of pipes with age, J. Ice, 7, 99–118, 1937.

CAC – Calibration Guidelines for Water Distribution System Modelling: Engineering Computer Application Committee, Proceedings of AWWA 1999 1mTech conference, American Water Works Association, 1999.

Gofman, E. and Rodeh, M.: Loop equations with unknown pipe characteristics, J. Hydraul. Div., 107, 1047–1060, 1982.

Goforth, G. and Floris, V.: OASIS: An intelligent water management system for South Florida, AI Appl., 5, 47–55, 1991.

Kapelan, Z.: Calibration of water distribution system hydraulic models, PhD thesis, University of Exeter, 2002.

Khosa, R., Parida, B., Singh, B., and Aggarwal, S.: Expert system for iit water supply, Master's thesis, Unpublished Undergraduate Project under Student Undergraduate Research Award(SURA), Indian Institute of Technology, Delhi, 1995.

Kulshrestha, S., Khosa, R., and Jha, K.: Roughness Calibration of a Water Distribution Network using Ant Colony Optimization with Prior Information, J. Water Resour. Pl.-ASCE, in review, 2001.

Lamont, P.: Common pipe flow formulas compared with the theory of roughness, J. Am. Water Work. Assoc., 73, 274–280, 1981.

Leon, C., Martín, S., Elena, J., and Luque, J.: EXPLORE- hybrid expert system for water networks management, J. Water Resour. Pl.-ASCE, 126, 65–74, 2000.

Machell, J., Mounce, S. R., and Boxall, J. B.: Online modelling of water distribution systems: a UK case study, Drink. Water Eng. Sci., 3, 21–27, doi:10.5194/dwes-3-21-2010, 2010.

Maier, H., Simpson, A., Zecchin, A., Foong, W., Phang, K., Seah, H., and Tan, C.: Ant colony optimization for design of water distribution systems, J. Water Resour. Pl.-ASCE, 129, 200–209, 2003.

Muttil, N. and Chau, K. W.: Machine-learning paradigms for selecting ecologically significant input variables, Eng. Appl. Artif. Intell., 20, 735–744, 2007.

Ormsbee, L.: Implicit network calibration, J. Water Resour. Pl.-ASCE, 115, 243–257, 1989.

Ormsbee, L. and Wood, D.: Explicit pipe network calibration, J. Water Resour. Pl.-ASCE, 112, 166–182, 1986.

Raghavendran, V., Gonsalves, T. A., Rani, U., Kumar, S., and Narasimhan, S.: Design and Implementation of a Network Management System for Water Distribution Networks, ADCOM'07 Proceedings of the 15th International Conference on Advanced Computing and Communications, IEEE Computer Society Washington, DC, USA, doi:10.11.09/ADCOM.2007.62, 2007.

Rahal, C., Sterling, M., and Coulbeck, B.: Parameter tuning for simulation models of water distribution networks, Proc. Inst. Eng., 69, 751–762, 1980.

Rossman, L.: EPANET water quality model, Technical report, PB-93-236552/XAB, Environmental Protection Agency, Cincinnati, OH (United States), Risk Reduction Engineering Lab., 1993.

Shepherd, A. and Ortolano, L.: Water-supply system operations: Critiquing expert-system approach, J. Water Resour. Pl.-ASCE, 122, 348–355, 1996.

Swamee, P. and Jain, A.: Explicit equation for pipe-flow problems, ASCE J. Hydraul. Div., 102, 657–664, 1976.

Todini, E. and Pilati, S.: Gradient Algorithm for the Analysis of Pipe Networks, Computer Applications in Water Supply, 1, Research Studies Press Ltd. Taunton, UK, 1988.

Walski, T.: Technique for calibrating network models, J. Water Resour. Pl.-ASCE, 109, 360–372, 1983.
Walski, T.: Tips for saving energy in pumping operations, J. Am. Water Work. Assoc., 85, 49–53, 1993.
Walski, T., Shields, F., and Sharp, W.: Predicting Internal Roughness in Water Mains, Miscellaneous Paper EL-88-2, US Army Engineer Waterways Experiment Station, Vicksburg, 1988.
Walski, T., Chase, D., Savic, D., Strafaci, A., and Dietrich, K.: Water Distribution Modelling, Haestad Press, Waterbury, CT, USA, 2001.
Wang, H. J., Zhang, J. P., Chau, K. W., and Anson, M.: 4D dynamic management for construction planning and resource utilization, Automat. Construct., 13, 575–589, 2004.
Zecchin, A., Simpson, A., Maier, H., and Nixon, J.: Parametric study for an ant algorithm applied to water distribution system optimization, IEEE T. Evol. Comput., 9, 175–191, 2005.

12

Development and evaluation of new behavioral indexes for a biological early warning system using *Daphnia magna*

T. Y. Jeong[1], J. Jeon[2], and S. D. Kim[1]

[1]Department of Environmental Science and Engineering, Gwangju Institute of Science and Technology (GIST), 1 Oryong-dong, Buk-gu, Gwangju 500-712, Republic of Korea
[2]Department of Environmental Chemistry, EAWAG, Swiss Federal Institute of Aquatic Science and Technology, Ueberlandstrasse 133, 8600, Duebendorf, Switzerland

Correspondence to: S. D. Kim (sdkim@gist.ac.kr)

Abstract. New behavioral indexes including combined index (CI), distribution index (DI), toxic index (TI), and altitude index (AI) for a biological early warning system (BEWS) were developed and evaluated using *Daphnia magna* in this study. The sensitivity and stability of each index were compared to evaluate the performance of the indexes through a real-time exposure test with a synthetic copper solution. The applicability of the CI to the field sample was evaluated through an effluent exposure test. The proportional relationship between toxicity level and magnitude of response was much lower in the effluent due to the complexity of water than in the copper solution. The results showed that the CI was most sensitive among the three indexes, while the DI was confirmed as the most useful index among the individual indexes. The combined index (CI) shows not only sensitivity but also stability in normal conditions below the statistically significant threshold ($p < 0.01$), whereas the individual indexes displayed unstable index values in normal conditions ($p > 0.01$). The CI improved performance of the BEWS in terms of sensitivity and stability, and it was confirmed as the higher correlation coefficient between the magnitude of the index and the toxicity level of the water sample.

1 Introduction

Because there is a need for the early detection of water contamination in the field of waste water treatment and drinking water supply, research has focused on a biological early warning system (BEWS) over the last few decades (Butterworth et al., 2001; Green et al., 2003). In order to apply the system to regional environmental conditions and achieve different monitoring aims, BEWS has been diversified through the employment of various test organisms such as water fleas, mussels, algae, and fish (Baldwin et al., 1994; Benecke et al., 1982; Borcherding et al., 1997; Hendriks et al., 1993). Among these organisms, *Daphnia magna*, which is a species of water flea, is widely used because of its usefulness in biomonitoring and high sensitivity to toxic chemicals (van der Schalie et al., 2001). Because of its high sensitivity, behavioral changes in living *Daphnia magna* are a powerful real-time indicator in the detection of toxicants in the warning system. Different types of behavioral parameters have been developed for various kinds of monitoring systems; the monitoring systems of Knie, Kerren Um-welttechnik, and BBE Moldaenke employ swimming activity, phototactic behavior and swimming speed, and so on (Baillieul et al., 1998; Kieu et al., 2001). When *Daphnia magna* faces a change in water quality or a toxic condition in the BEWS, its behavior changes markedly, and the level of abnormality is determined from the recorded behavioral data. If the toxic level calculated from the measured data via the specific algorithm increases to above an intentionally fixed threshold, an alarm is triggered by a BEWS (Jeon et al., 2008; Lechlet et al., 2000).

Even if the design of BEWS is appropriate for its goal, the possibility exists that the system will trigger false alarms during its practical operation. Problems can be also caused by uncontrolled variables such as improperly defined thresholds or machine error (Butterworth et al., 2001). In addition to these problems, inherent limitation exists due to behavioral deviations in living test organisms during the operation period and between individual organisms. To minimize the possibility of false alarms caused by these biological variations, the simultaneous utilization of a sufficient number of behavioral parameters is needed to supplement the weakness of individual parameters. The Daphnia-Toximeter, produced by BBE Moldaenke, is an example of a system that reduces the frequency of false alarms by simultaneously employing various parameters: velocity, fractal dimension, swimming height, distance between water fleas, and the number of moving water fleas (Baillieul et al., 1998; Michels et al., 1999; Lechelt et al., 2000).

In a previous study, Jeon et al. (2008) developed a new BEWS, which is equipped with a grid counter device (GC) that identifies the position of the test organism at the grid level instead of at the pixel level. In addition, this system adopted a new algorithm for the alarm function, achieving not only cost effectiveness but also high performance. Accordingly, this study was conducted to enhance the power of the system developed by Jeon et al. The objectives of the present study are as follows: (1) develop a new, reasonable behavioral index that can be applied to a GC device-equipped BEWS; (2) improve the performance of the BEWS by applying various parameters at the same time. To reach these goals, a new type of index called the distribution Index (DI) was developed from newly suggested behavioral parameters, and the altitude Index (AI) was additionally generated by modifying the swimming height parameter of the Daphnia Toximeter (Green et al., 2003; Lechelt et al., 2000). Finally, the combined index (CI) was derived by combining three behavioral indexes, including the toxic index (TI) that was developed by Jeon et al. The CI performance was evaluated by real-time exposure tests using synthetic water and effluent.

2 Materials and methods

2.1 System design and organism preparation

This study employed a BEWS equipped with a grid counter device. Compositions of the system were the same as in the previous study (Jeon et al., 2008), except that the number of chambers and test organism in each chamber was adjusted from six and one to two and ten, respectively. It can be assumed that there were no differences between the TI value from the previous and recent systems except the increase of the statistical power by enlarging total number of the test organism used for the system, because calculation and recording method were identical. All methods for culturing *Daphnia magna* were conducted according to the USEPA guidelines (Weber, 1993). A constant temperature of $22.5 \pm 1\,°C$ and 16 h of light with 8 h of dark were maintained during the culturing and the exposure test. A synthesized freshwater was prepared within the pH range of 8.0 ± 0.2, with hardness of $170 \pm 10\,mg\,L^{-1}$ and alkalinity of $110 \pm 10\,mg\,L^{-1}$ as CaCO3 ($MgSO_4$, 9.98×10^{-4} M; $CaSO_4$, 6.98×10^{-4} M; $NaHCO_3$, 2.28×10^{-3} M; KCl, 1.07×10^{-4} M).

2.2 Design of indexes

2.2.1 Distribution index

The DI is a newly developed index generated from the swimming range parameter. This behavioral parameter is based on the homeostasis of *Daphnia*'s swimming range. It is well known that the movement of *Daphnia* is predominantly affected by the degree of light intensity and food abundance (Young et al., 1993). Hence, if the food distribution and light intensity are maintained regularly, the swimming range will be consistent. Previous studies of the behavior of *Daphnia* observed typical changes in activity mass resulting from the stimulus of environmental condition change (Butterworth et al., 2001; Shimizu et al., 2002). This finding motivated our study because activity mass is proportional to movement range. The previous research found that stimulus expressed as avoidance of chemicals or retardation by adverse effects affects the swimming activity, making it greater or lesser than in normal conditions. This means that the movement range could be widened or narrowed, according to the type of behavior elicited by the stimulus. Additionally, spinning behavior, which is an inherent habit of *Daphnia* in stressful conditions (Dodson et al., 1995), could also contribute to dramatic changes in movement ranges. We, therefore, assumed that the range of swimming distribution could be differentiated from normal conditions according to these activity patterns – wide, narrow, and extremely narrow ranges – and developed a new behavioral index based on this assumption.

In this study, with the GC device, a rectangular chamber containing water fleas was divided into 100 parts. Each section was marked for the location of water fleas according to the combination of numbers on the x and y axes; the sections were numbered from 0 to 99 (Fig. 1a). A camera was used to record the spatial data of individual water fleas, and the information was transferred to a computer. When a water flea migrated to another section, a new section number was recorded in the same manner. The data of the locations of the water fleas were shown in detail. The frequencies of appearance at each section were obtained at one minute intervals (Fig. 1b), and the distribution level (DL) was calculated (Fig. 1c). If the distribution range of the water fleas was concentrated in a specific, narrow region of the chambers, only a few sections showed a high frequency of appearance, so the value of DL increased. On the other hand, if each section showed a similar frequency of appearance, it implied that the water fleas moved in a wide range around the test chamber, and the value of DL decreased.

The equation for the DL is

$$\mathrm{DL} = \frac{\sigma_{\mathrm{Dis}}}{\mu_{\mathrm{Dis}}}, \qquad (1)$$

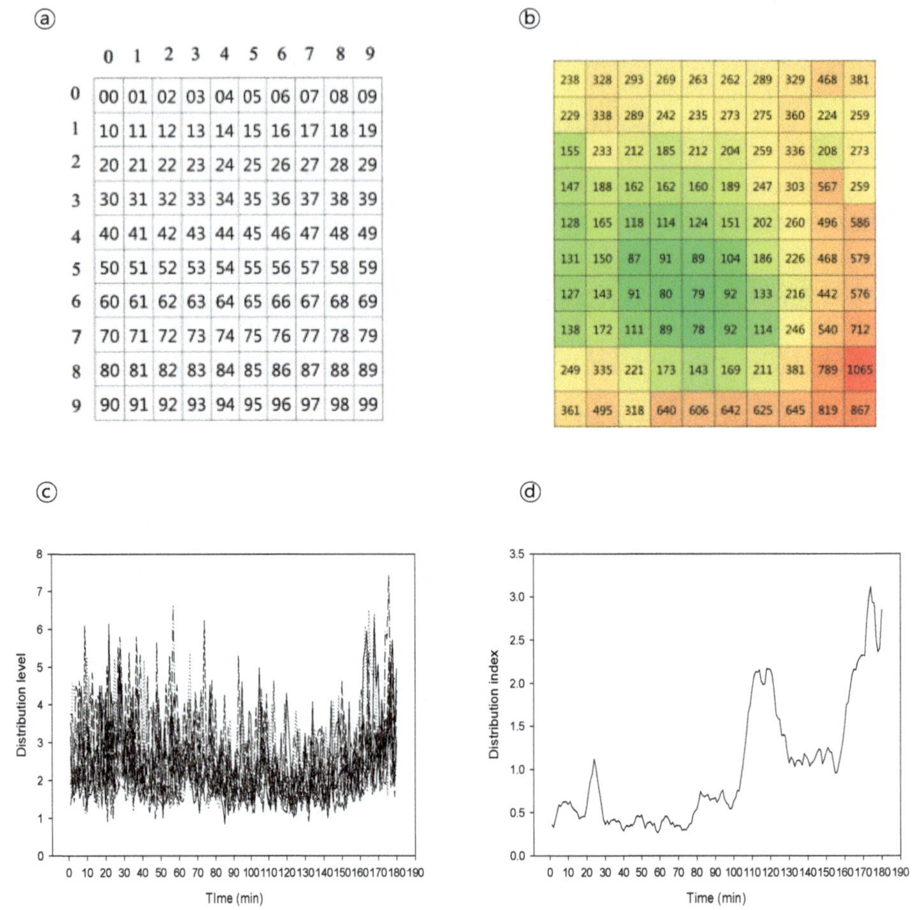

Figure 1. Procedure for generating the distribution index. **(a)** Numbering sections 00 to 99 for recognition of each section. **(b)** Calculation of appearance frequency of *Daphnia magna* at each section; this image was colored to easily view the frequency level of occupation. (Red: highest, Yellow: medium, Green: lowest). **(c)** Plots of 10 distribution levels derived from 10 *Daphnia magna*. **(d)** A Plot of a DI derived from 10 distribution levels.

where σ_{Dis} is the standard deviation of every observed appearance frequency for 1 min, and μ_{Dis} is the average of every observed appearance frequency for 1 min. After this process, the deviation ratio (*Za*) for DL was calculated to detect whether a considerable change of DL had occurred. The equation for the *Za* calculation is

$$Za = \frac{X_i - \mu_{\text{i}}}{s_i}, \tag{2}$$

where X_i is the observed DL at the time (i), μ_i is the average accumulated DL until the time (*i*), and s_i is the standard deviation of the accumulated DL until time (i). In terms of statistics, to determine whether a significant difference was observed, a student's *t* test was conducted between *Za* calculations in unexposed and exposed periods to the toxicant. The *p* values derived from the student's *t* test were then transformed to the logarithmic value as a log $(1/p)$, and the moving average of the log $(1/p)$ for 5 min was simultaneously calculated. The function of this moving average was to reduce the noise in the time series data. Finally, the DI value was calculated from the chamber as shown in Fig. 1d.

2.2.2 Altitude index

The AI is another index based on the swimming height parameter of the Daphnia Toximeter (Lechelt et al., 2000). The concept of swimming height refers to the average swimming position of *Daphnia* in terms of swimming altitude. *Daphnia* usually maintain a specific swimming altitude in normal conditions because they prefer to swim in an appropriate light range where they cannot be detected by predators (Young et al., 1993). When they are exposed to a toxic condition, this habit is broken. Consequently, when the toxicant stimulates *Daphnia*, the average swimming position moves to an abnormal height, which makes the AI differ from normal conditions.

We generated the AI using a grid counter device. As mentioned in the case of the DI, each chamber was divided into

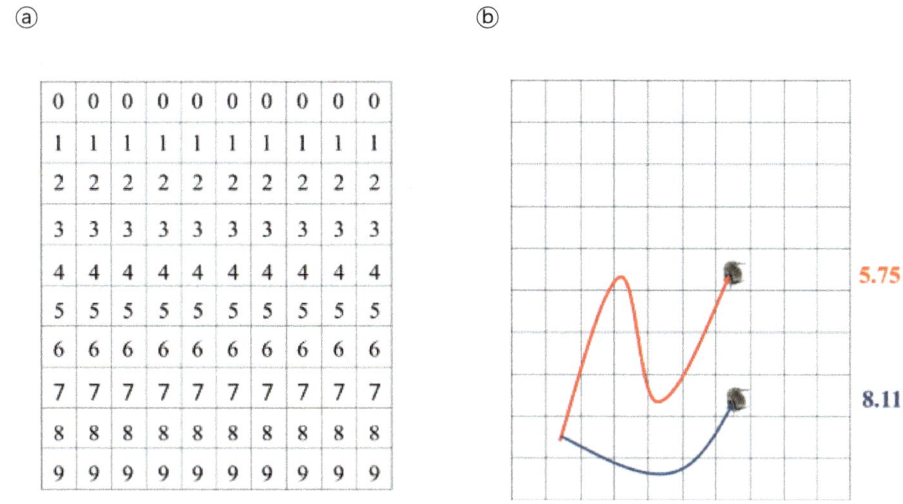

Figure 2. Recognition of swimming altitude of *Daphnia magna*. **(a)** Numbering sections 0 to 9 for recognition of each altitude. **(b)** Example of altitude recognition; the average altitudes of *Daphnia magna* colored red and blue are calculated.

100 parts, and each section was numbered. The altitude of *Daphnia magna* was thus recognized according to section numbers from 0 to 9 along the *y* axis from the top to the bottom, as shown in Fig. 2a. The method used to record the altitude was the same as the method used with DI. The average altitude, called altitude level (AL), was calculated from the recorded altitudes (Fig. 2b). The equation for AL is

$$\mathrm{AL} = \mu_{\mathrm{Alt}} \tag{3}$$

where μ_{Alt} is the average of every observed altitude for 1 min. After this AL calculation, the *Za* for AL and the log $(1/p)$ for *Za* using a student's *t* test were derived in the same way as for DI. Finally, the AI was produced in real time by adding a step to reduce noise at an interval of 1 min.

2.2.3 Toxic index and combined index

The TI previously developed by Jeon et al. (2008) was also employed in this study for the derivation of the CI and the comparison with the new indexes. All calculation methods were identical to those in that previous study except for the transformation procedure. The raw data on swimming distance were directly used to calculate *Za* in this study. First, the number of events counted (NC) was measured to indicate the swimming distance of *Daphnia* for 1 min, and the deviation ratio (*Za*) for NC was calculated to detect whether a considerable change in NC had occurred. *Za* were compared between the normal condition and the toxic condition using a student's *t* test, and the log $(1/P)$ was derived. The TI was then calculated with a moving average using the same methods as used for the DI and AI at 1 min intervals. Finally, three parameters were combined into the CI by averaging the values of DI, AI, and TI. This CI was derived every 1 min and was compared to the DI, AI, and TI through a real-time exposure test.

2.3 Experiment methods

2.3.1 Preparation for tests

The experiment was composed of two different real-time exposure tests. The copper exposure test was conducted on synthetic water samples, and the effluent exposure test used effluents sampled from three separate regions. The performance of the indexes was compared, and the applicability of the CI, which combined three indexes, to the field sample was evaluated using the results of the copper and effluent exposure tests. The age of *Daphnia magna* used in the tests ranged from 96 h to 120 h for the investigation of behavioral changes. If the test organism is too young, it could be excessively sensitive, or it could die soon after exposure to a toxic chemical. The test organisms were selected from one stock culture to reduce variance in the behavioral character among individuals and the third brood neonates were used. Every test was conducted without food; the absence of food does not affect the behavior of *Daphnia magna* (Demeester, 1992).

In the copper exposure test, copper (CuSO4•5H2O) was used as the representative toxic chemical because of *Daphnia magna*'s high sensitivity to it and its wide use in various industries. Notably, South Korea uses 400 Gg Cu $\mathrm{yr^{-1}}$ of copper annually and is the third highest copper-producing country in Asia, making copper an important toxicant and thus relevant for use in this study (Graedel et al., 2004). Before the test, a preliminary acute toxicity test for determining LC50 of copper was conducted over 24 h using *Daphnia magna* in a static condition. Because 24 h LC50 is typically chosen as the toxic level of one toxic unit (TU), the determination of a middle concentration of 75 ppb for the main test was based on the 24 h LC50 value of the pre-test (54.371 to 70.717 ppb). The minimum concentration of 18.75 ppb was determined as the basis for the lowest observed effect concentration (12.5 ppb) in the pre-test to investigate the effects at different concentrations both above and below 24 h LC50. In the main test, we prepared copper solution daily with hard water at concentrations of 0ppb, 18.75 ppb, 37.5 ppb, 75 ppb, 150 ppb, and 300 ppb. These concentrations are relevant for the BEWS study; it was reported that copper was present in a range of 0.2–11 ppb in the Yeongsan River in South Korea (Kim et al., 2011). The copper exposure tests were triplicated for every test set.

The effluent exposure test sampled three different effluents from waste water treatment plant in Iksan, Yesan, and Yeosu. The effluent from Yesan was treated with extended aeration process and the effluents from Iksan and Yeosu were treated with conventional activated sludge process. The parameters estimated at the point of exposure study were descripted in Table 1 in the Supplement. The toxicity level of each effluent was respectively confirmed as 0 TU, 50.50 TU, and 1.23 TU by the preliminary 24 h whole effluent toxicity test. The nontoxic effluent from Iksan was used for demonstrating the stability of parameters. Others, including the diluted samples, were used to estimate the early warning capability of parameters in the field water samples. The effluent from Yesan was diluted to 0.3, 0.7, 1, 1.6, 2.6, 3.3, 3.4, and 4.3 TU. The main contaminants in the sample waters from Yesan and Yeosu were not revealed; however, the characteristics of the two water samples were quite different. The sample from Yesan was of fresh water and that from Yeosu was of salt water (Kim et al., 2009). The ion concentration estimated by Kim et al. was shown in Table 2 in the Supplement. For the effluent, every exposure test was not replicated to simulate real spillage incident in the area of receiving water and water supply source.

2.3.2 Test procedures

The exposure test was conducted for 180 min. Before the exposure, we carried out a 120 min feeding time, 180 min adaptation time, and 180 min unexposed period. During the feeding time, we fed *Daphnia magna* a mixture of Pseudokirchneriella subcapitata and yeast, CEROPHYLL, and trout chow (YCT). Because the copper exposure test was conducted in the absence of food, *Daphnia magna* needed a full feeding. To avoid the effect of environmental conditions such as light, temperature, and food level, we did not support any nutritive element and maintained constant light and temperature conditions in the isolated monitoring room after feeding (Dodson, 1989). In the next step, the adaptation time, *Daphnia magna* were transferred into two test chambers; each chamber contained 10 *Daphnia magna*. *Daphnia magna* were allowed to swim in the chamber with water flux for 180 min, because an adaption period of at least 2 h is required for *Daphnia magna* to obtain a constant swimming pattern by acclimating to water flow after being transferred (Wolf et al., 1998; Untersteiner et al., 2003). In the subsequent step, the unexposed period, the behaviors of *Daphnia* were recorded for 180 min. These data were used as a control when the student's *t* test was conducted with the data on the exposed period. In the last step, the exposure period, the influent sample to chambers was replaced with a copper solution or effluent, and this condition was maintained for 180 min. Every pump was stopped during the replacement period. Data obtained for 30 min before and after stopping the pump was excluded to avoid exceptional behavioral changes caused by this procedure. Because the detection time of toxicant exposure by BEWS is not lengthy, we limited the maximum exposure time to 180 min to evaluate whether our system could detect differences in behavior in the early phase of contamination.

2.4 Data analysis

A student's *t* test was utilized to compare differences between the behavioral index data during the unexposed and exposed periods. Matlab R2009b, Microsoft Excel 2010, and Sigmaplot 10.0 were used for every data treatment introduced in the above sections, including the student's *t* test and graph illustration.

3 Results and discussion

3.1 Copper exposure test

The results of the indexes in the normal condition (0 ppb of copper) are expressed as graphs in Fig. 3. All individual indexes (DI, AI, and TI) displayed statistically significant differences in index values ($p < 0.01$, index > 2) in the uncontaminated condition. In contrast, the result of the CI was stable at below 2, a threshold implying a probability of 0.01 % for significant behavioral difference. This result proves the usefulness of combining indexes at the same time in terms of data stability. The raw data from the copper exposure test are summarized in Fig. S1 in the Supplement. Although fluctuations in the time series data are apparent, all indexes show the increasing tendency with increased concentration. For the quantitative evaluation of the power of the indexes to reflect changes in toxicity levels, the averages of every calculated index value during exposure time at each concentration were derived as a representative value. Then, linear regressions between the copper concentration and the average values were conducted, as shown in Fig. 4. A comparison of the correlation coefficients (r^2) showed that the CI had the highest one and the DI showed the highest value among the individual indexes. This result means that the CI is most capable of performing accurate early water contamination warning, and the DI best reflects the differences in toxicity levels among the individual indexes derived from the three different behavioral parameters.

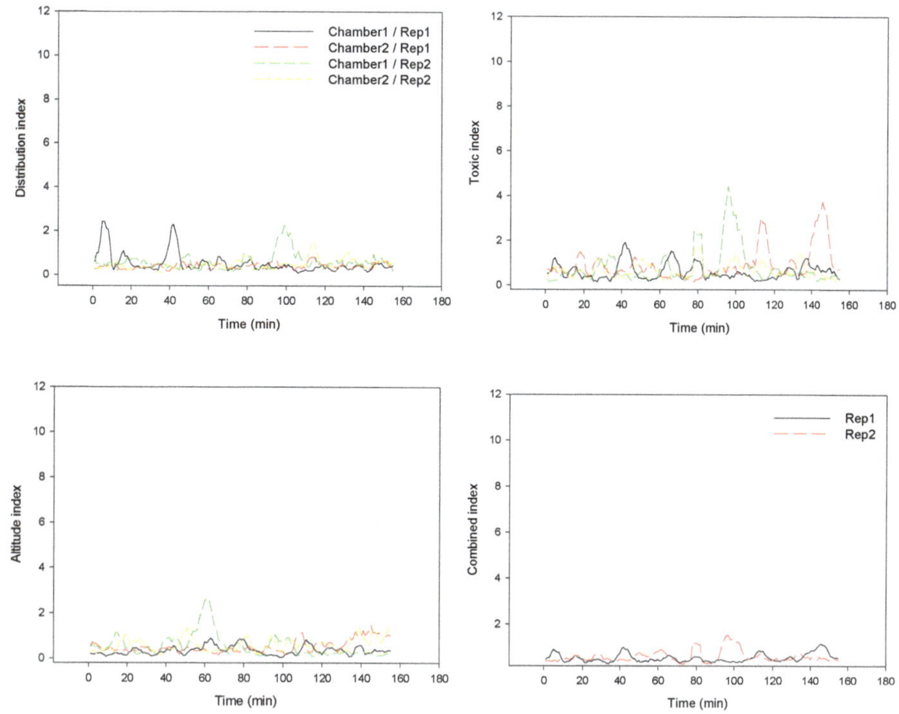

Figure 3. Results of indexes (DI, TI, AI, and CI) in normal condition. The Rep# refers replication number of test.

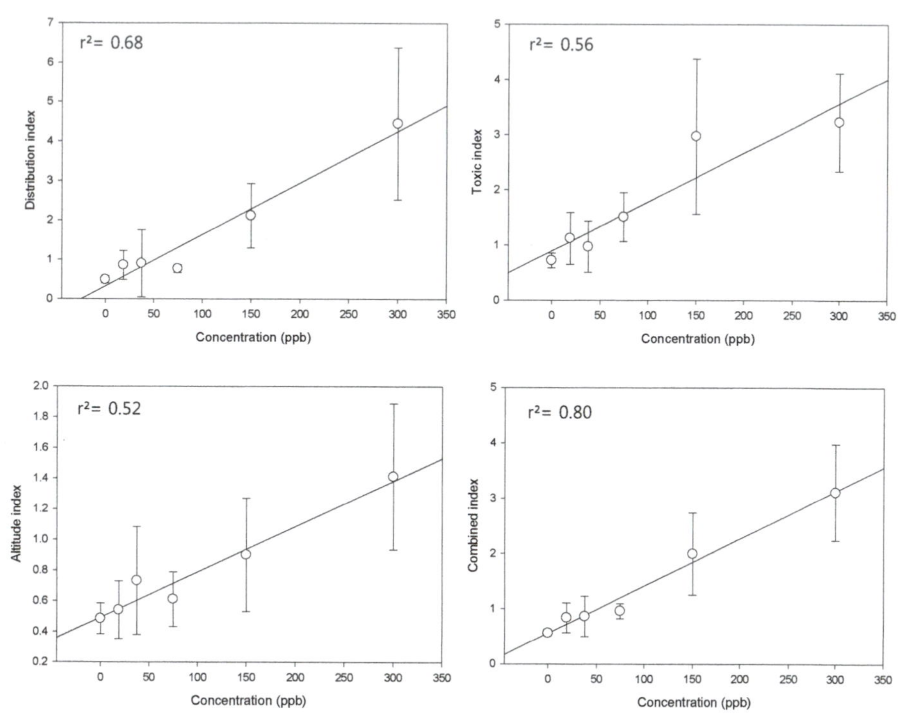

Figure 4. Linear regressions between representing value (average) of index and copper concentration. The line is regression line and r^2 is correlation coefficient value.

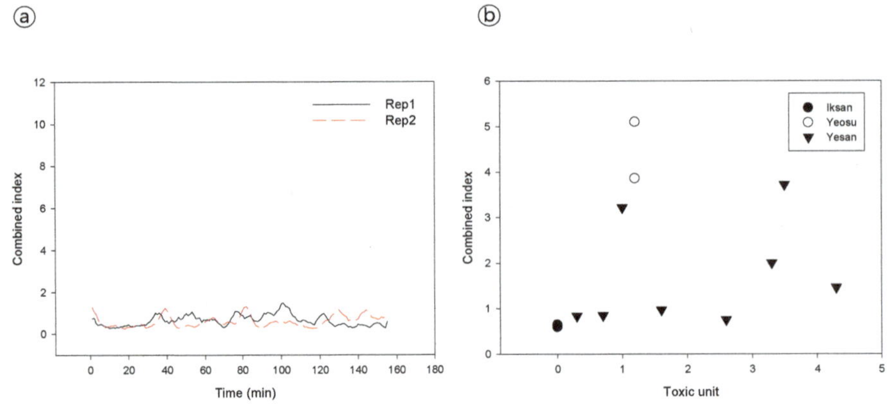

Figure 5. **(a)** Results of Iksan effluent (0TU). Rep# refers replication number. **(b)** Result of Iksan, Yeosu, and Yesan effluents.

3.2 Effluent exposure test

The results of the CI in the real-time effluent exposure test are expressed as graphs in Fig. 5, where (a) and (b) display the results of the non-toxic effluent from Iksan and different effluents, respectively. The exposure to non-toxic effluent shows stable index values during the test period at below 2 ($p > 0.01$, index < 2), which proves the stability of the CI. In (b), which shows the results of tests using effluent from Yeosu and Yesan, the CI and the TU are not correlated with a huge deviation. This result was expected from the deviation of data from the copper exposure test. These fluctuating data confirmed that the system is insufficiently powerful to detect precise toxicity levels of the field samples without replication. The raw data from the effluent exposure tests are summarized in Fig. S2 in the Supplement. This result shows that data from the Yeosu effluent have a higher index value than even the water sample containing a higher number of toxic units, which implies that the different characteristics of water samples can affect the magnitude of index values.

3.3 Discussion

In this study, we suggested a new concept of behavioral parameters related to changes in swimming range. Using this new parameter, the DI and CI were derived, and the utility of combining indexes and the value of the DI for generating the CI were confirmed. The DI showed the best correlation between the magnitude of the index and the toxicity level of the water sample among all individual indexes derived from three different behavioral parameters. The results showed that the swimming range parameter is a novel indicator in terms of both sensitivity and stability. The performance of the DI was better than that of TI, which is based on the velocity parameter, the best behavioral parameter for monitoring water quality among various parameters (Watson et al., 2007). In addition, the CI was more stable in the normal condition and more sensitive to toxicity change than the individual indexes were. Because no CI data from the non-toxic water exceeded 2 CI, a statistically meaningful threshold at $p = 0.01$; this study shows the possibility of developing a powerful index that can indicate whether a water sample is contaminated without an intentionally fixed threshold. Although the CI failed to reflect toxicity levels successfully according to the differentiated TU of the effluent samples without replication of the exposure test, the contribution of this study was in upgrading system stability and sensitivity, which showed the potential to improve BEWSs. Since false alarms are still a major obstacle for accurate BEWSs, this achievement is important in the field.

It is clear that the performance of the BEWS was improved by adopting a new type of behavioral parameter; however, problems in the BEWS occurred. The two major problems were (a) the great deviation in time series data at the same toxicity level, particularly at the lower concentration, and (b) the impact of different water characteristics on index magnitude. Although the problems were complemented by replicated exposure tests in the case of the copper test, it is not realistic to apply the system when there is no chance to replicate early warnings for water contamination. The first problem should be solved by trials to reduce biological deviation, for example, studies strictly selecting homologous organisms (Keui et al., 2001), developing new behavioral parameters and a hardware system (Gerhardt et al., 1998; Michels et al., 1999; Jeon et al., 2008), and improving the performance of hardware and data treatment methods (Li et al., 2008). Regarding the second problem, more studies analyzing relationships between regional water and the behavior of test organisms are required at the monitoring site where the system is set (Soldán., 2011). Additionally, test using synthetic water with a variety of chemical composition, physico-chemical parameters should also be conducted (Gerhardt et al., 2006; Green et al., 2003; Ren et al., 2007, 2009).

4 Conclusions

The new behavioral indexes DI and CI that can be utilized for BEWS to detect water contamination in drinking water resource and receiving water were developed from swimming behavior of *Daphnia magna*. The conclusions derived from this study are as follows.

1. The DI derived from the new behavioral parameter shows the best performance among the three indexes employed in the copper exposure test.
2. The CI improved performance of the BEWS in terms of sensitivity and stability, and it was confirmed as the higher correlation coefficient between the magnitude of the index and the toxicity level of the water sample.
3. Large deviations in the index data and unexpected changes in behavioral responses to water samples containing different characteristics occurred in the present study; therefore, further studies to enhance the accuracy and applicability of BEWSs in the field are needed.

Acknowledgements. This study (2013R1A2A2A03014187) was supported by Mid-career Researcher Program through NRF grant funded by the MEST.

References

Baillieul, M. and Scheunders, P.: On-line determination of the velocity of simultaneously moving organisms by image analysis for the detection of sublethal toxicity, Water Res., 4, 1027–1034, 1998.

Baldwin, I. G., Harman, M. M. I., and Neville, D. A.: Performance characteristics of a fish monitor for detection of toxic substances, Water Res., 28, 2191–2199, 1994.

Benecke, G., Falke, W., and Schmidt, C.: Use of algal fluorescence for an automated biological monitoring system, Bull. Environ. Contam. Toxico., 28, 385–395, 1982.

Borcherding, J. and Jantz, B.: Valve movement response of the mussel Dreissena polymorpha – the influence of pH and turbidity on the acute toxicity of pentachlorophenol under laboratory and field conditions, Ecotoxicology, 6, 153–165, 1997.

Butterworth, F. M., Gunatilaka, A., and Gonsebatt, M. E. Biomonitors and biomarkers as indicators of environmental change 2: A handbook, Plenum Publishers, New York, 2001.

Demeester, L.: The phototactic behaviour of male and female *Daphnia magna*, Anim. Behav., 43, 696–698, 1992.

Dodson, S. I.: Predator-induced reaction norms, Bioscience, 39, 447–452, 1989.

Dodson, S., Hanazato, T., and Gorski, P. R.: Behavioral responses of *Daphnia pulex* exposed to carbaryl and Chaoborus kairomone, Environ. Toxicol. Chem., 14, 43–50, 1995.

Gerhardt, A. and Clostermann, M.: A new biomonitor system based on magnetic inductivity for freshwater and marine environments, Environ. Int., 24, 699–701, 1998.

Gerhardt, A., Ingram, M. K., Kang, I. J., and Ulitzur, S.: In situ online toxicity biomonitoring in water: recent developments, Environ. Toxicol. Chem., 25, 2263–2271, 2006.

Graedel, T. E., Vanbeers, D., Bertram, M., Fuse, K., Gordon, R. B., Gritsinin, A., Kapur, A., Klee, R. J., Lifset, R. J., Memon, L., Rechberger, H., Spatari, S., and Vexler, D.: Multilevel cycle of anthropogenic copper, Environ. Sci. Technol., 38, 1242–1252, 2004.

Green, U., Kremer, J. H., Zillmer, M., and Moldaenke, C.: Detection of Chemical threat agents in drinking water by an early warning real-time biomonitor, Environ. Toxicol., 18, 368–374, 2003.

Hendriks, A. J. and Stouten, M. D. A.: Monitoring the response of microcontaminants by dynamic Daphnia magna and Leuciscus idus assays in the Rhine delta: biological early warning as a useful supplement, Ecotoxicol. Environ. Saf., 26, 265–279, 1993.

Jeon, J. H., Kim, J. H., Lee, B. C., and Kim, S. D.: Development of a new biomonitoring method to detect the abnormal activity of *Daphnia magna* using automated grid counter device, Sci. Total Environ., 389, 545–556, 2008.

Kieu, N. D., Michels, E., and Meester, L. D.: Phototactic behavior of *Daphnia* and the continuous monitoring of water quality: Interference of fish kairomones and food quality, Environ. Toxicol. Chem., 5, 1098–1103, 2001.

Kim, S. D., Ra, J. S., Kim, K. T., Kim, J. Y., Park, J. E., Kim, H. D., and Kim, E. Y.: Identification of ecotoxicity from wastewater treatment facilities and investigation of sources of toxicity, Ministry Of Environment, Republic of Korea, 2009.

Kim, S. D., Lee, S. H., Kim, H. Y., Kim, J. Y., Jeong, T. Y., Yoon, S. H., Jung, Y. J., and Jung, D. E.: Monitoring and risk assessment of the potentially hazardous chemicals in the Yeongsan and the Seomjin rivers. Yongsan River Environment Research Center, Republic of Korea, 2011.

Lechelt, M., Blohm, W., Kirschneit, B., Pfeiffer, M., Gresens, E., Liley, J., Holz, R., Lüring, C., and Moldaenke, C.: Monitoring of surface water by Ultrasensitive Daphnia Toximeter, Environ. Toxicol., 15, 390–400, 2000.

Li, Y., Seo, D. H., and Lee, W. D.: New classifier applied to biological early warning systems for toxicity detection, Applications of Digital Information and Web Technologies, 360–365, 2008.

Michels, E., Leynen, M., Cousyn, C., Meester, L. D., and Ollevier, F.: Phototactic behavior of *Daphnia* as a tool in the continuous monitoring of water quality: Experiments with a positively phototactic *Daphnia magna* clone, Water Res., 2, 401–408, 1999.

Ren, Z., Zha, J., Ma, M., Wang, Z., and Gerhardt, A.: The early warning of aquatic organophosphorus pesticide contamination by on-line monitoring behavioral changes of *Daphnia magna*, Environ. Monit. Assess., 134, 373–383, 2007.

Ren, Z., Li, Z., Ma, M., Wang, Z., and Fu, R.: Behavioral responses of *Daphnia magna* to stresses of chemicals with different toxic characteristics, Bull. Environ. Contam. Toxicol., 82, 310–316, 2009.

Shimizu, N., Ogino, C., Kawanishi, T., and Hayashi, Y.: Fractal analysis of *Daphnia* motion for acute toxicity bioassay, Environ. Toxicol., 17, 441–448, 2002.

Soldán, P.: Possible way to substantial improvement of early warning system in the International Odra (Oder) River Basin, Environ. Monit. Assess., 178, 349–359, 2011.

Untersteiner, H., Kahapka, J., and Kaiser, H.: Behavioural response of the cladoceran Daphnia magna to sublethal copper stress-validation by image analysis, Aquat. Toxicol., 65, 435–442, 2003.

van der Schalie, W. H., Sheddb, T. R., Knechtgesb, P. L., and Widder, M. W.: Using higher organisms in biological early warning systems for real-time toxicity detection, Biosens. Bioelectron., 16, 457–465, 2001.

Watson, S. B., Jüttner, F., and Köster, O.: *Daphnia* behavioural responses to taste and odour compounds: ecological significance and application as an inline treatment plant monitoring tool, Water Sci. Technol., 55, 23–31, 2007.

Weber, C. I.: Methods for measuring the acute toxicity of effluents and receiving waters to freshwater and marine organisms, U.S. Environmental Protection Agency, Cincinnati, 1993.

Wolf, G., Scheundersb, P., and Selens, M.: Evaluation of the swimming activity of *Daphnia magna* by image analysis after administration of sublethal cadmium concentrations, Comp. Biochem. Physiol., 120, 99–105, 1998.

Young, S. and Watt, P.: Behavioral mechanisms controlling vertical migration in *Daphnia*, American Society of Limnology and Oceanography, 39, 70–79, 1993.

13

Dynamic hydraulic models to study sedimentation in drinking water networks in detail

I. W. M. Pothof[1,2] and E. J. M. Blokker[3]

[1]Deltares/Industrial Hydrodynamics, Delft, The Netherlands
[2]Delft University of Technology/Water management, Delft, The Netherlands
[3]KWR Watercycle Research Institute, Nieuwegein, The Netherlands

Correspondence to: I. W. M. Pothof (ivo.pothof@deltares.nl)

Abstract. Sedimentation in drinking water networks can lead to discolouration complaints. A sufficient criterion to prevent sedimentation in the Dutch drinking water networks is a daily maximum velocity of 0.25 m s^{-1}. Flushing experiments have shown that this criterion is a sufficient condition for a clean network, but not a necessary condition. Drinking water networks include many locations with a maximum velocity well below 0.25 m s^{-1} without accumulated sediments. Other criteria need to be developed to predict which locations are susceptible to sedimentation and to prevent sedimentation in future networks. More distinctive criteria are helpful to prioritise flushing operations and to prevent water quality complaints.

The authors use three different numerical modelling approaches – quasi-steady, rigid column and water hammer – with a temporal discretisation of 1 s in order to assess the influence of unsteady flows on the wall shear stress, causing resuspension of sediment particles. The model predictions are combined with results from flushing experiments in the drinking water distribution system of Purmerend, the Netherlands. The waterhammer model does not result in essentially different flow distribution patterns, compared to the rigid column and quasi-steady modelling approach. The extra information from the waterhammer model is a velocity oscillation of approximately 0.02 m s^{-1} around the quasi-steady solution. The presence of stagnation zones and multiple flow direction reversals seem to be interesting new parameters to predict sediment accumulation, which are consistent with the observed turbidity data and theoretical considerations on critical shear stresses.

1 Introduction

The goal of drinking water companies is to supply their customers with good quality drinking water 24 h per day. With respect to water quality, the focus has for many years been on the drinking water treatment. Recently, interest in water quality in the drinking water distribution system (DWDS) has been growing. On the one hand, this is driven by customers who expect the water company to ensure the best water quality by preventing such obvious deficiencies in water quality as discolouration and (in many countries) by assuring a sufficient level of chlorine residual. On the other hand, since "9/11" there is a growing concern about (deliberate) contaminations in the DWDS.

Sedimentation in drinking water networks may lead to discolouration complaints. A sufficient criterion for Dutch DWDS, consisting of PVC, AC and lined CI mains, to prevent sedimentation is a daily maximum velocity of 0.25 m s^{-1} (Blokker et al., 2010a). Flushing experiments have shown that this criterion is a sufficient condition for a clean network, but not a necessary condition.

Transient models, including pressure wave propagation, are used for waterhammer analysis and for the evaluation of valve operations, pump switches and the design of control systems. More recently, transient models have been applied in DWDS for the prediction of a number of water quality parameters, such as chlorine decay or intrusion volumes during low pressure transients (Ebacher et al., 2011).

Figure 1. Purmerend DWDS and selected test area (grey rectangle). Source: Blokker et al. (2010a).

In this paper, we investigate whether more detailed hydrodynamic models will result in more accurate criteria for the prediction, efficient mitigation and ultimately prevention of sedimentation in the DWDS. We have used three different numerical modelling approaches: (1) the traditional quasi-steady model, as implemented in EPANET-based models; (2) a rigid column model, in which the inertia of the water mass in all pipes is taken into account and (3) the complete waterhammer model, including liquid compressibility and pipe stiffness so that the propagation of pressure waves is correctly simulated (Wylie and Streeter, 1993). The quasi-steady modeling results were obtained with EPANET (Rossman, 2000). The Rigid Column (RC) and waterhammer results were obtained with WANDA, developed and validated by Deltares (Deltares, 1993–2011).

2 Approach

2.1 Network selection

Ideally, we would investigate a DWDS with loops and a single water source in which sedimentation has been measured in all pipes and in which the velocity time series between two consecutive flushing procedures has been measured in all pipes. Turbidity measurements during well-defined flushing procedures provide a reasonable spatial distribution of the sediment load in all flushed pipes. Obviously, the second criterion is not practically feasible in any DWDS. However, if the network layout (pipe length, material, internal diameter, wall roughness) is known and sufficient demographic information is available on the inhabitants, then a reasonable assessment of the time series of the velocities can be computed from detailed stochastic water demand model simulations with SIMDEUM (Blokker et al., 2010b).

The turbidity was measured during flushing procedures in the Purmerend DWDS (the Netherlands). Furthermore, a SIMDEUM model of the Purmered DWDS is available. The Purmerend DWDS and the flushing procedure are described by Blokker (2010) and in two papers, presented at the 2011 CCWI conference (Blokker et al., 2011; Schaap and Blokker, 2011). An area within the Purmerend DWDS has been selected based on the availability of accurate sedimentation data obtained via flushing procedures. Furthermore, the test area includes sections with a lot of accumulated sediment and other similar sections without. The test area is shown in Fig. 1. The test area includes approximately 200 house connections and 450 pipes. The water demands of the individual households are a realization of the stochastic water demand model SIMDEUM (Blokker, 2010). The water demands have a temporal resolution of 1 s. The simulations cover a period from 05:00 a.m. until 11:00 a.m., so that the minimum and maximum water demands are included.

Table 1. Pipe properties for waterhammer model for pipes with a pressure rating of 6 barg.

Pipe material	Young's modulus [Gpa]	Internal diameter [mm]	Wall thickness [mm]
		19.6	1.2
		25	1.2
PVC	3	44.2	2.0
		59	2.0
		90	2.7
		100	10
		150	10
AC	30	200	11
		250	12
		400	18

2.2 Rigid column and waterhammer model

The Rigid column model does not need any additional information in comparison with the EPANET model. The only difference is the extension of the momentum equation with the inertia term:

$$\Delta H = \frac{\lambda L}{D}\frac{v^2}{2g} + \frac{L}{g}\frac{\mathrm{d}v}{\mathrm{d}t} \quad (1)$$

where ΔH [m] is the differential head of an individual pipe in the DWDS, λ [–] is the quasi-steady friction factor according to White-Colebrook, L the pipe length, D [m] is the internal pipe diameter, v [m s^{-1}] is the pipe velocity and g [m s^{-2}] is the constant of gravitational acceleration.

The more advanced waterhammer model takes the pipe elasticity and water compressibility into account, so that the effects of pressure waves in the network are computed. This requires additional information on the pipe material and wall thickness. The pipe materials are shown in Fig. 1 and the applied wall thickness values and Young's moduli are listed in Table 1.

These data result in typical acoustic wave speeds of 350 m s^{-1} in PVC pipes and 1000 m s^{-1} in AC pipes. Pipes with a length of less than 2 m have been modelled as rigid column pipes, in order to prevent a time step of less than 0.002 s. The timestep of the waterhammer model is 0.003 s.

Due to the fact that the test area includes two loops, the rigid column and waterhammer models may lead to a different pressure and flow distribution than the EPANET model, due to the presence of the inertia term in the momentum balance. Both modelling approaches have been modeled in WANDA (Deltares, 1993–2011). All boundary conditions are identical for the three different modeling approaches.

2.3 Sedimentation and resuspension

The typical particle size ($d < 25\,\mu$m $= 0.025$ mm) (Vreeburg, 2007) and particle density ($\rho_s = 1200$ kg m^{-3}) of material in

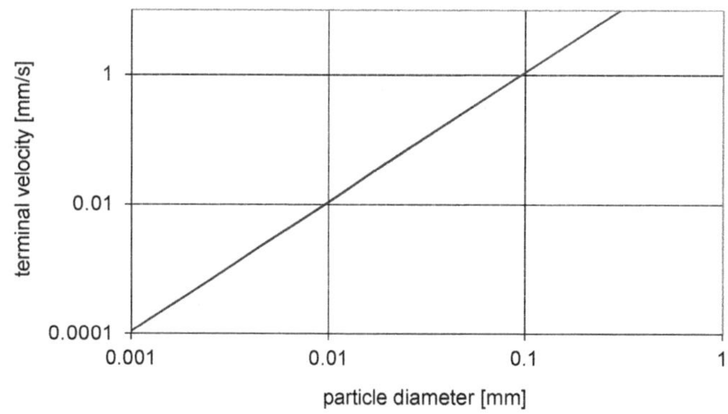

Figure 2. Particle terminal falling velocity as a function of particle size; $\rho_s = 1200$ kg m^{-3}.

drinking water networks are so small that the terminal velocity can be determined with Stokes' law. The terminal particle velocity v_t [m s^{-1}] follows directly from:

$$v_t = \frac{(\rho_s - \rho_f)}{\mu_f}\frac{gd^2}{18} \quad (2)$$

The terminal velocity v_t is only 0.07 mm s^{-1} at particle diameter $d = 0.025$ mm and $\rho_s = 1200$ kg m^{-3} and fluid density $\rho_f = 1000$ kg m^{-3} (Fig. 2); μ_f [Pa s] is the dynamic viscosity of water. A particle of this size needs 12 min to drop 50 mm. If these particles do settle at all, they will easily be resuspended at the so-called critical shear stress. Settled particles reside in the laminar sublayer of a distribution pipe. Therefore, Soulsby's model for the critical shear stress is applied (Soulsby, 1997). Soulsby has developed his model for non-cohesive particles. For particles smaller than 100 μm the dimensionless critical shear stress θ_{cr} tends to 0.3, but experimental evidence is limited in this particle range. The shear stress τ_{cr} [Pa] then becomes:

$$\tau_{cr} = \theta_{cr}(\rho_s - \rho_f)gd = 0.015\,\text{Pa} \quad (3)$$

where a maximum particle size of $d = 0.025 \times 10^{-3}$ m was substituted. The critical shear stress may increase if the particles exhibit cohesive behaviour.

The laminar wall shear stress $\tau_{w,lam}$ [Pa] is a known function of the average pipe velocity U [m s^{-1}] and pipe radius R [m].

$$\tau_{w,lam} = \frac{R}{2}\frac{\mathrm{d}p}{\mathrm{d}x} = \frac{4\mu_f U}{R} \quad (4)$$

If the flow becomes turbulent ($Re > 2300$), then a typical friction factor is $\lambda = 0.03$ for pipes with diameter $D = 0.1$ m. In this case the wall shear stress $\tau_{w,tur}$ [Pa] is computed as

$$\tau_{w,tur} = \frac{\lambda}{8}\rho_f U^2 = 3.75U^2 \quad (5)$$

The critical shear stress for resuspension and the steady wall shear stress in a pipe with $D = 0.1$ m have been

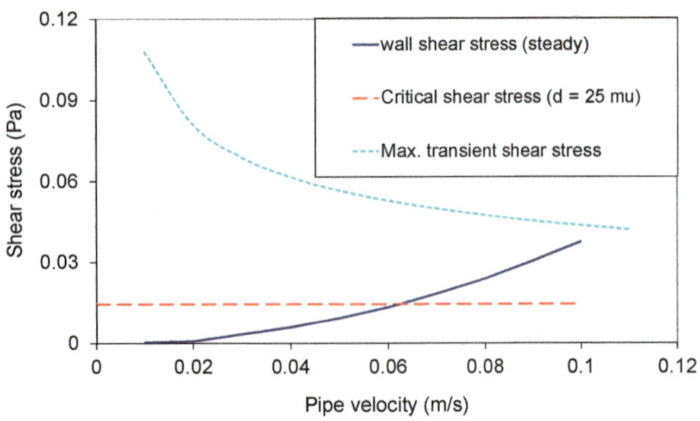

Figure 3. Wall shear stress (Eqs. 4 and 5), critical shear stress for resuspension (Eq. 3) and maximum unsteady shear stress (Eq. 6).

plotted in Fig. 3, showing that the larger particles ($\rho_s = 1200\,\mathrm{kg\,m^{-3}}$ and $d = 25\,\mu\mathrm{m}$) will move if the water velocity $U > 0.06\,\mathrm{m\,s^{-1}}$. The critical shear stress for resuspension increases linearly with the particle diameter and density, so that the critical water velocity for other particles can be derived from Fig. 3.

Due to acceleration and deceleration of the flow, the velocity profile does not vary in a quasi-steady manner. Therefore, the unsteady wall shear stress may contribute significantly to the total wall shear stress. The modelling of these unsteady friction phenomena has not yet led to a generally accepted modelling approach. Brunone et al. (2000) has proposed a model that is based on instantaneous accelerations. Others have extended unsteady friction models for laminar flows (Vardy and Brown, 2003), based on Zielke (1968). Pothof (2008) has developed a model in which the unsteady shear stress model is based on a decelerating turbulent flow. Vardy and Brown (2003) have derived a maximum unsteady wall shear stress, $\tau_{\mathrm{wu,max}}$ [Pa]:

$$\tau_{\mathrm{wu,max}} = \frac{\rho_w \sqrt{C^*} D \mathrm{d}U/\mathrm{d}t}{2} \qquad (6)$$

where C^* [–] is a function of the Reynolds number

$$C^* = 12.86/Re^{\kappa}$$
$$\kappa = \log\left(15.29/Re^{0.0567}\right) \qquad (7)$$

The transient simulation with the waterhammer model shows typical velocity decelerations of $2\,\mathrm{cm\,s^{-2}}$, independent of the water velocity. This information can be combined with Eq. (6) to obtain the maximum unsteady shear stress in a pipe with $D = 0.1$ m as a function of the water velocity (Fig. 3). Figure 3 shows that particles of $d = 25\,\mu\mathrm{m}$ may be easily resuspended by flow acceleration or decelerations at velocities well below $U = 0.06\,\mathrm{m\,s^{-1}}$.

This analysis suggests that sedimentation can only occur in stagnation zones. We will therefore focus on the stagnation zones in the network loops.

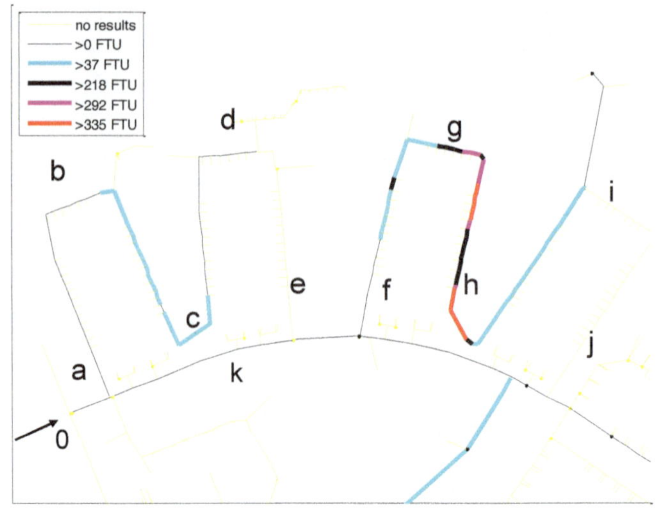

Figure 4. Spatial distribution of turbidity measurement in the Purmerend test area.

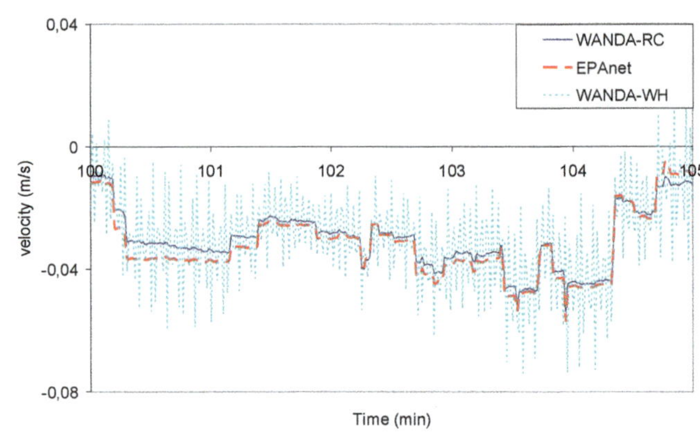

Figure 5. 5-min detail of the three modelling approaches in pipe e (100 = 06:40 a.m.).

3 Results

3.1 Turbidity measurements

The turbidity time series measurements have been translated to a spatial distribution of turbidity (Fig. 4). For each flushing action, the measured turbidity [FTU, Formazin Turbidity Unit] of the 1st turnover was linked to the location in the stretch of pipes from which the particles originated. This was done by converting the measurement time [s] to the flushed pipe length [m] with the help of the flushing flow [$\mathrm{m^3\,s^{-1}}$] and pipe diameter [m] (Blokker et al., 2010a). For some pipe stretches no results are shown in Fig. 4, because the pipe was not flushed, or the conversion to the pipe length was too inaccurate in case of short flushing times, leaking valves or high turbidities in the 2nd turnover.

Both loops in the test area (a-b-c-d-e and f-g-h-i-j) are very similar with respect to pipe diameter and spatial distribution

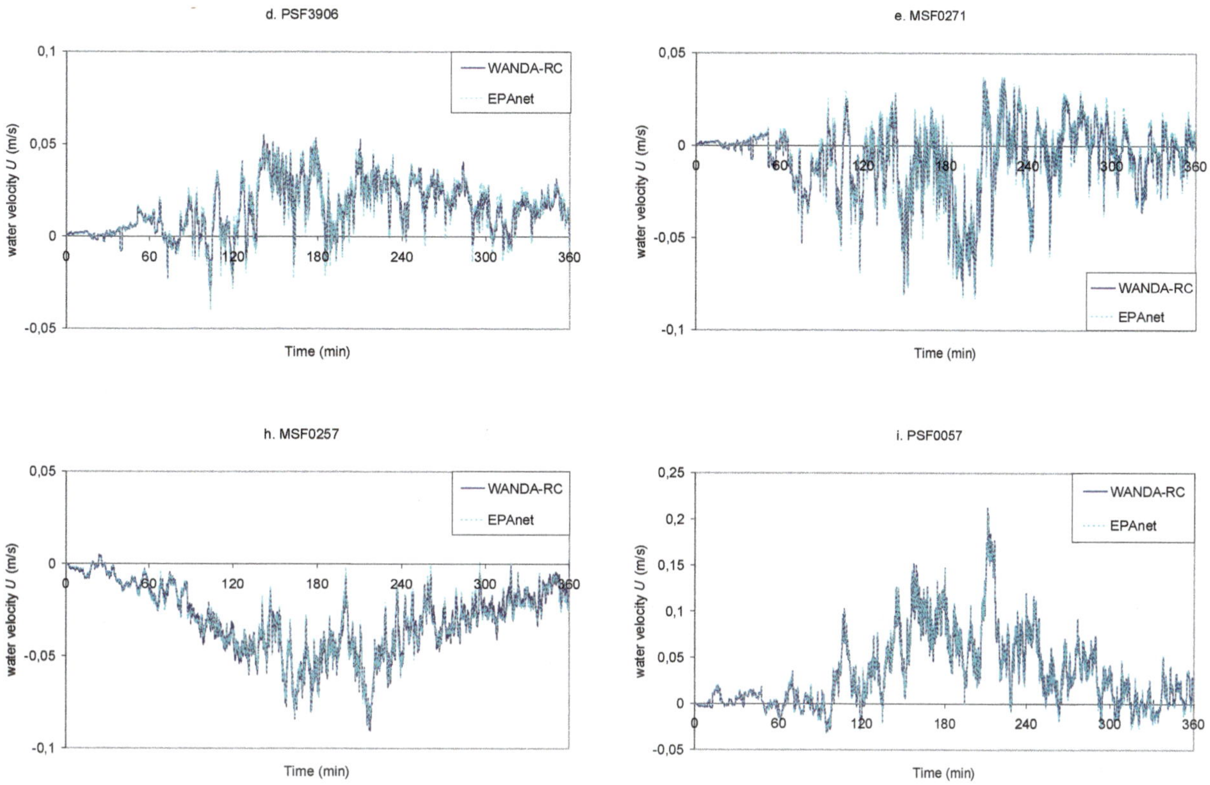

Figure 6. Velocities in reaches d, e., h and i (labels in Fig. 4); start time corresponds to 05:00 a.m.

of homes. However, the first loop hardly contains any sediment, whereas a fair amount of sediment was found in the second loop. Following the reasoning in Sect. 2, a stagnation zone should be absent in loop (b-c-d) and present in loop (g-h-i). The presence of stagnation zones will be discussed in Sect. 3.2.

3.2 Simulation results

The rigid column model is practically identical to the EPANET model, even at the temporal resolution in demands of 1 s, as illustrated in Figs. 5 and 6. The waterhammer model shows typical velocity oscillations of $0.02\,\mathrm{m\,s^{-1}}$ and typical accelerations of $0.02\,\mathrm{m\,s^{-2}}$ around the EPANET solution (Fig. 5). Since the magnitude of velocity oscillations are independent of the absolute velocities, these simulations suggest that these oscillations are generally valid at the local network level, but this statement requires further substantiation with local measured flow data at the appropriate temporal scale of 1 s as applied in these simulations. These kind of flow data were not available for the Purmerend test area.

The stochastic nature of the drinking water demand is responsible for spatial variation of the stagnation point. In fact, there is no stagnation zone, but there are pipes where many flow reversals occur. The velocity time series at location e contains most flow reversals; this is the connection with the large AC pipe, marked k in Fig. 4. The simulated flow is uni-directional most of the time between pipes a and d, so that stagnation does not occur in these pipes. Since sediment was hardly measured between pipes a. and d., the presence of stagnation zones or the number of flow direction reversals may correlate with the sediment load. In the second loop (f-g-h-i-j) the stagnant zone is located between pipes h. and i. (Fig. 6) and most sediment was measured near pipe h. and between pipes g. and h. The presence of a stagnant zone, or equivalently many flow direction reversals, in the loop (f-g-h-i-j) may be an indication for the presence of sediment. The match between the sediment concentration and the number of flow direction reversals is not perfect, because of a number of inherent uncertainties associated with DWDS modelling and the turbidity data processing. First, the actual water demand distribution over the years may differ somewhat from the simulated demand distribution of only one day. More simulations can show the variability of the flow distributions and the location of the stagnation zones. Secondly, the turbidity data analysis assumes that the sediment bed erodes completely and instantaneously during the flushing procedure. In reality, the bed may not erode instantaneously due to cohesive behaviour of the sediment. This uncertainty would cause a shift in the locations where sediment has accumulated. Cohesive sediment behaviour would also increase the critical bed shear stress for erosion. Therefore, it is recommended to characterise the cohesive properties of sediment in the DWDS.

4 Conclusions and recommendations

Detailed hydraulic simulations have been performed with a temporal resolution of 1 s and with three modelling approaches: an EPANET model (quasi steady state), a Rigid Column model and a waterhammer model. We have investigated whether the more advanced hydraulic modelling approaches provide necessary conditions or unambiguous criteria for the presence of sedimentation in a DWDS. The detailed simulation results have been combined with turbidity measurements during flushing procedures in order to identify promising sedimentation criteria, which are summarised hereafter.

The Rigid Column simulation is practically identical with the EPANET simulation. The water hammer simulation shows velocity oscillations of approximately $2\,\mathrm{cm\,s^{-1}}$ and $2\,\mathrm{cm\,s^{-2}}$ around the EPANET solution, independent of the magnitude of the velocity. It is recommended to verify these transient velocities with flow measurements with sufficient temporal resolution. The more detailed simulations do not lead to different flow distributions in the Purmerend DWDS.

The presence of stagnation zones and multiple flow direction reversals may serve as alternative parameters to predict sediment accumulation, which are consistent with theoretical considerations on critical shear stresses and with the observed turbidity data. The analysis of critical shear stress, including unsteady shear stresses, confirms that sediment in the DWDS must exhibit cohesive behaviour to accumulate any material. It is recommended to determine the cohesive properties of sediment in DWDS. A direct consequence of the analysis in this paper states that sediment accumulation will not occur in branched distribution networks, because of the low critical shear stress for resuspension and low terminal velocity of typical particles. It suggests that branched distribution networks will be self-cleaning if the daily maximum velocity exceeds $0.06\,\mathrm{m\,s^{-1}}$ (Fig. 3), assuming that all particles have a diameter $d < 25\,\mu\mathrm{m}$ and density $\rho_s < 1200\,\mathrm{kg\,m^{-3}}$ and exhibit non-cohesive behaviour.

Acknowledgements. The authors acknowledge PWN Water Supply North Holland for their co-operation and permission to publish the turbidity data.

References

Blokker, E. J. M.: Stochastic water demand modelling for a better understanding of hydraulics in water distribution networks, Civil Engineering and Geosciences, Ph.D. thesis, 205, 978-90-8957-015-4, 2010.

Blokker, E. J. M., Vreeburg, J. H. G., Schaap, P. G., and van Dijk, J. C.: The self-cleaning velocity in practice, Water Distribution System Analysis, Tucson, AZ, 12–15 September, 2010a.

Blokker, E. J. M., Vreeburg, J. H. G., and van Dijk, J. C.: Simulating residential water demand with a stochastic end-use model, J. Water Res. Pl.-ASCE, 136, 19–26, 2010b.

Blokker, E. J. M., Schaap, P. G., and Vreeburg, J. H. G.: Comparing the fouling rate of one drinking water distribution systems in two different flow configurations, Computing and Control in the Water Industry (CCWI), Exeter, September, 2011.

Brunone, B., Karney, B. W., Mercarelli, M., and Ferrante, M.: Velocity profiles and unsteady pipe friction in transient flow, J. Water Res. Pl.-ASCE, 126, 236–244, 2000.

Deltares: WANDA, pipeline simulation software, Deltares Software, 1993–2011.

Ebacher, G., Besner, M. C., Lavoie, J., Jung, B. S., Karney, B. W., and Prévost, M.: Transient Modeling of a Full-Scale Distribution System: Comparison with Field Data, J. Water Res. Pl.-ASCE, 137, 173–182, 2011.

Pothof, I. W. M.: A turbulent approach to unsteady friction, J. Hydraul. Res., 46, 679–690, 2008.

Rossman, L. A.: EPAnet 2 User Manual, 2000.

Schaap, P. G. and Blokker, E. J. M.: Carefully designed measurements provide insight into sediment build-up in drinking water distribution systems, Computing and Control in the Water Industry (CCWI), Exeter, UK, September, 2011.

Soulsby, R. L.: Dynamics of marine sand, Thomas Telford, 1997.

Vardy, A. E. and Brown, J. M. B.: Transient turbulent friction in smooth pipe flows, J. Sound Vib., 259, 1011–1036, 2003.

Vreeburg, J. H. G.: Discolouration in drinking water systems: a particular approach, PhD, Civil Engineering and Geosciences, Delft University of Technology, Delft, 183 pp., 2007.

Wylie, E. B. and Streeter, V. L.: Fluid Transients in Systems, Prentice-Hall, New York, 1993.

Zielke, W.: Frequency dependent friction in transient pipe flow, ASME J. Basic Engineering, 90, 109–115, 1968.

14

Effect of fouling on removal of trace organic compounds by nanofiltration

S. Hajibabania[1], A. Verliefde[2,3], J. E. Drewes[4], L. D. Nghiem[5], J. McDonald[6], S. Khan[6], and P. Le-Clech[1]

[1]UNESCO Centre for Membrane Science and Technology, School of Chemical Engineering, The University of New South Wales, Sydney 2052, Australia
[2]Particle and Interfacial Technology Group, Faculty of Bioscience Engineering, Ghent University, Coupure Links 653, 9000 Gent, Belgium
[3]Delft University of Technology, Faculty of Civil Engineering and Geosciences, Department of Sanitary Engineering, P.O. Box 5048, 2600 GA Delft, The Netherlands
[4]Environmental Science and Engineering Division, Colorado School of Mines, Golden, CO 80401-1887, USA
[5]School of Civil Mining and Environmental Engineering, The University of Wollongong, Wollongong, NSW 2522, Australia
[6]Water Research Centre, The University of New South Wales, Sydney 2052, Australia

Abstract. The fate of chemical of concern is not yet fully understood during treatment of impaired waters. The aim of this paper is to assess the impact of different organic-based fouling layers on the removal of a large range of trace organics. Both model and real water samples (mixed with trace organic contaminants at environmental concentration of $2\,\mu g\,l^{-1}$) were used to simulate fouling in nanofiltration under controlled environment. The new and fouled membranes were systematically characterised for surface charge, hydrophobicity and roughness. It was observed that fouling generally reduced the membrane surface charge; however, the alterations of the membrane hydrophobicity and surface roughness were dependent on the foulants composition. The rejection of charged trace organics was observed to be improved due to the increased electrostatic repulsion by fouled membranes and the adsorption of the trace organic chemicals onto organic matters. On the other hand, the removal of nonionic compounds decreased when fouling occurred, due to the presence of cake enhanced concentration polarization. The fouling layer structure was found to play an important role in the rejection of the trace organic compounds.

Keywords. Nanofiltration, Fouling, Trace organic compounds, Characterisation

1 Introduction

Water recycling, including direct/indirect potable reuse, has been identified as efficient use of available water resources and thereby providing some relief to water scarcity. However, a wide range of trace organic contaminants such as endocrine disrupting compounds (EDCs), pharmaceutically active compounds (PhACs) and pesticides are known to persist (in $\mu g\,l^{-1}$ to $ng\,l^{-1}$ concentrations) in wastewater-treated effluents (Le-Minh et al., 2010; Watkinson et al., 2009). The negative effects of these contaminants on human health as the result of their long term exposure are generally unknown. Therefore, effective removal of trace organic compounds is required to prevent any potential risks of human and environmental exposure (Ongerth and Khan, 2004).

High pressure membrane processes such as nanofiltration (NF) has been increasingly used for treatment of surface water in both direct and indirect potable water recycling applications (Schäfer et al., 2003; Van der Bruggen and Vandecasteele, 2003). The rejection of some trace contaminants trough NF was reported to be incomplete (Van der Bruggen et al., 2008; Van der Bruggen and Vandecasteele, 2003).

Several studies have been conducted over the last decade to identify the different factors involved in NF rejection mechanisms (Bellona et al., 2004; Bolong et al., 2009; Nghiem and Schäfer, 2006; Radjenovic et al., 2008). It is generally believed that a given solute can be rejected by NF membranes by size exclusion and charge repulsion, and also by adsorption to the membrane surface. All of these mechanisms are largely dependent on both solute (Verliefde et al., 2009) and membrane properties (Boussu et al., 2008). However, detailed understanding of the rejection mechanisms of trace organics by NF, as well as the interactions between trace contaminants and the membrane polymer, fouling layers and other solutes in the solution is still incomplete (Van der Bruggen et al., 2008).

Fouling is an inevitable phenomenon in membrane filtration systems. Besides affecting the permeate flux, fouling alters membrane surface properties. In a previous study, it was reported that membrane surface charge decreased, while contact angle increased, during fouling with surface water (Verliefde et al., 2009). In another example, significant increase of membrane surface roughness was observed as a result of protein fouling (Bowen et al., 2003).

Several studies have investigated the effects of organic and colloidal fouling on membrane performance, with emphasis on permeate flux decline, salt rejection. A number of papers have also focused on the effect of fouling on removal of trace organics (Elimelech and Ng, 2004; Hajibabania et al., 2011; Nghiem and Hawkes, 2009; Verliefde et al., 2009; Xu et al., 2006; Zazouli et al., 2009). It has been observed that the rejection of trace organic compounds was enhanced when membrane was fouled with humic acid, due to pore restriction (Nghiem et al., 2008). In another study, an increase in rejection of hydrophilic compounds was reported, also due to pore restriction, following NF fouling with varying concentrations of alginate (Xu et al., 2006). Moreover, it was noted that the magnitude of this rejection enhancement was more pronounced for membranes with large pore sizes. A significant increase in rejection of nonionic solutes after fouling with secondary treated wastewater effluent was also reported. This was due to the formation of a fouling layer which isolated the solute-membrane interactions and increased adsorbed mass of hydrophobic nonionic compounds on the fouling layer.

Yangali-Quintanilla et al. reported that the rejection of both charged and neutral compounds decreased due to the occurrence of cake enhanced concentration polarization (CECP) when alginate fouling was established on the membrane surface (Yangali-Quintanilla et al., 2009). It has also been reported that the rejection of compounds by NF membranes fouled with filtered and raw lake waters increased significantly (Comerton et al., 2009). This increased rejection of trace organics was attributed to the higher trace organic-organic complexes in the presence of calcium.

The seemingly contradictory findings in these studies are difficult to untangle since NF rejection mechanisms are affected by a large number of variables such as membrane type, feed water composition (feed concentration of trace organics, pH, and salt balance) and operating conditions. Moreover, detailed characterisation of fouled membranes should be considered as a prerequisite for any investigation aiming to better understand the effects of fouling on membrane surface properties and on resulting rejection behaviour of trace organics.

Table 1. Selected properties for the virgin NF270 membrane.

Membrane	MWCO ($g\,mol^{-1}$)	Membrane thickness (nm)	Pure water permeability ($l\,m^{-2}\,h^{-1}\,bar^{-1}$)	NaCl rejection (%)
NF270	270–300	15–40 (Freger, 2003)	13.6	40–60

Most previous studies have investigated removal of trace organics at high feed concentrations ($mg\,l^{-1}$) in order to facilitate detection of the trace chemicals in membrane feed and permeate samples. Also, most studies were carried out under constant transmembrane pressure (TMP) since this is often a simpler protocol to implement for laboratory-scale studies. As a result, previous research failed to accurately represent full-scale plants, which usually operate under constant flux and treat lower trace organics concentrations in the feed ($\mu g\,l^{-1} - ng\,l^{-1}$). It has been observed that filtration at constant flux may result in significantly lower fouling rates, compared to studies conducted at constant TMP (Defrance and Jaffrin, 1999). In this research, the phenomena occurring in NF operation will be studied using environmental concentrations of Organic matter (OM) and trace organics in constant flux mode, as a more realistic approach.

In the current study, the effect of membrane fouling on the rejection of trace organics in conjunction with solute and membrane physicochemical characteristics is assessed. Various types of OM are used to foul the membrane, to investigate the influence of OM nature on the rejection of trace organics. This study will also address the influence of the foulants nature on the temporal changes and properties of the fouled membrane, and their interactions with the physicochemical properties of trace organics.

2 Materials and methods

2.1 NF Membrane and filtration protocol

A commercially available NF membrane, NF270 (Dow Chemicals) was used in this study. The membrane properties are shown in Table 1. The molecular weight cut off (MWCO) and salt rejection values were provided by the membrane manufacturer. The membrane was rinsed with Milli-Q water for duration of 1 h prior to use. Then, the virgin membrane

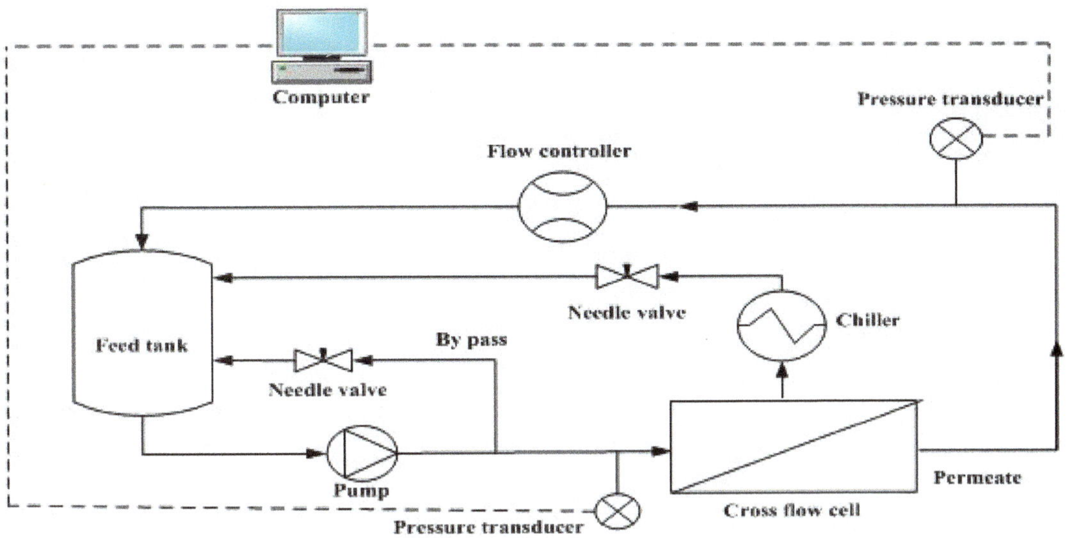

Figure 1. Schematic illustration of the experimental set up.

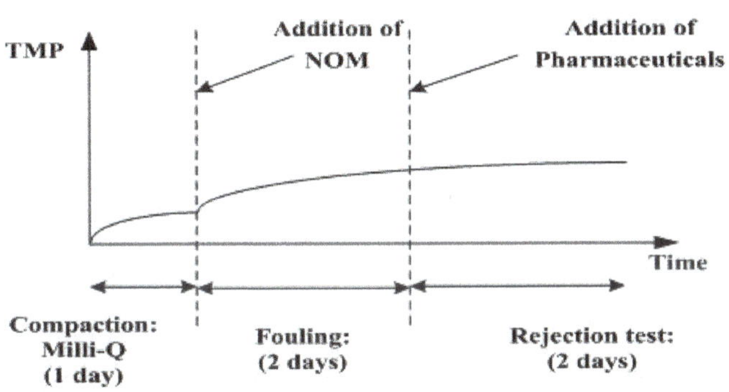

Figure 2. Filtration protocol for compaction, fouling and rejection test.

was characterised in terms of pure water permeability with Milli-Q water.

A bench-scale NF set up was used in this study (Fig. 1). A flat sheet membrane cell with channel dimensions of 12.2 cm in length, 3.8 cm in width and 0.3 cm in height was used to accommodate the membrane active area of 46.4 cm^2. In all experiments, the permeate and concentrate were recirculated back to the feed vessel, except when samples were collected for analysis. The feed temperature was maintained at 25 ± 1 °C using a stainless steel water cooling system (Grant thermostat).

All experiments in this study were carried out in 3 steps as shown in Fig. 2. In the first step, the membrane was compacted with Milli-Q at 15 bar for 24 h, to allow TMP stabilisation. Then, the membrane was characterised for salt rejection using a 500 mg l^{-1} NaCl-solution to ensure that the specimen was not damaged or ruptured before or during compaction. In the second step, organic fouling was developed by introducing a synthetic feed solution containing organic foulant, surface water, or MBR effluent to the feed reservoir. Fouling experiments were conducted at applied pressure of 5.5 bar, constant cross-flow velocity of 0.2 m s^{-1} and constant flux of 20 l m^{-2} h^{-1} over 48 h. The increase in TMP was monitored to evaluate membrane hydraulic performances. In the third step, the concentrated stock solution of trace organics was spiked in the feed solution to obtain a concentration of 2 μg l^{-1} of each individual compound. Triplicate samples of both feed and permeate streams were taken (150 ml) to determine the statistical variance on the results. Rejection (R in %) was defined as indicated in Eq. (1):

$$R\,\% = 100 \cdot \left(1 - \frac{C_P}{C_F}\right) \tag{1}$$

Where C_P and C_F are permeate and feed concentration, respectively. After each experiment, the membrane sample was carefully preserved in dry conditions for subsequent surface characterisation.

2.2 Membrane characterisation

Membrane contact angles were measured using a Rame-Hart goniometer (Model 200-F1). Contact angle measurements were performed using the sessile drop technique and Milli-Q water. The average of at least 10 measurements on different locations of the membrane was calculated. Membrane roughness was measured by atomic force microscopy (AFM) with tapping mode in air using a DI3000 microscope (Veeco), also on 10 different locations across the sample (2 × 2 μm scanning area). Membrane charge (zeta potential) was determined at pH 6.5 in a 1 mM KCl (Aldrich) background solution using streaming potential equipment (SurPass, Anton

Table 2. Properties of selected trace organic compounds.

Compounds	Abbreviation	MW ($g\,mol^{-1}$)	Length (nm)	Width (nm)	Height (nm)	pKa[a]	Log *D* (at pH 6.5)[b]
Hydrophilic nonionic							
Trimethoprim	TTP	290	1.05	0.76	0.42	7.2	−0.105
Paracetamol	PCL	151	0.75	0.61	0.41	9.9	0.49
Caffeine	CFN	193	0.75	0.67	0.41	14.39	−0.07
Sulfamethoxazole	SMX	253	1.03	0.59	0.41	10.39	−0.105
Ionic							
Diclofenac	DFN	318	0.89	0.83	0.42	4.18	1.75
Ibuprofen	IBF	206	0.90	0.61	0.42	4.41	1.64
Naproxen	NPN	230	0.99	0.62	0.41	4.84	1.33
Hydrophobic nonionic							
Carbamazepine	CBZ	236	0.87	0.70	0.35	13.49	1.89
Risperidone	RPD	410	1.47	0.72	0.42	7.89	1.475
Fluoxetine	FLX	309	1.12	0.70	0.42	10.05	1.17
Bisphenol A	BPA	228	0.92	0.66	0.41	9.73	3.63

[a] Available data calculated using Sparc calculator.
[b] Available data calculated using for acids, and bases $\log D_{(pH)} = \log P - \log(1 + 10^{|pH-pKa|})$.

Paar). Membrane samples were dried in a dessicator 24 h before roughness, streaming potential and contact angle measurements.

2.3 Trace organic compounds and analysis

Target organic solutes were selected to represent a wide range of physicochemical properties and were categorised according to their acid dissociation constants (pKa) and Log octanol-water distribution ratio (Log *D*) values. The physicochemical properties of the selected compounds are presented in Table 2. The hydrophobicity of solutes can be correlated and quantified with the logarithm of the octanol-water partitioning coefficient (Log K_{ow}). The Log *D* value is similar to Log K_{ow}, but also considers the ionic speciation of the compound (Nghiem and Coleman, 2008). In previous studies, compounds with Log *D* smaller than 1 and greater than 3 have been defined as highly hydrophilic and hydrophobic, respectively. However, uncertainties exist for compounds with Log *D* values between 1 and 3 (Wells, 2006). In this research, solutes with Log *D* below 1 were ranked as hydrophilic, between 1 and 3 as intermediate hydrophobic and above 3 as highly hydrophobic compounds.

All of the trace organic compounds were of analytical grade and were purchased from Supelco. A concentrated stock solution was prepared in methanol and was stored at 4 °C in an amber glass container. The concentrations of trace organic solutes in feed and permeate samples were analysed by high performance liquid chromatography/tandem mass spectrometry (HPLC-MS/MS) using a method adapted from the literature (Vanderford and Snyder, 2006). These analyses were undertaken with an Agilent HPLC 1200 series coupled to an Applied Biosystems API 4000 triple quadrupole mass spectrometer. Before the analysis, solid phase extraction (SPE) was carried out using 1 g of hydrophilic-lipophilic balance (HLB) solid phase material (Bakerbond Polar Plus) filled in a 6 ml polyethylene cartridge. Total organic carbon (TOC) analysis was conducted using a Shimatzu800 TOC analyser. The confidence level of the trace organics analysis was assessed using three replicates of filtration of L-lysine. The confidence level (i.e., the relative error on the results) of the NF experiments is reflected as standard deviation (STD). The calculated standard error (STD %) from triplicate measurements were used for all experiments in this study.

2.4 Feed water characteristics and analysis

The fouling experiments were carried out with different model organic foulants, including humic acid (Aldrich), L-lysine (Aldrich), alginic acid (Ajax Chemicals) and Suwannee River Humic Acid (International Humic Substances Society). The feed solution also contained a buffer solution containing KH_2PO_4 0.1 M (Aldrich) and NaOH 0.1 M (Aldrich) to maintain the solution pH at 6.5. Sodium azide (Riedel-de Haën) was added at a concentration of 0.02 % by weight to prevent growth of bacteria, which could result in the development of biofilm on the membrane surface. As a result, the potential impact of biofilm in NF rejection was not considered within the scope of this study. The chemicals used for this project were of analytical grade without any further modification. Milli-Q water was provided via Milli-Q system by Millipore and water quality was greater than 18 $M\Omega\,cm^{-1}$.

Table 3. Characterisation of the feed waters by LC-OCD.

	Humic acid	Alginic acid	L-Lysine	SRHA	Surface water	MBR permeate
Organic carbon fraction (%)						
Bio-polymers[a]	n.q.	100	100	n.q.	3.1	11.1
Humic Subst. (HS)	67.0	n.q.	n.q.	80.2	33.1	32.4
Building Blocks[b]	14.8	n.q.	n.q.	8.0	17.3	16.6
Neutral LMW organics	18.2	n.q.	n.q.	11.8	12.5	39.9
Acidic LMW organics Acids	n.q.	n.q.	n.q.	n.q.	34.0	0.0
Average molecular weight ($g\,mol^{-1}$)						
	961	8291	554	1033	484	599

[a] Biopolymers = polysaccharides, proteins, aminosugars; [b] Building Blocks = breakdown products of humics; n.q. Not quantifiable (<1 ppb calculated)

Each fouling solution was prepared by dissolving individual foulant in the electrolyte solution described above to make up a total organic carbon (TOC) concentration of $5\,mg\,l^{-1}$. Since the experiments were conducted in constant flux mode, the same amount of TOC was delivered to the membrane surface for a given volume of permeate in each experiment. Hence, 15 l of solutions filtered in recirculation mode during 96 h, resulted in TOC delivered of 0.01 mg per cm^2 of membrane, calculated using Eq. (2).

$$\text{Delivered TOC}\left(\frac{\text{mg}}{\text{cm}^2}\right)=V\cdot\frac{C}{A} \quad (2)$$

Where V (l) is the permeate volume collected during filtration, C ($mg\,l^{-1}$) is the feed TOC concentration and A (cm^2) is the membrane surface area. However, it is noteworthy that a similar amount of DOC delivered to the membrane for different foulants does not necessarily result in similar fouling extents.

In addition to the synthetic solutions described above, untreated surface water (Woronora dam, NSW) and an anaerobic membrane bioreactor (MBR) permeate (pilot plant, UNSW Water Research Centre) were used. The TOC and pH of these feed waters were adjusted to match the model solutions. Sodium azide was also added to minimise the growth of biofilm. The relative impact of the presence of inorganics in the effluent was not studied in this work.

The dissolved organics in both model and real feed solutions were characterised by dissolved organic carbon (DOC) Labor Model 8 liquid chromatography-organic carbon detection (LC-OCD). During LC-OCD analysis, the dissolved organics were separated into the chromatographable dissolved organic carbon (CDOC) and non chromatographable dissolved organic fractions (NDOC), depending on the affinity of the DOC for the resin (HW50S). The NDOC fractions may be approximately described as the lipid based compounds greater than 1 µm in size. The CDOC portion was fractionated into five major fractions, as shown in Table 3. The organics were characterised based on the MW, also taking the ionic and hydrophobic interactions into account. Separation was based on the steric interaction of the DOC with the column over the wide range of MW.

3 Results and discussions

3.1 Rejection of trace organic compounds by virgin membrane

The rejection of the mixture of selected pharmaceuticals in Milli-Q water was determined in order to obtain the baseline rejection values of the virgin membrane during 48 h filtration. In this initial experiment, the impact of the presence of trace organic compounds on the membrane hydraulic performances was also assessed.

In constant flux operation, an increase in TMP generally indicates the occurrence of fouling. During filtration of the solution containing Milli-Q water and trace organics, no significant TMP increase was observed. This indicates that TMP reached a steady-state during the initial 24 h compaction period, and the stable trend of TMP after the compaction suggests that the trace organics did not foul the membrane. This was expected because of the very low concentration of the trace organics used in these experiments. It was also observed that the rejection of all trace organics reached a steady-state value within the first 4 h of filtration.

3.1.1 Adsorption onto NF membrane

For most of the trace organic chemicals in this study, the rejection levels by the virgin membrane reached stable value within the first hour of filtration. The concentrations of these compounds in the feed solution and permeate did not change significantly during the remaining of the filtration and thus, little adsorption on NF was expected. Amongst the hydrophobic nonionic trace organic solutes in this study, only bisphenol A was observed to exhibit different rejection behaviour. Figure 3 shows the rejection of bisphenol A by the virgin membrane during 48 h of filtration. Bisphenol A was

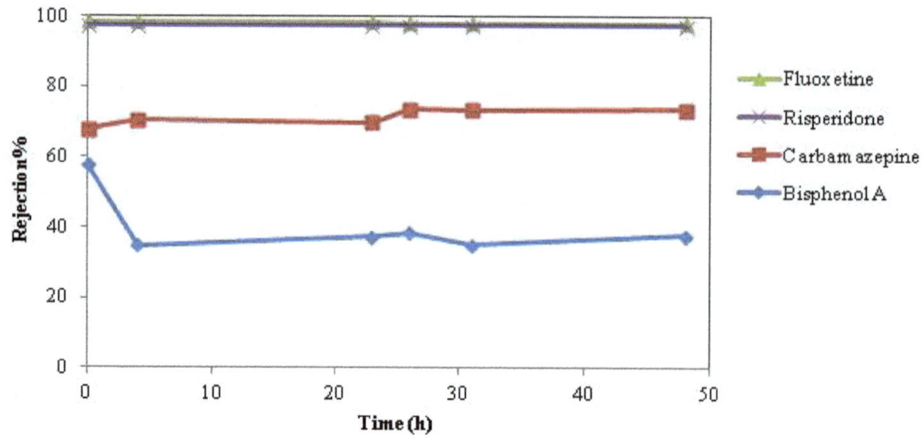

Figure 3. Temporal changes for rejection of fluoxetine, risperidone, carbamazepine and bisphenol A rejection by new NF270 membrane.

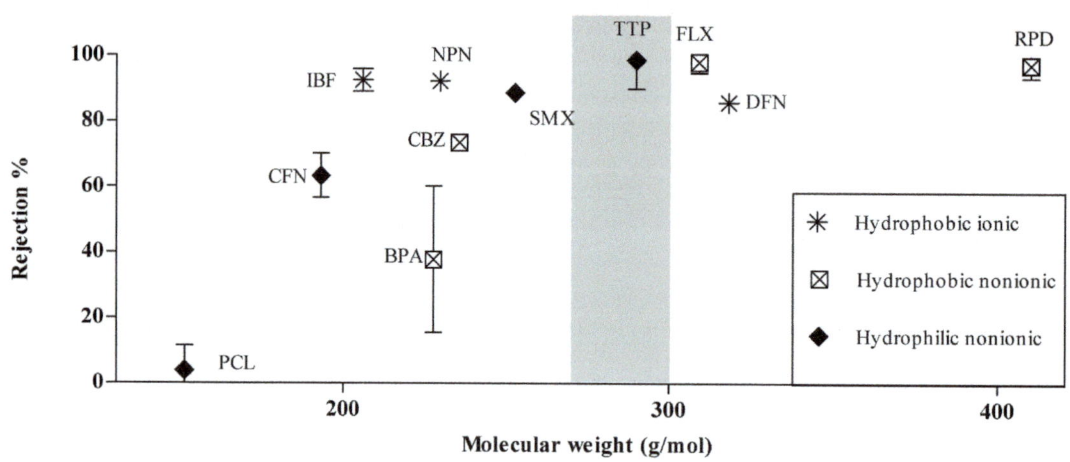

Figure 4. Rejection of trace organic compounds vs. their molecular weights. The high-lighted area indicates the MWCO of the NF270 membrane.

rejected at 60 % in the beginning of filtration. The high initial retention of bisphenol A was then followed by a sharp decline to 35 %. The rejection of bisphenol A stabilised during the first 4 h of the experiment. In a previous study, it was hypothesised that high initial rejections of hydrophobic solutes followed by a sharp decline to a much lower stabilised rejection value were due to adsorption on the membrane (Kimura et al., 2003a). The adsorption of trace organics to the membrane was reported to be mainly affected by hydrophobic interactions for aromatic pesticides (Kiso et al., 2001). In this experiment, bisphenol A featured the highest Log *D* value amongst the hydrophobic solutes and thus, exhibited a higher propensity to adsorb onto the membrane. Indeed in another study, considerable static adsorption of both bisphenol A and risperidone onto NF270 was already reported. However, because of its large MW (410 g mol^{-1}), risperidone was not expected to partition as easily as bisphenol A (228 g mol^{-1}). This was confirmed by the stable rejection values observed for risperidone.

3.1.2 Rejection of trace organics by virgin NF membrane

The steady-state rejection values of the three groups of compounds after 48 h filtration by the virgin NF270 membrane are shown in Fig. 4. The trace organics in each group were plotted according to their molecular weight.

3.1.3 Negatively charged ionic compounds

The negatively charged ionic solutes were efficiently removed at more than 90 %, as the result of electrostatic repulsion with the virgin membrane. According to Yangali-Quintanilla et al., the observed retention of naproxen and ibuprofen by a NF200 membrane was due to the combined influence of steric hindrance and electrostatic repulsion (Yangali-Quintanilla et al., 2009). However, in the current study, no significant relationship between the compounds MW and their rejection was observed. As such, it can be hypothesised that steric hindrance was not significant, and the main rejection mechanism for the negatively charged

Table 4. Summary of rejection mechanisms observed for the trace organic compounds (MW < 270 g mol^{-1}) by virgin NF270.

Type of compounds	Steric hindrance	Electrostatic repulsion	Solute membrane affinity	dominant mechanism
Ionic compounds	–	Higher rejection independent of the solutes MW	Lower rejection of the more hydrophobic solutes	Electrostatic repulsion
Hydrophilic nonionic compounds	Partial rejection depending on solutes MW	–	–	Steric hindrance
Hydrophobic nonionic compounds	Partial rejection depending on solutes MW	–	Lower rejection of solutes with higher Log *D*	Steric hindrance/ solute-membrane affinity

ionic compounds by the virgin membrane used in this study (NF270) was electrostatic repulsion. One notable exception was diclofenac, which was slightly less well rejected than the other negatively-charged solutes. This can be explained by the physicochemical properties of diclofenac, which features the highest Log *D* value in this group. As mentioned previously, solutes with high Log *D* are considered more hydrophobic and thus tend to adsorb and partition into the membrane matrix. As a result, lower rejection values are expected, since the compounds are eventually released into the permeate (Kimura et al., 2003b). In the case for diclofenac, the non-electrostatic affinity between the compound and the membrane offsets its potential rejection by electrostatic repulsion.

3.1.4 Nonionic compounds

Hydrophobic compounds

Rejection of hydrophobic nonionic compounds by the NF270 membrane varied significantly from 38 to 98 %. As expected, compounds with larger MW were more efficiently removed (>90 %), compared to the smaller molecules (Fig. 4).

Although fluoxetine and risperidone seem to be neutral, according to their single pKa value at pH 6.5, their full association indicated that they can form positively charged ionic species as well. On the other hand, both risperidone and fluoxetine are relatively large compounds and thus, the overwhelming effect of size exclusion overcomes the less significant sorption and electrostatic attraction between them and the membrane.

It was also observed that the difference between the rejection of compounds with similar MW and below the MWCO of the membrane was significant. For example, bisphenol A and carbamazepine rejection values were expected to be in the same range, based on their MW, whereas they were rejected at 38 and 73 %, respectively. As can be seen in Table 2, bisphenol A features higher Log *D* value and thus more hydrophobic than carbamazepine (3.63 compared to 1.89). The results indicate that compounds with higher Log *D* values, but which are also too small to be mainly rejected via steric hindrance, feature lower rejection levels. This is due to the higher affinity between the more hydrophobic compounds and the hydrophobic membrane matrix: hydrophobic solutes partition into the membrane, and consequently diffuse more readily through the membrane. Once again, there was no significant relationship between the rejection of large solutes (i.e., MW above 300 g mol^{-1}) and their Log *D* value.

Nonionic hydrophilic compounds

Rejection values of non-ionic hydrophilic solutes by the virgin membrane ranged from 4 to 99 %, and were clearly dependent on their MW. Trimethoprim, the solute with the largest MW (290 g mol^{-1}) in this group, was removed almost completely (99 %). Paracetamol, on the other hand, was only slightly rejected by the NF270 membrane, due to the small MW of this compound (151 g mol^{-1}). These results suggest that a major rejection mechanism for nonionic hydrophilic compounds remained size exclusion. This is consistent with the findings of Yangali-Quintanilla et al., who attributed rejection of nonionic compounds to the sieving phenomenon (Yangali-Quintanilla et al., 2009). Table 4 shows the summary of the observed rejection mechanisms of the target compounds by the virgin membrane.

3.2 Rejection of trace organic compounds by fouled membranes

3.2.1 Characterisation of the fouled membrane

Membrane fouling experiments were conducted using different organic foulants, under constant flux operation and constant TOC delivery rate. The characteristics for the new and fouled membrane are presented in Table 5.

TMP increase and conductivity rejection for new and fouled membranes were used to confirm the occurrence of fouling and concentration polarization within the fouling layer. The changes in membrane performance after fouling varied noticeably for the different macromolecules despite

Table 5. Characteristics for compacted and fouled NF270 in terms of TMP increase, salt rejection, roughness, contact angle and zeta potential.

Experiments	TMP increase (%)	Conductivity rejection (%)	Roughness (nm)	Contact angle (°)	Zeta potential at pH 6.5 (mV)
New membrane (compacted)	–	59.2	20 ± 4	58 ± 3	−21
HA	18	56.3	25 ± 1	57 ± 5	−17
L-Lysine	10	67.6	61 ± 1	60 ± 5	−22
Alginate	13	60.5	13 ± 6	36 ± 4	−48
SRHA	5	53.7	27 ± 3	35 ± 3	−18
Surface water	6	30.9	31 ± 4	78 ± 4	−13
MBR permeate	7	24.6	10 ± 3	66 ± 2	−22

the constant amount of TOC delivered to membrane in all experiments. For the real water types, the changes in conductivity rejection were even more significant. The normalized TMP generally increased by 5 to 18 % after 96 h of filtration for the different foulants. This indicates that the extent of fouling was significant but not severe.

Increased TMP and improved conductivity rejection after fouling was observed for the model foulants humic acid, L-lysine and alginate. This indicated that a compacted fouling layer was formed by these foulants, acting as an additional separation barrier.

On the other hand, for some foulants, fouling resulted in a small TMP increase and a considerable decrease in conductivity rejection. This was probably due to the formation of a rather porous cake layer by these foulants. The decrease in salt rejection in the presence of a porous cake layer is probably caused by CECP: the concentration polarization within a porous fouling is higher than in the bulk solution (for a virgin membrane) due to entrapment of small solutes within the cake layer, which hinders their back-diffusion into the bulk solution. This was already observed in previous studies, where it was reported that membrane fouling by NOM resulted in low hydraulic resistance and thus porous cake layer (Li and Elimelech, 2004; Nghiem and Hawkes, 2009; Hoek and Elimelech, 2003).

The characterisation of the fouled membranes confirmed that their surface properties were significantly different, compared to the virgin membrane. The virgin membrane was negatively charged (−21 mV) at pH 6.5, whereas fouled membranes generally demonstrated significant variations from that value (−13 to −48 mV). The change in membrane hydrophobicity was dependent on the properties of the fouling material. The roughness data indicated that the smooth NF270 membrane (20 nm) became noticeably rougher after fouling (25 to 61 nm), except when being fouled with alginate and MBR permeate (13 and 10 nm, respectively). Alginate and MBR permeate both created a "gel-like" fouling layer (i.e., tipically formed by extracellular polymeric substances). Therefore, the roughness for this specimen measured with AFM (in dry mode) might not represent the real surface morphology of alginate and MBR fouling in wet conditions.

3.2.2 Rejection of ionic compounds by fouled membranes

The temporal changes in rejection of diclofenac, naproxen and ibuprofen during the different fouling runs were monitored. The rejection values of all three ionic compounds were observed to be relatively stable and constant during filtration, indicating that the ionic solutes did not significantly adsorb onto the fouling layers. This is probably due to the electrostatic repulsion between the ionic solutes and the negatively-charged membrane surface. As these solutes cannot approach the membrane surface, hydrophobic interactions (adsorption) between the membrane and these solutes are limited (Verliefde et al., 2009).

The rejection levels for the negatively-charged hydrophobic compounds by fouled membranes are shown in Fig. 5. Generally, the rejection of these compounds did not vary significantly when the membrane was fouled, except for naproxen with the SHRA- and surface water-fouled NF270. Rejection of naproxen by these membranes significantly decreased, compared to the clean membrane.

Since the main rejection mechanism for ionic solutes was expected to be electrostatic repulsion, no significant correlation between the rejection and the MW of these solutes was expected. Fouling generally reduced the membrane surface charge (except for the alginate-fouled membranes), and thus electrostatic repulsions between the fouled membranes and the ionic solutes was expected to be lower. This was, however, not consistent with the slight increase in rejection of diclofenac when the membranes became fouled, nor with the stable rejections of the other ionic solutes.

The higher rejection of diclofenac after fouling can be explained by the increased overall charge density of the fouled membranes. Even though the membrane charge decreased, the roughness of the fouled membranes was higher (i.e., larger membrane surface area). The combined effect of surface morphology and membrane charge, on the removal of

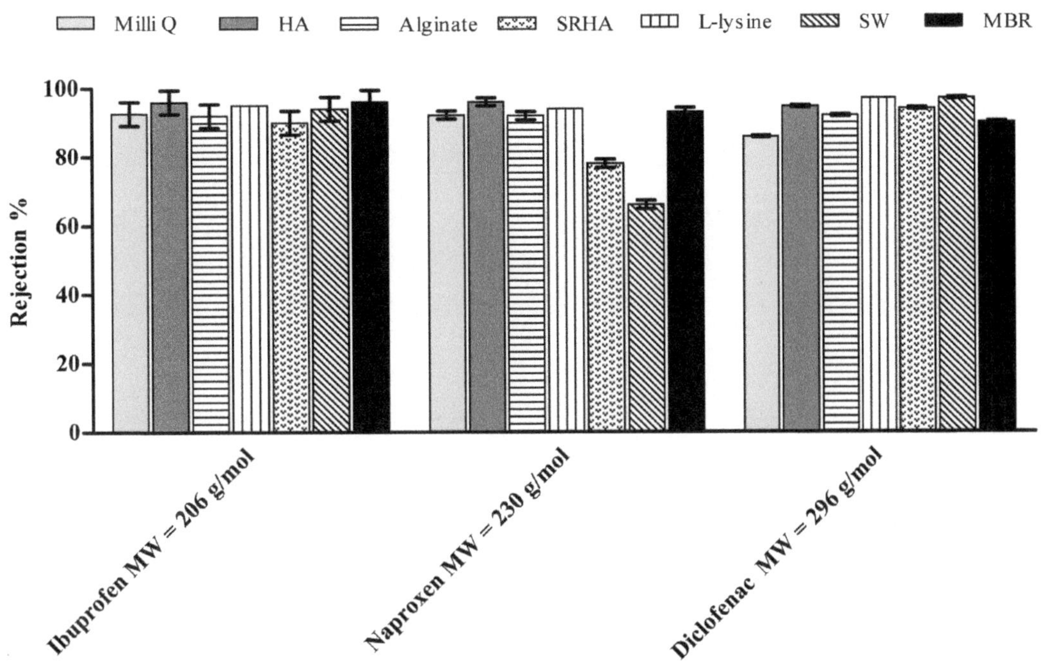

Figure 5. Rejection of hydrophobic ionic compounds by membranes fouled by various organic matters.

ionic trace organics has been reported in a previous study (Verliefde et al., 2009). For rough membranes, negatively-charged solutes that are forced into the crevices of the rough fouling layer, are surrounded by negative charges, resulting in an increase in electrostatic repulsion, even though the surface charge is lower. As discussed previously, protein- and surface-water-fouled membranes were the roughest, which enhanced the removal of diclofenac for these membranes more significantly.

For the other two ionic compounds, however, this increase of rejection with increasing surface roughness was not observed. In addition, no correlation between rejection and the characteristics of fouled membranes was observed.

3.2.3 Nonionic hydrophobic compounds

The rejection levels of the hydrophobic solutes by new and fouled membranes after 48 h of filtration time are shown in Fig. 6. Risperidone, carbamazepine and bisphenol A were rejected in increasing order, according to their MW. This was already observed in the experiment with Milli-Q water.

Fouling generally reduced the removal of the non-ionic hydrophobic trace organics. For risperidone and fluoxetine, the fouled membrane surface charge can also be responsible for the decreased rejection, since the negatively charged compounds removal indicated a higher negative charge density on the membrane. Theoretically, higher negative charge leads to a decrease in rejection of positively charged solutes (Verliefde et al., 2009).

As shown in Fig. 6, the extent of the decrease in rejection of hydrophobic nonionic compounds was different for each foulant. The variation in rejection by different foulants was due to the different characteristics of these foulants such as organic carbon fractions with different size, hydrophobicity and charge. According to the results of the LC-OCD characterisation, the DOC in the surface water comprised organic matter with a vast range of MW, preventing efficient stacking of the foulant, and thus resulting in a relatively loose cake layer. The gradual increase in concentration of solutes in the porous cake layer leads to a decrease in their removal (Elimelech and Ng, 2004).

To investigate the hydrophobic interactions between the solutes and the different foulants, the rejection of the solutes was plotted against the fouled membrane contact angle values (e.g., for risperidone in Fig. 7). Rejection of hydrophobic nonionic compounds by the fouled membranes generally correlated positively with the contact angle of these fouled membranes, except for carbamazepine. This is in contradiction with the expected lower rejection where the compounds presented higher affinity with the surface. Interestingly in this case, the membranes were fouled and the measured contact angle represented the hydrophobicity of the foulant, rather than those of the membrane surface. Therefore, it can be hypothesised that although fouling generally decreased the rejection level of hydrophobic compounds, the extent of this decrease was less severe for more hydrophobic solutes. This was most likely due to the adsorption of hydrophobic solutes onto the foulants in the feed. In a previous study, CECP and adsorption (to the foulants) were found to be two counteractive mechanisms influencing the rejection of hydrophilic nonionic trace organics (Hajibabania et al., 2011).

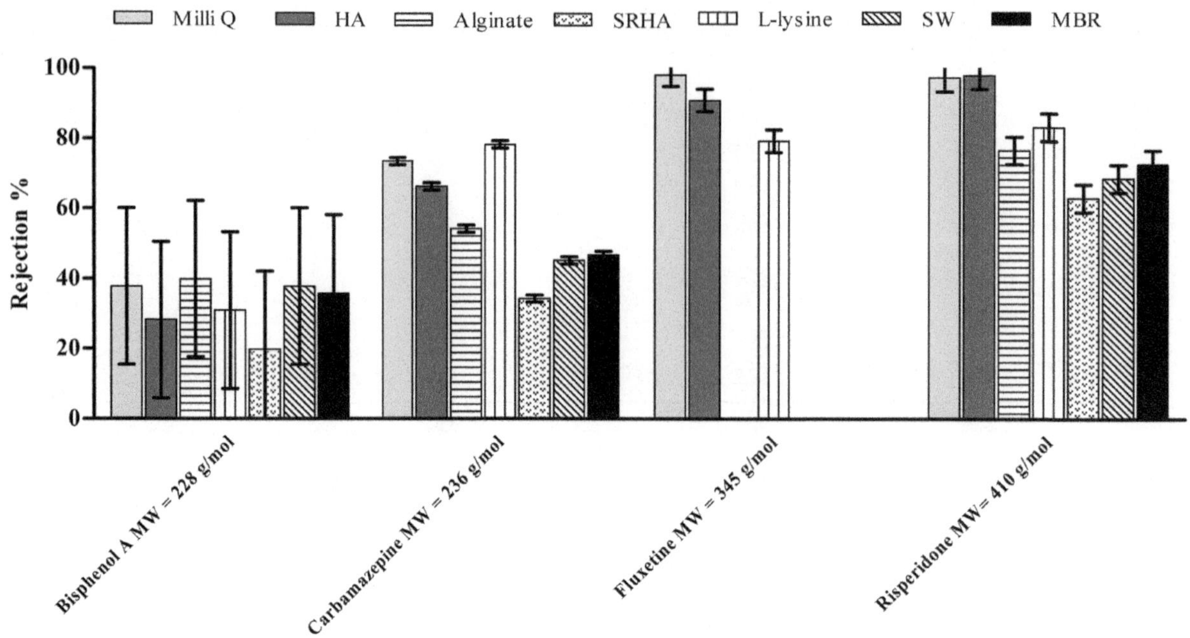

Figure 6. Rejection of hydrophobic nonionic compounds by different fouled membranes.

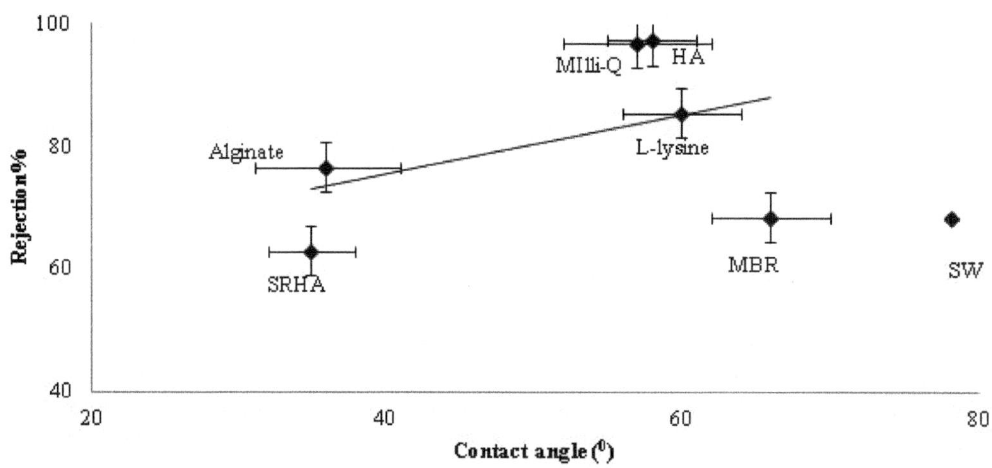

Figure 7. Rejection of risperidone by HA, alginate, lysine, SRHA and surface water fouled NF270 vs. fouled membranes contact angle.

However, the rejection of risperidone and surface water fouled membrane contact angle were an anomaly which might be due to more severe CECP. No significant relationship between rejection and surface roughness of the fouled membrane was observed.

3.2.4 Nonionic hydrophilic compounds

The effect of fouling on rejection efficiencies of these compounds was more pronounced compared to the hydrophobic solutes. The rejection efficiencies of the nonionic hydrophilic solutes by new and fouled membranes are shown as a function of their MW in Fig. 8. As can be seen, fouling generally reduced the removal of trace organics, except for paracetamol. The striking increase in paracetamol rejection with alginate fouled membrane observed here was not confirmed by previous studies (Hajibabania et al., 2011), and can be the result of an unexpected manipulation error.

Paracetamol thus showed different rejection behaviours compared to the other solutes, which again was due to its small size. For paracetamol alone, fouling enhanced size exclusion, with the extent of the increase in rejection varying depending on the foulant. The largest increases in rejection for paracetamol were observed for the surface water and MBR permeate-fouled membranes, i.e., the foulants with the largest fraction of low MW organic carbon fractions.

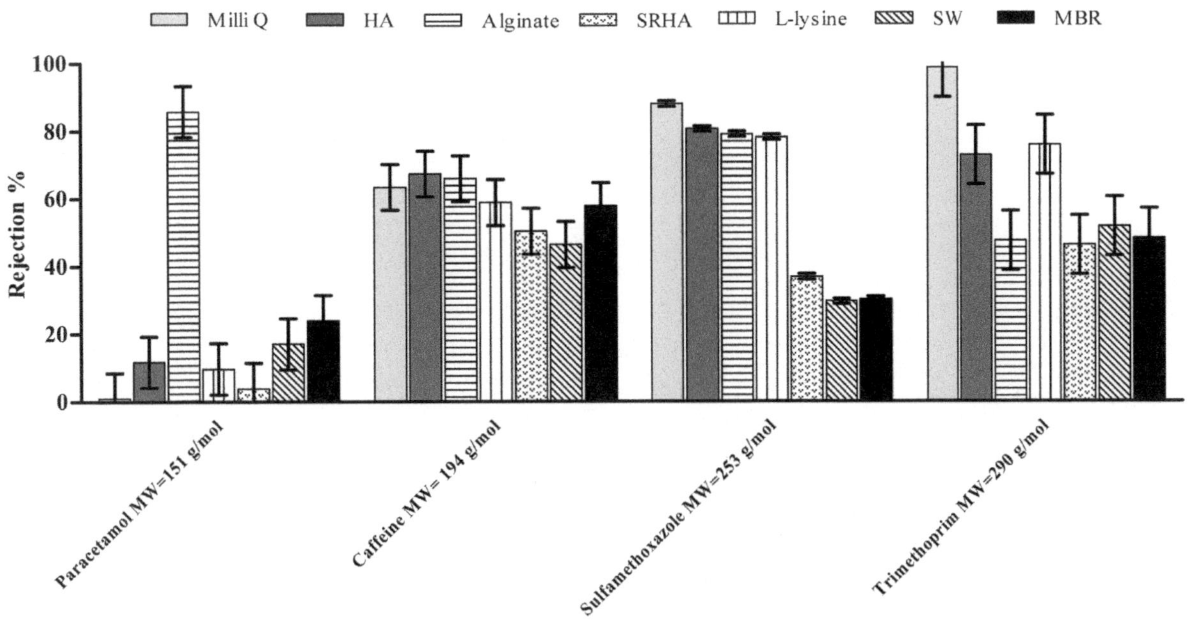

Figure 8. Rejection of hydrophilic nonionic compounds by new and fouled membranes.

Trimethoprim, which was the largest solute in this group, experienced a significant decrease in rejection from 98 to 46 % after membrane fouling. This indicates that the effect of CECP on rejection was more pronounced for larger solutes (since these solutes experienced a more significant hindered back-diffusion).

Porous fouling structures enhanced the occurrence of CECP, and consequently lead to increased transport of solutes through the membrane. This decrease due to severe CECP was most significant for SRHA and surface water fouling layers. Humic acid and lysine fouling layers, on the other hand, decreased the rejection of trace organics less than the other OM used in this study. Both foulants resulted in a higher TMP increase and conductivity rejection, which indicates that they form closely-stacked fouling layers. Apparently, the further size exclusion (sieving effect) was more pronounced than the CECP effect for these foulants.

To investigate the effect of fouled membrane characteristics on removal of nonionic hydrophilic compounds, the rejection of these solutes was plotted against the roughness and contact angles of the fouled membranes. There was no linear relationship observed between the rejection of hydrophilic nonionic compounds and membrane surface roughness and hydrophobicity (data not presented here). Therefore, size exclusion still seems to be the most important mechanism for rejection of hydrophilic solutes, with CECP playing an important role in the rejection of trace organics by fouled membranes.

4 Conclusions

In this study, both synthetic and real waters containing organic matters were used to simulate membrane fouling and to assess the impact of fouling on the rejection of trace organics by NF membrane. The well-controlled environment used in this study allowed a detailed assessment of the relative contribution of different organic foulants. It can be concluded that the effect of fouling on removal of trace organics was more pronounced for nonionic compounds, as they can approach and enter the fouling layer, in contrast to ionic solutes, which tend to be repelled by the negatively-charged fouling.

Porous fouling layers resulted in more severe CECP and lower rejections of nonionic solutes. The similar rejection trends obtained for hydrophobic and hydrophilic compounds confirmed the influence of CECP. No clear correlation between rejection by fouled membranes and parameter used to describe surface properties was observed. However, the combination of two or more parameters (such as roughness and charge and the concept of charge density) could be used to explain the observed variations in rejection.

Acknowledgements. The authors would like to thank ARC for the funding of this project. Dow Water & Process Solutions is also thanked for providing membrane samples.

References

Bellona, C., Drewes, J. E., Xu, P., and Amy, G.: Factors affecting the rejection of organic solutes during NF/RO treatment a literature review, Water Res., 38, 2795–2809, 2004.

Bolong, N., Ismail, A. F., Salim, M. R., and Matsuura, T.: A review of the effects of emerging contaminants in wastewater and options for their removal, Desalination, 239, 229–246, 2009.

Boussu, K., Vandecasteele, C., and Van der Bruggen, B.: Relation between membrane characteristics and performance in nanofiltration, J. Membrane Sci., 310, 51–65, 2008.

Bowen, W., Doneva, T., and Stoton, J.: Protein deposition during cross-flow membrane filtration: AFM studies and flux loss, Colloid. Surface. B, 27, 103–113, 2003.

Comerton, A. M., Andrews, R. C., and Bagley, D. M.: The influence of natural organic matter and cations on the rejection of endocrine disrupting and pharmaceutically active compounds by nanofiltration, Water Res., 43, 613–622, 2009.

Defrance, L. and Jaffrin, M. Y.: Comparison between filtrations at fixed transmembrane pressure and fixed permeate flux: application to a membrane bioreactor used for wastewater treatment, J. Membrane Sci., 152, 203–210, 1999.

Elimelech, M. and Ng, H.: Influence of colloidal fouling on rejection of trace organic contaminants by reverse osmosis, J. Membrane Sci., 244, 215–226, 2004.

Freger, V.: Nanoscale heterogeneity of polyamide membranes formed by interfacial polymerization, Langmuir, 19, 4791–4797, 2003.

Hajibabania, S., Verliefde, A., McDonald, J. A., Khan, S. J., and Le-Clech, P.: Fate of trace organic compounds during treatment by nanofiltration, J. Membrane Sci., 373, 130–139, doi:10.1016/j.memsci.2011.02.040, 2011.

Hoek, E. M. V. and Elimelech, M.: Cake-enhanced concentration polarization: a new fouling mechanism for salt-rejecting membranes, Environ. Sci. Technol., 37, 5581–5588, 2003.

Kimura, K., Amy, G., Drewes, J., and Watanabe, Y.: Adsorption of hydrophobic compounds onto NF/RO membranes: an artifact leading to overestimation of rejection, J. Membrane Sci., 221, 89–101, 2003a.

Kimura, K., Amy, G., Drewes, J., Heberer, T., Kim, T., and Watanabe, Y.: Rejection of organic micropollutants (disinfection byproducts, endocrine disrupting compounds, and pharmaceutically active compounds) by NF/RO membranes, J. Membrane Sci., 227, 113–121, 2003b.

Kiso, Y., Sugiura, Y., Kitao, T., and Nishimura, K.: Effects of hydrophobicity and molecular size on rejection of aromatic pesticides with nanofiltration membranes, J. Membrane Sci., 192, 1–10, 2001.

Le-Minh, N., Khan, S., Drewes, J., and Stuetz, R.: Fate of antibiotics during municipal water recycling treatment processes, Water Res., 44, 4295–4323, 2010.

Li, Q. and Elimelech, M.: Organic Fouling and Chemical Cleaning of Nanofiltration Membranes: Measurements and Mechanisms, Environ. Sci. Technol., 38, 4683–4693, 2004.

Nghiem, L. and Coleman, P.: NF/RO filtration of the hydrophobic ionogenic compound triclosan: Transport mechanisms and the influence of membrane fouling, Sep. Purif. Technol., 62, 709–716, 2008.

Nghiem, L. D. and Schäfer, A. I.: Critical risk points of nanofiltration and reverse osmosis processes in water recycling applications, Desalination, 187, 303–312, 2006.

Nghiem, L. D., Vogel, D., and Khan, S.: Characterising humic acid fouling of nanofiltration membranes using bisphenol A as a molecular indicator, Water Res., 42, 4049–4058, 2008.

Nghiem, L. D. and Hawkes, S.: Effects of membrane fouling on the nanofiltration of trace organic contaminants, Desalination, 236, 273–281, 2009.

Ongerth, J. and Khan, S.: Drug Residuals: How Zenobiotics Can Affect Water Supply Sources, J. Am. Water Works Ass., 96, 94–101, 2004.

Radjenovic, J., Petrovic, M., Ventura, F., and Barceló, D.: Rejection of pharmaceuticals in nanofiltration and reverse osmosis membrane drinking water treatment, Water Res., 42, 3601–3610, 2008.

Schäfer, A. I., Fane, A. G., and Waite, T. D.: Nanofiltration: principles and applications, New York, Elsevier Advanced Technology, 2003.

Van der Bruggen, B. and Vandecasteele, C.: Removal of pollutants from surface water and groundwater by nanofiltration: overview of possible applications in the drinking water industry, Environ. Pollut., 122, 435–445, 2003.

Van der Bruggen, B., Mänttäri, M., and Nyström, M.: Drawbacks of applying nanofiltration and how to avoid them: A review, Sep. Purif. Technol., 63, 251–263, 2008.

Vanderford, B. and Snyder, S.: Analysis of pharmaceuticals in water by isotope dilution liquid chromatography/tandem mass spectrometry, Environ. Sci. Technol., 40, 7312–7320, 2006.

Verliefde, A. R. D., Cornelissen, E. R., Heijman, S. G. J., Petrinic, I., Luxbacher, T., Amy, G. L., Van der Bruggen, B., and van Dijk, J. C.: Influence of membrane fouling by (pretreated) surface water on rejection of pharmaceutically active compounds (PhACs) by nanofiltration membranes, J. Membrane Sci., 330, 90–103, 2009.

Watkinson, A. J., Murby, E. J., Kolpin, D. W., and Costanzo, S. D.: The occurrence of antibiotics in an urban watershed: From wastewater to drinking water, Sci. Total Environ., 407, 2711–2723, 2009.

Wells, M.: Log D OW: Key to Understanding and Regulating Wastewater-Derived Contaminants, Environ. Chem., 3, 439–449, 2006.

Xu, P., Drewes, J. E., Kim, T. U., Bellona, C., and Amy, G.: Effect of membrane fouling on transport of organic contaminants in NF/RO membrane applications, J. Membrane Sci., 279, 165–175, 2006.

Yangali-Quintanilla, V., Sadmani, A., McConville, M., Kennedy, M., and Amy, G.: Rejection of pharmaceutically active compounds and endocrine disrupting compounds by clean and fouled nanofiltration membranes, Water Res., 43, 2349–2362, 2009.

Zazouli, M. A., Susanto, H., Nasseri, S., and Ulbricht, M.: Influences of solution chemistry and polymeric natural organic matter on the removal of aquatic pharmaceutical residuals by nanofiltration, Water Res., 43, 3270–3280, 2009.

15

How much are households willing to contribute to the cost recovery of drinking water supply? Results from a household survey

S. Tarfasa

Department of Economics, Hawassa University, Hawassa, Ethiopia

Correspondence to: S. Tarfasa (gutamajale@gmail.com)

Abstract. Financial resources are crucial to improve existing urban drinking water supply in developing countries typically characterized by low cost recovery rates and high and rapidly growing demand for more reliable services. This study examines the willingness to pay for improved urban drinking water supply employing a choice model (CM) in an urban context in Ethiopia, Hawassa, with a household survey of 170 respondents. The design of the choice model allows the estimation of the values of two attributes of urban drinking water service (extra day water delivery per week and safer water). The findings indicate that households are willing to pay up to 60 % extra for improved levels of water supply over and above their current water bill. Especially those households living in the poorest part of the city with the lowest service levels demonstrate that they are willing to pay more despite significant income constraints they are facing. Women value the improvement of water quality most, while a significant effect is found for averting behavior and expenditures. The estimated economic values can be used in policy appraisals of investment decisions.

1 Introduction

Financing domestic water supply is important to ensure water access for the urban poor and to broaden livelihood options. Urban drinking water is generally supplied publicly or under regulation. In the world, of every 10 people, 2 lack accesses to safe water supply, 5 have inadequate sanitation, and 9 do not have their wastewater treated to any degree (World Bank, 2004). The low public investment and insufficient tariffs in urban water systems in Ethiopia resulted in a low level of service, rationing and unscheduled disruptions, a long wait of months or even years for customers who want new connections due to lack of pipes and meters among other facilities (World Bank, 2004). In Ethiopia, for instance, 44 % of the financial requirement of the entire water supply program was provided by government and the consumers, and the rest by external support agencies. This reflected the huge gap between the finance needed to maintain and operate the existing water supply system in Ethiopia. It represents a financial burden for the local and central governments which are already under considerable pressure from other investment activities and the fast growing urban population (Rahamato, 1999).

The main objective of this study is to investigate urban households demand for improved water supply and to identify their willingness to pay in developing countries context, Ethiopia.

Economists typically use two approaches to measure the economic value of the quality and availability of non-market commodities. In the revealed preference (RP) approach the values of non-market environmental goods and services are inferred from other market transaction. Included under this approach are aversion cost, travel cost method and hedonic pricing. The other approach of stated preference (SP) asks individuals hypothetical questions about their willingness to pay. Contingent valuation (CV) and choice modeling are common variants of stated preference. In implementing a CV survey, respondents are presented with questions on whether or not they are willing to pay/be reimbursed for a change of some characteristics of the commodity in question. Choice modeling (CM) is a generalization of the CVM in that it gives respondents a list of alternative scenarios characterized by several attributes, from which they have to choose (Adamowicz et al., 1998). Among stated preference (SP) methods, the contingent valuation of public programs is the most frequently employed valuation tool in environmental economics (Bateman et al., 2003). A limited number of studies have been conducted to investigate the demand for a domestic water supply service in rural and urban Ethiopia using contingent valuation (see Fissiha, 1997; Dhunfa, 1998; Alebel, 2004; Kinfe and Berhanu, 2007). However, the contingent valuation method is inadequate to value a single attribute[1] of a multi-attribute non market goods and services, such as a domestic water supply service. An appropriate alternative tool is Choice Modeling (Louviere et al., 2000) which allows the investigation of an attribute of a good and also attempts to model the decision process of an individual or a household in a particular hypothetical situation.

The contributions of this study are that it adds to the limited study in this area as this choice modeling is the first study applied to the topic of urban domestic water services in Ethiopia and to inform policy makers on the provision of reasonable urban domestic water supply. In this study, a more advanced stated preference Choice Modeling (e.g. Blamey et al., 1999; Scarpa et al., 2007), is applied where households are asked to choose between different policy scenarios of improved water supply services at different water price levels. In the design of the Choice Modeling, a distinction is made between improved water supply delivery in terms of extra day per week and water quality where there is no boiling of water for anyone at all. The limited number of Choice Modeling conducted in this area in the developed world focused on WTP to avoid water restrictions, for instance due to droughts (Hensher et al., 2006).

2 Methodology

2.1 Choice modeling

Choice Models rely on the theory of consumer demand, and are based specifically on Lancaster's characteristics theory of value, which assumes that consumers derive satisfaction not consuming the good per se but from the characteristics composing the good (Hanley et al., 2001). Employed in this analysis is the Random Parameter Logit (RPL) model which does not require the assumption of independence from irrelevant alternatives (IIA) and which can also account for unobserved, unconditional heterogeneity in preferences across respondents, unlike the Multinomial Logit model (MNL) that suffers from the IIA assumption and which treats preferences across respondents as being constant (Hausman and McFadden, 1984).

The random utility function in the random parameter logit (RPL) model is given by:

$$U_{ijt} = V_{ijt} + \varepsilon_{ijt} = Z_{ijt}(\beta + \eta_i) + \varepsilon_{ijt} \quad (1)$$

where respondent i receives utility U by choosing alternative j from a choice situation t. The utility is decomposed into a deterministic component V_{ijt} and a stochastic term ε_{ijt}. Indirect utility is assumed to be a linear function of the choice attributes Z_{ijt} (as well as the social, economic and demographic characteristic, if included in the model) and parameter, β, which due to preference heterogeneity may vary between respondents by a random component, η_i. Assuming the error term follows an IID extreme value distribution of type I, the probability of choosing j in each of the choice sets can be derived (Revelt and Train, 1998).

Recent applications of the RPL model have shown that this model is superior to the MNL model in terms of overall fit and welfare estimates (Breffle and Morey, 2000). But, even if unobserved heterogeneity can be accounted for in the RPL model, the model fails to explain the sources of heterogeneity (Boxall et al., 1996). One solution to detecting the sources of heterogeneity while accounting for unobserved heterogeneity could be to include interactions of household characteristics with choice specific attributes in the utility function. When the interaction terms are included, the indirect linear utility function (V_{ijt}) that is estimated becomes:

$$V_{ijt} = \beta_0 + \beta_1 Z_1 + \beta_2 Z_2 + \text{----} + \beta_n Z_n + \delta_1 S_1 + \delta_2 S_2 + \text{---} + \delta_m S_m \quad (2)$$

where n is the number of urban drinking water attributes considered and the vector of utility parameters β_1 toβ_n are attached to the vector of attributes Z_1 to Z_n. In this specification, m is the number of respondent specific household characteristics that explain the choice of the improved drinking water, and the vector of coefficients δ_1 to δ_m correspond to the vector of interaction terms S that influence utility.

2.2 Economic welfare measurement

The compensating surplus (CS) is a measure of the change in utility arising from a change in a good or service, and in this particular case a change in a domestic water supply service. In this section, the compensating surplus (CS) is calculated following Rolfe et al. (2000) and Bateman et al. (2003). It measures the change in income that would make an individual indifferent between the initial (status quo) and a subsequent situation (improved water supply) assuming the individual has the right to the status quo. This change in income reflects the individual's willingness to pay (WTP) to obtain an improved water supply. Assessment of economic welfare involves an investigation of the difference between the well-being (or utility) achieved by the individual under the status quo (or constant base) alternative and some other alternatives. It is therefore, a matter of considering the marginal

[1] In Lancasterian microeconomic approach (Lancaster, 1966), individuals derive utility from the characteristics, also called attributes, of the goods rather than directly from the goods themselves. The set of attributes used in this study are (1) extra day water supply per week (2) water quality and (3) an increase in the households monthly water bill in Birr.

Table 1. An example of choice task.

	Status Quo	Policy Option 1	Policy Option 2
Water Supply(Extra days per week)	0	2	1
Water Quality(dummy: no boiling at all)	For infants	N0	For infants
Increase in monthly water bill	0	Birr 10	Birr 3
I prefer	□	□	□

value of a change away from the status quo. First, the values of the attributes that are associated with the status quo are substituted into the equation that estimates the indirect utility associated with that option. If socio-economic variables are included in that equation, the values to be substituted are the sample mean (or the individual specific welfare measures can be computed). Note that the monetary attribute is assigned a value of zero for this stage. Next, the values of the attributes that are associated with an alternative allocation of resources are substituted into the equation that relates to the relevant change attribute. The value of the relevant alternative specific constant (ASC) should be included in this calculation. Socio-economic variables are treated the same as for the status quo option and again the monetary attribute is set at zero. The utility associated with the change alternative (V_1) is then subtracted from the utility associated with the status quo option (V_0). If the model is linear (in the monetary attribute) this 'indirect utility difference' is then divided by the negative of the coefficient associated with the monetary attribute (β):

$$\text{Compensating Surplus} = \frac{-(V_0 - V_1)}{\beta_{\text{Monetary attribute}}} \quad (3)$$

A negative value for this surplus estimate would indicate that the respondents are willing to pay the amount of the surplus in order to experience an improvement in their well-being caused by the reallocation of resources from the status quo to the change alternative. By setting up multiple scenarios of alternative resource allocation (by varying the values the levels of attributes can take) and repeating this arithmetic exercise, an array of values associated with the scenarios can be estimated. Note that these results apply only when all attributes enter in a linear fashion (Bennett and Blamey, 2001).

2.3 The experimental design

The set of attributes and levels used in this study are: (i) water supply (1, 2, 3 extra day per week), (ii) water quality (dummy: no boiling at all), and (iii) an increase in the households monthly water bill in Birr[2] (3, 5, 10, 15 and 20) (see Table 1).

Paired choice sets were created using the fractional factorial orthogonal design procedure in SPSS, enabling the capture of main effects plus two-way interactions which produced 12 paired choice sets. A status quo alternative was added in all sets whose inclusion is instrumental to achieving welfare measures that are consistent with demand theory (Louviere et al., 2000; Bateman et al., 2003). Each choice task asked the respondent to hypothetically choose one of the two available options, in addition to the status quo option. If neither of the two options was found satisfactory, the respondent could choose the "opt-out option". After collecting socio-economic data, each respondent was introduced to the type of choice task required. The results of 25 respondents were discarded because of inconsistencies. Thus, a total of 1740 (12×145) observations were obtained.

In June of 2010, a household survey was carried out with about 170 households in Hawassa city, about 300km south of the capital city of Ethiopia, Addis Ababa, with a total population of about 160 000 (CSA, 2007). Because no recent statistical information about household characteristics was available, the survey was conducted in three zones identified by Hawassa Water Supply and Sewerage Service (HWSSS) namely, Misrak Wukro, Manaharia and Mahal Piasa. Misrak Wukro is generally considered the poorest part of the city, whereas households in Manaharia and Mahal Piasa are better off. In selecting respondents from each zone, first households with private compounds (i.e., relatively poor households in the study area usually live in houses with no compound) were identified, and secondly from these a household was chosen at random. The third criterion was that respondents were split as evenly as possible between male and female in order to allow for testing of gender related differences. Trained interviewers were used in the survey.

3 Results

3.1 Sample characteristics

The sample characteristics are shown in Table 2.The average monthly household income is USD 145. Given the average household size of 5.6, this gives a monthly per capita net income of USD 26. A larger proportion of the respondents (63 %) were female and respondents are, on average, 34 yr old. Consumption varies between 10 and 800 L per day per household. On average, a household has access of 4 days per week to drinking water supply and pays USD 2.2 per month for their water bill. Most households (58 %) also spend a

[2]Birr is Ethiopia's national currency. At the time of the study, 1 Birr was equal to approximately 0.06 USD.

Table 2. Sample characteristics.

Variable	Mean	Std.Er	Min	Max
Monthly household income (in Birr)	2388.42	1665	50	8000
Proportion of female respondents	0.633	0.48	0	1
Age	33.917	6	16	80
Family size	5.617	3.25	1	21
Monthly water bill	36.199	31.22	2.0	210.0
Amount of water consumed per household/month	99.158	87.3	10	800
Aversion cost	32.092	57.2	0	252
Water supply per week per household	4.095	1.68	1	7

considerable amount of money every month on substitutes such as bottled water. The result also shows that a household, on average, consumes about 3000 L (3 m^3) per month per household, which is about 100 L per day per household, according to this study. Given the average family size, per capita water consumption per day is 18 L, which is far below the world standard WHO (2004) of 45 L per capita per day.

3.2 The random parameter logit model

The results of RPL model are presented in Table 3. All the attribute parameters are highly significant and have the expected signs. The positive sign of the values for all non-monetary parameters suggests that improvements in all the non-monetary attributes are more likely to bring about a positive utility among individuals. The results also show that the coefficient for the monetary attribute is negative, as expected, implying that households demand less drinking water as its monthly bill increases. The estimated standard deviations are also significant and sizeable, indicating that we have captured unobserved heterogeneity with the random parameter specification (see Eq. 1). The alternative specific constant (combined into one as the experiment was generic) is also positive and statistically significant at 95 per cent, indicating that respondents receive more utility from the improvement than from the current water supply, ceteris paribus. Also, since the experiment was generic, this indicates that factors other than attribute levels affect behavior.

Results of the random parameter logit model with interaction given in Table 4 also show that out of the five socio-demographic variables, four of them were significant. These are: sex of the household head, income, the zone in which the respondents live and households' aversion behavior. The sex of household head being positive and significant implies that women prefer improvement in the domestic water supply compared to male household heads. The positive sign of income indicates that the higher the income of a household the more the household head is willing to pay for the proposed change. Households in the study area, especially those living in the poorest part of the city with the lowest service levels, are willing to pay up to 60 % extra for improved levels of water supply over and above their current water bill. A

Table 3. Random parameter logit model.

Variable	Coefficient	Standard error
ASC	1.162***	0.168
Water Bill	−0.196***	0.012
Water Quality	2.933***	0.302
Water Supply	0.765***	0.091
Derived standard deviations		
Water quality	2.933***	0.303
Water Supply	0.765***	0.091
Log likelihood function	−1198.995	
McFadden Pseudo R^2	0.390	
Number of respondents	145	

Note: *** denotes significance at 1 %.

significant effect is found for averting behavior[3] and expenditures. The explanatory power of the model is increased by about 28 % as a result of including the socio-demographic characteristics. The models are estimated with a simulated maximum likelihood using Halton draws of 500 replications.

3.3 Welfare analysis

As indicated (see Eq. 3), the compensating surplus for the change from the status quo to the new scenario is estimated by calculating the difference between the values of the two scenarios and multiplying by the negative inverse of the coefficient for the monetary attribute (that is the water bill). Assuming that the status quo is water supply per week is 4 days, on average, and boiling for infants only is needed, the mean WTP for three scenarios are presented in Table 5. These three scenarios are: (1) availability of water per week is 5 days and boiling for infants only is needed; (2) availability of water per week is 4 days and boiling is not needed;

[3]A possible averting behavior is for households in the survey to switch from the existing urban drinking water to bottled water when the quality of the water deteriorates. The cost (expenditure) of bottled water is an estimate of the maximum total willingness to pay for improving urban drinking water quality (Prato, 1998).

Table 4. Results of random parameter logit model with interactions.

Mean Parameters	Coefficient	Standard Error
ASC	1.136***	0.185
Water Bill	−0.136***	0.013
Water Quality	1.826***	0.937
Water Supply	0.763***	0.495
Water Quality X Female respondents	0.889**	0.281
ASC X Income	0.0001*	0.0001
Water supply X living in Misrak Wukro	0.296**	0.145
Water quality X aversion costs(Birr/month)	0.011***	0.005
Derived Standard deviation		
Water Quality	2.867***	0.310
Water Supply	0.715***	0.079
Log likelihood function	−1176.864	
McFadden Pseudo R^2	0.381	
Number of respondents	145	

Note: ***, **, and * denote significance at 1 %, 5 %, and 10 %, respectively.

Table 5. Mean WTP per month per household in USD.

	With Boiling		Without Boiling	
Water supply per week(in days)	4	5	4	5
WTP	0	0.66	1.12	1.36

(3) availability of water per week is 5 days and boiling is not needed.

It can be seen from the results that as the supply of water increases, the WTP also increases while the quality of the water is kept constant. When boiling for infants only is needed, WTP increases from 0 to USD 0.66 per month as the availability of water increases per week by one day. If the quality of water improves to the point where no boiling at all is needed, households are willing to pay USD 1.36. If it is supposed that, at least all of the households with private connection in the city, roughly 12 500, would be willing to pay this amount, then a total of USD 17 000 per month could be generated. The present value of these total benefits could then be compared with the present value of capital costs of this "ideal" option in order to calculate the net benefits. This implies that households prefer an improvement in the urban water service in terms of both the attributes of quality and supply of water. Thus, there is room for policy intervention to improve urban drinking water supply in the study area.

4 Conclusions

Choice Modeling can be used to investigate urban households demand for improved drinking water supply in terms of their willingness to pay in developing countries such as Ethiopia. From the results of the model, it can be concluded that households in Hawassa city support an improvement in domestic water service in terms of the attributes supply and water safety. Therefore, in an attempt to improve and expand urban drinking water supply in Hawassa city and similar cities in the country, the attributes can be targeted and additional revenue to minimize the operating and maintenance cost can be generated as the households show a positive willingness to pay for the improvement plan. Moreover, the present value of the total benefits could then be compared with the present value of capital costs of this "ideal" option in order to calculate its net benefits. This economic value can be used in policy and project appraisals of improved drinking water investment decisions.

Acknowledgements. I would like to acknowledge Hawassa University Research and Development Directorate for financially supporting this research.

References

Adamowicz, W., Boxall, P., Williams, M., and Louviere, J.: Stated preferences approaches to measuring passive use values, Am. J. Agr. Econ., 80, 64–75, 1998.

Alebel, B.: Affordability and Willingness to Pay for Improved Water Supply in Urban Ethiopia: The Case of Adama Regional City, Unpublished M.Sc. Thesis, Addis Ababa University, 2004.

Bateman, I. J., Carson, R. T., Day, B., Hanemann, W. M., Hanley, N.,Hett, T., Jones-Lee, M., Loomes, G., Mourato, S., Özdemiroğlu, E., Pearce, D. W., Sugden, R., and Swanson, S.: Guidelines for the Use of Stated Preference Techniques for the Valuation of Preferences for Non-market Goods, Edward Elgar, Cheltenham, 2003.

Bennett, J. and Blamey, R.: The Choice Modeling Approach to Environmental Valuation, Edward Elgar publishing, Inc, UK, 93–104, 2001.

Blamey, R., Gordon, J., and Chapman, R.: Choice modeling: assessing the environmental values of water supply options, Aust. J. Agr. Resour. Ec., 43, 337–357, 1999.

Boxall, P. C., Adamowicz, W. L., Swait, J., Williams, M., and Louviere, J.: A comparison of stated preference methods for environmental valuation, Ecol. Econ., 18, 243–253, 1996.

Breffle, W. and Morey, E.: Investigating Preference Heterogeneity in Repeated Discrete-Choice Recreation Demand Model of Atlantic Salmon fishing, Marine Resource Economics, 15, 1–20, 2000.

Central Statistical Agency of Ethiopia (CSA): Census Tables, Addis Ababa, 2007.

Dhunfa, L.: The Willingness to Pay of Rural Households for Improved Rural water Provision in Ada'a Liban District, Central

Ethiopia: Unpublished M.Sc thesis, Addis Ababa University, 50–57, 1998.

Fissiha, A.: Estimating Willingness to Pay for Water: Contingent Valuation Study on Meki Town, Ethiopia, Unpublished M.Sc. thesis, Department of Economics, Addis Ababa University, 55–60, 1997.

Hanley, N., Mourato, S., and Wright, R.: Choice Modeling Approaches: A Superior Alternative for Environmental Valuation, Journal of Economic Surveys, 15, 435–462, 2001.

Hausman, J. and McFadden, D.: Specification tests for the multinomial logit model, Econometrica, 52, 1219–1240, 1984.

Hensher, D. A., Shore, N., and Train, K.: Water Supply Security and Willingness to Pay to Avoid Drought Restrictions, The Economic Record, 82, 56–66, 2006.

Kinfe, G. and Berhanu, A.: Valuing Water Supply Service Improvements in Addis Ababa, Ethiopian, Journal of Economics, 17, 50–72, 2007.

Louviere, J., Hensher, D., and Swait, J.: Stated Choice Methods, Analysis and Application in Marketing, Transportation and Environmental Valuation, Cambridge University Press, 2000.

Rahamato, D.: Forum for Social Studies: Discussion Papers on Ethiopia's water resources and the strategy of water development, Addis Ababa, Ethiopia, 69–90, 1999.

Revelt, D. and Train, K.: Mixed logit with repeated choices; households' choice of appliance efficiency level, Review of Economics and Statistics, 53, 647–657, 1998.

Rolfe, J. C., Bennett, J. W., and Louviere, J. J.: Choice Modelling and its Potential Application to Tropical Rainforest Preservation, Ecol. Econ., 35, 289–302, 2000.

Scarpa, R., Willis, K. G., and Acutt, M.: Valuing Externalities from Water Supply. Status Quo, Choice Complexity and Individual Random Effects in Panel Kernel Logit Analysis of Choice Experiments, Journal of Environmental Planning and Management, 50, 449–466, 2007.

World Bank Groups Program for Water Supply and Sanitation report: Water Supply and Sanitation Sector Board, 6–8, 2004.

World Health Organization: The World Health Organization Report, Geneva: WHO, Switzerland, 2004.

16

Accumulation and modeling of particles in drinking water pipe fittings

K. Neilands, M. Bernats, and J. Rubulis

Riga Technical University, Department of Water Engineering and Technology, Latvia

Correspondence to: K. Neilands (kaspars.neilands@rtu.lv)

Abstract. The effect of pipe fittings (mainly T-pieces) on particle accumulation in drinking water distribution networks were shown in this work. The online measurements of flow and turbidity for cast iron, polyethylene and polyvinyl chloride pipe sections were linked with analysis of pipe geometry. Up to 0.29 kg of the total amount mobilized in T-pieces ranging from DN 100/100–DN 250/250. The accumulated amount of particles in fittings was defined as J and introduced into the existing turbidity model PODDS (prediction of discoloration in distribution systems) proposed by Boxall et al. (2001) which describes the erosion of particles leading to discoloration events in drinking water network viz sections of straight pipes. However, this work does not interpret mobilization of particles in pipe fittings which have been considered in this article. T-pieces were the object of this study and depending of the diameter or daily flow velocity, the coefficient J varied from 1.16 to 8.02. The study showed that pipe fittings act as catchment areas for particle accumulation in drinking water networks.

1 Introduction

The accumulation of particles in drinking water distribution networks is commonly observed. A considerable number of studies have been done on explaining fundamentals of processes (van Thienen et al., 2011a; Boxall et al., 2003), a mathematical description (Ryan et al., 2008; Boxall et al., 2001) and management advices (Vreeburg et al., 2004; Matsui et al., 2007; Saldarriaga et al., 2010; Husband et al., 2010) regarding water discoloration in a water supply network has been achieved.

A detailed work (Ryan et al., 2008) was carried out in order to explain the mechanisms responsible for particle accumulation in drinking water networks. However, in the proposed particle sediment model (PSM) only the mechanism of gravitational force of particles where the threshold value for deposition is set as $0.07\ \mathrm{m\,s^{-1}}$ and for resuspension $0.2\ \mathrm{m\,s^{-1}}$ (in 100 mm pipe) was included. As a second mechanism, the deposition of particles on the pipe walls was described (Ryan et al., 2008), which could affect the deposition by the van der Walls forces which are known to be weak; in this case the small distance between the particles and pipe surface is important. Fick's law was used to describe diffusion of particles from bulk flow to pipe wall. Although this mechanism was not included in the model, from the experimental results it was apparent that the particles (specific gravity 1.64) precipitated better on polyvinyl chloride (PVC) than on cement lined metal pipe. The concentration of particles on PVC pipe wall is more sensitive to flow velocity, which do not include region of gravitational settling (0.07–$0.3\ \mathrm{m\,s^{-1}}$) (Ryan et al., 2008). The amount of particles deposited on the wall of the PVC pipes was shown to be larger than the amount deposited on the wall of the cement lined metal pipes, for the same working conditions (velocity, initial concentration). Although the lab-scale experiments were done in pipe loop (length = 41 m) consisting of four 90° bends, there was no information on any influence of loop geometry on the accumulation results (Ryan et al., 2008).

The application of the stochastic water demand model SIMDEUM (SIMulation of Demand, an End-Use Model) together with turbidity measurements from a uni-directional flushing (UDF) program showed that maximum daily velocities in range 0.05–$0.1\ \mathrm{m\,s^{-1}}$ give the highest concentrations of particles in a network (Blokker et al., 2010). While it has been concluded that maximum daily velocity is not the only parameter responsible for particle accumulation, the role of radial transport processes can be explained by, for example, turbophoresis and turbulent diffusion, which was based on the laboratory-scale experiments (van Thienen et al., 2011b) in which the deposition of particles on the entire wall circumference at velocity $0.14\,m\,s^{-1}$ with particles of specific gravity 1.2 was demonstrated. To describe these processes the virtual mass effect which refers to inertia of particles in water together with the Magnus and the Saffman forces, which describe the rotation of a particle in water, was discussed regarding the particle accumulation process in pipe sections. The authors have not taken a closer look at the geometry; remarked that the internal irregularities in the distribution pipe system may be catchment areas for particles. It should be noted that particles from wash-out during first flushing turnover mobilized along the entire pipe length can be called locally settled sediments (LSS) (Blokker et al., 2010).

Boxall et al. (2001), Boxall and Saul (2005) proposed the PODDS (prediction of discolouration in distribution systems) model by involving the term "cohesive transport model" where discoloration material is assumed to be homogenously distributed around the pipes circumference in cohesive layers of particulate matter. The layer strength is a function of the maximum daily shear stress. Material erosion and regeneration processes can be modelled through calibration. Proposed empirical equations for the description of erosion have been verified in lab-scale polyethylene (PE) pipe loop (length = 600 m) with an elliptical form (Husband et al., 2008), and in both plastic and rough corroded metal pipes with different diameters in field conditions (Husband and Boxall, 2010). The shear stress and subsequent erosion of particles increased with the flow. It is considered that the accumulation of particles occurs in layers around the complete pipe circumference which is conditioned via daily hydraulic flow. The accumulation mechanisms via corrosion (direct) and the background concentration of material from upstream or the treatment plant (indirect) have been discussed (Husband and Boxall, 2010). The origin of the particles in distribution network is has been widely described (Gauthier et al., 1999; Verberk et al., 2006; Husband et al., 2008; Vreeburg, 2007; Vreeburg et al., 2008); although , this is not the subject of the current paper. There is some criticism expressed towards the PODDS model, e.g. that (1) the formulations of the equations which predict the accumulation rate of particles should be subordinated to the fundamental laws (van Thienen et al., 2011a) and (2) the parameter that the rate of change of the velocity or, subsequently, unsteady shear stress is applied during the early stage of hydraulic transient event (e.g. fast valve opening/closing) plays a significant role for the first seconds of erosion of the particles (Aisopou et al., 2010). It should be mentioned that PODDS supports the cohesive layer assumptions in the approach considering the entire pipe length irrespective of the topographical features, while afterwards, some irregularities in washed out particle concentration obtained during the trunk main flushing within section consisting of four 90° bends and depressions in the pipe topography have been noted (Husband et al., 2010).

The particle accumulation and the subsequent erosion within the drinking water pipe networks has been broadly described within the past few years. However, there is a lack of information regarding the particle concentration in locations where water flow changes due to geometry of conventional fittings (T-pieces, valves, bends etc.). In a recent study, we reported the analysis of detailed pipe geometry (topology) and online turbidity curves obtained from particle erosion during UDF via one and two hydrants simultaneously, and demonstrated that these spikes corresponded to location of network fittings (Rubulis and Neilands, 2010).

In this work the influence of pipe geometry in accumulation process of particles in drinking water distribution networks has been investigated. The major objectives of this study were to (1) analyze UDF data from field events with different diameter T-pieces and other fittings; (2) link UDF data (shear stress, turbidity and mass) with equations of PODDS by using MS Excel to model the erosion of particles; (3) explain the field measurements using an approach from tracer studies regarding the particle transport in other industrial applications.

2 Materials and methods

2.1 Description of study site

The UDF operation was performed in a town with population of about 9000 inhabitants. The overall network consists of 13 km of cast iron (CI) material, of 4 km of PVC and 3 km of polyethylene (PE) pipes (Fig. 1).

The CI pipes were mainly constructed between the years 1970–1990, while PVC and PE pipes were built during the years 2000–2011. For the supply of drinking water networks the treatment with aeration followed by rapid sand filtration of $700\,m^3\,day^{-1}$ has been applied. The water quality outlet of the plant from the grab sample program summarized in Table 1 is characteristic of corrosive water with a high salt concentration. Neilands et al. (2009) proved that the application of the online monitoring system reveals the problems with the particle load in the network due to a deficient operation of rapid sand filters during the flushing process in the morning hours (Fig. 2).

The network had not been flushed before. The accumulation of sediments in the network was assessed with the resuspension potential method (RPM) proposed by Vreeburg et al. (2004) which revealed considerable risk of discoloration (Table 2).

2.2 Experimental design of field measurements

In general, the online turbidity data from UDF were linked with a careful analysis of technical drawings of each flushed

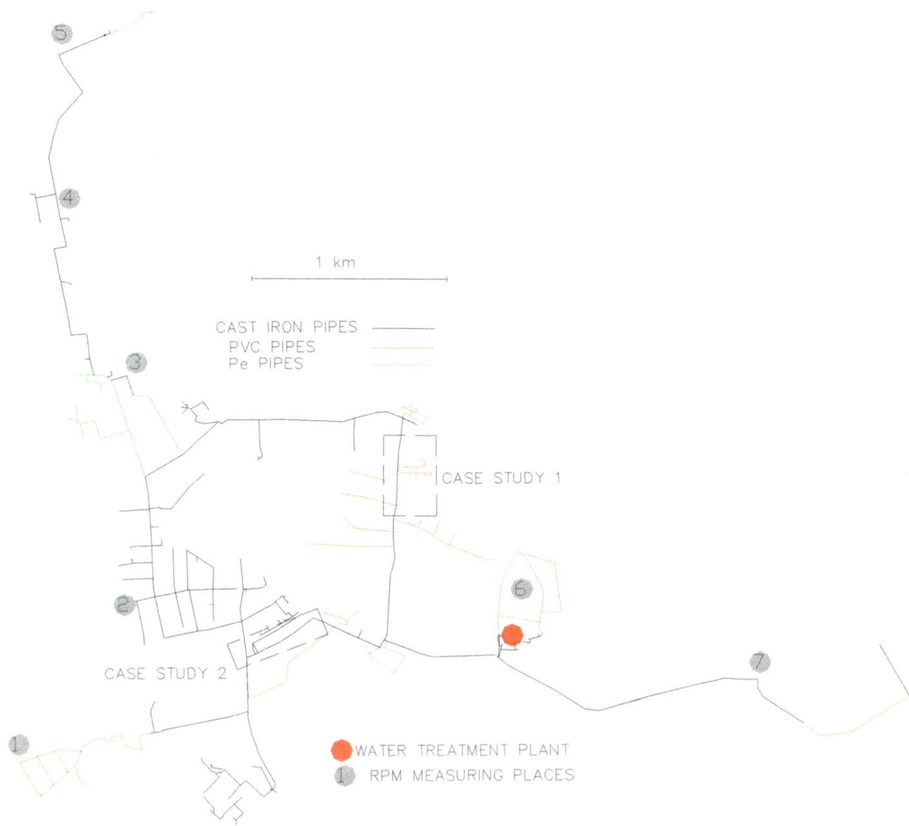

Figure 1. The diagram of the drinking water network indicating numerated circles for locations of resuspension potential method. Flushing sections are indicated with a dashed line. CI pipes are shown in blue, the PVC pipes are shown in orange and the PE pipes are shown in green.

Table 1. The water quality in water works (treatment plants). The average results with standard deviations and coefficient of variation (in brackets) are shown.

Parameter, units	Average results
Conductivity, $\mu S\ cm^{-1}$	$739 \pm 16(20)$
pH	$6.97 \pm 0.38(18)$
Iron (Fe tot), $mg\ l^{-1}$	$0.05 \pm 0.02(4)$
Iron (FeII), $mg\ l^{-1}$	$0.01 \pm 0.01(4)$
Turbidity, NTU	$0.344 \pm 0.32(94)$
Chlorine (tot) $mg\ l^{-1}$	$0.038 \pm 0.033(17)$
Chlorine (free), $mg\ l^{-1}$	$0.012 \pm 0.021(8)$
Langelier Index	-1.23
Red-ox, mV	$204 \pm 10(17)$

pipe section while paying particular attention to the location of the fitting. In total, 67 isolated pipe sections were flushed. The turbidity data during the flushing process were recorded for at least every minute (in several cases even 5 s) using calibrated online photometer ULTRATURB plus™ equipped with a controller sc 100™ (Dr. Bruno Lange GmbH&Co, Düsseldorf, Germany) installed into a flushing vehicle equipped with a manometer and the Siemens Sitrans FM MAG 8000 online flow meter (Siemens Energy & Automation Inc., Germany). The standard deviation for measurements was ±30 s, depending on how fast the necessary flow rate was reached during the opening of the fire hydrant. Mechanical cleaning of the measuring chamber in addition to the automatic cleaning of the measuring chamber of ULTRATURB plus™ was routinely performed.

The clear PVC pipe segment (inner Ø 75 mm) in the flushing vehicle was connected with a 1 m long fire hose (inner Ø 75 mm), which was connected directly to the standpipe of the hydrant. The construction of the flushing vehicle was done assuming that the water flow from the standpipe would have the shortest possible path till the online turbidity meter. The online turbidity measurements during UDF were controlled with simultaneous turbidity measurements in grab samples performed with a portable manual turbidity meter Hach 2100P (Dr. Bruno Lange GmbH&Co, Düsseldorf, Germany). This was done in a synchronized manner with all the measuring equipment at the start of the flushing.

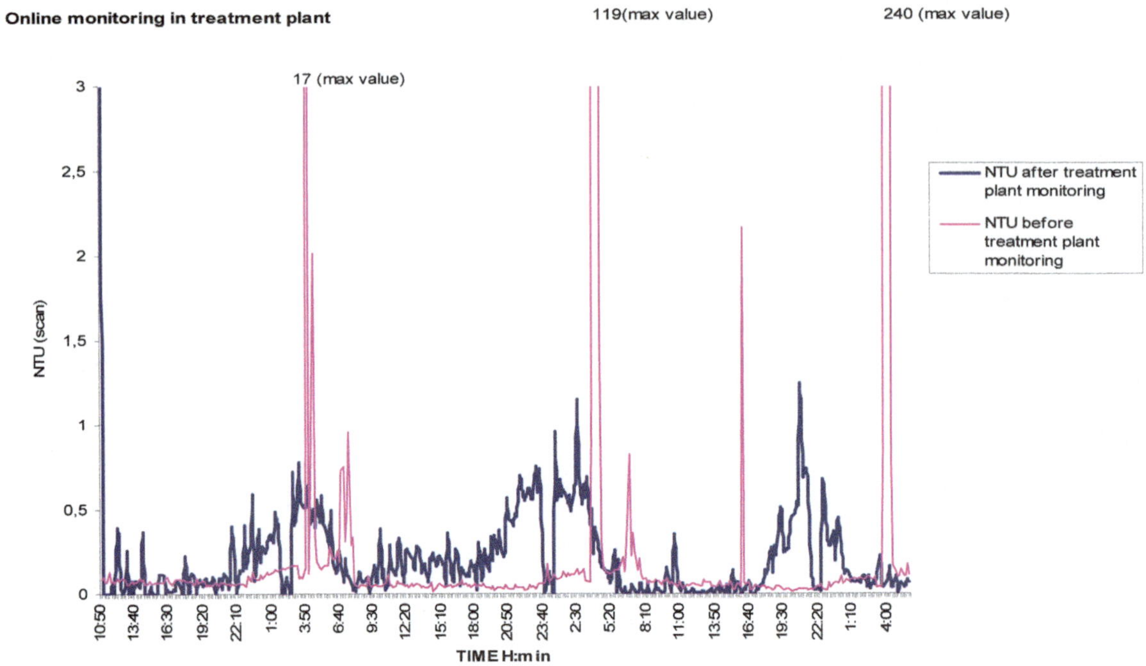

Figure 2. The online measurements of turbidity in the outlet of the WTP of the distribution network before and after the repair of the rapid sand filters. Values on the top of figure show the peaks of turbidity. Modified from Neilands et al. (2009).

Table 2. The results of RPM ranking measured with Ultraturb™ equipment in the field study place (Fig. 1).

Nr.	Diameter (mm)	Material	Section length (m)	RPM	Average turbidity (NTU)
1	100	PVC	326	14	114.6
2	150	CI	326	14	36.0
3	150	CI	397	7	6.9
4	150	CI	423	5	3.8
5	150	CI	401	10	10.7
6	200	PE	324	15	105.8
7	100/200	PVC/CI	456	6	6.8

2.3 Hydraulic model

The pipe hydraulic model (1073 pipes and 986 nodes) was built up using the EPANET 2.0 system (Rossman, 2000) and the hydraulic head loss was calculated with the Darcy-Weisbach equation and top-down demand approach. The time step for the hydraulic head loss calculation was set to 1 s. The node demands were collected from a 6 month data survey for every connection. The daily patterns of consumption for inhabitants was set to 630 m^3 day^{-1}. The daily pattern for companies and municipal organizations were set according to the information from the municipality of registered drinking water consumptions. There were two organizations in municipality, working constantly day and night, but the total amount of water consumption is only about 0.5 m^3 h^{-1}, thus, they did not influence the data generated by the model. The verification of water residence time was conducted by tracer studies (Fig. 3) since each well in the water treatment plant had a different electric conductivity (EC) in the range of 500–900 $\mu S\ cm^{-1}$ (25 °C).

All field measurements were recorded for several hours and were made from hydrants or at the taps in the buildings belonging to the drinking water supplier. For the field measurements, online instruments 3400sc™ with controller sc 100™ (Dr. Bruno Lange GmbH&Co, Düsseldorf, Germany) and condu::lyser™ with controller con::stat™ (s::can Meßtechnik GmbH, Vienna, Austria) supplemented with portable manual EC meter Cond 315i, (WTW, Germany) were used. The results of model verification are not shown in this study.

2.4 Analysis of turbidity curves

In order to generalize the trends of particle erosion observed during field experiments/UDF the following assumptions were made:

1. The hydraulic circumstances of the particle disturbance were based on the assumption that the accumulated particles erode and are transported by a plug flow. The distance (L, m) from hydrant outlet to location where erosion of particles starts was calculated with the formula below:

$$L = \frac{4 \cdot Q \cdot t}{\pi \cdot D^2} \tag{1}$$

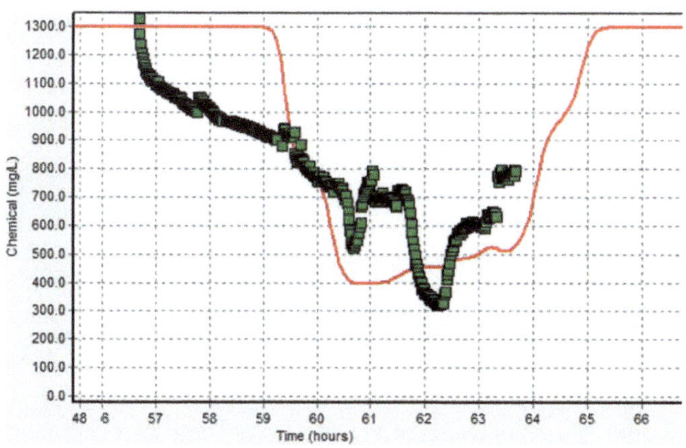

Figure 3. Measured (green squares) and modeled (red line) concentrations of the tracer near RPM measurement location No. 3 (see Fig. 1).

where t is time (s), Q – flushing flow [$m^3\,s^{-1}$] and D – diameter of pipe [m]. The same assumption has been used previously and in addition a term "locally settled sediment" (LSS) has been introduced (Blokker et al., 2010).

2. By using the principle of back-tracing, the location in networks corresponding to the spikes of 30 NTU and 26 NTU in Fig. 4 were found using Eq. (1) and an estimates of flow.

3. The maximum value (30 NTU and 26 NTU) of each spike was divided by the turbidity value just before spike, which means the average turbidity in straight section is (17 NTU of turbidity from beginning) $Turb_{max}/Turb_f = J$ coefficient. This value was used for the calculation of next peak magnitude, in case the upflow had some other fittings. Hence, a comparison of the difference between the particles originating from a linear pipe section and the particles which accumulated in the fittings could be compared. If a section is without fitting, spike, which comes with clear front (turnover), is not considered. As many data as possible are needed to create a database for several situations, according theoretical mass-balance from a treatment station.

A large number of data from flushing studies is needed to calculate a fitting coefficient for various conditions in order to predict the fouling of pipes with particles at the locations of T-pieces and valves.

2.5 Modeling of particle erosion in drinking water network including fittings

The stability of deposit layers is, according to the PODDS model, based on the hydraulically induced shear of the bulk water on the deposit surfaces. The daily maximum occurring

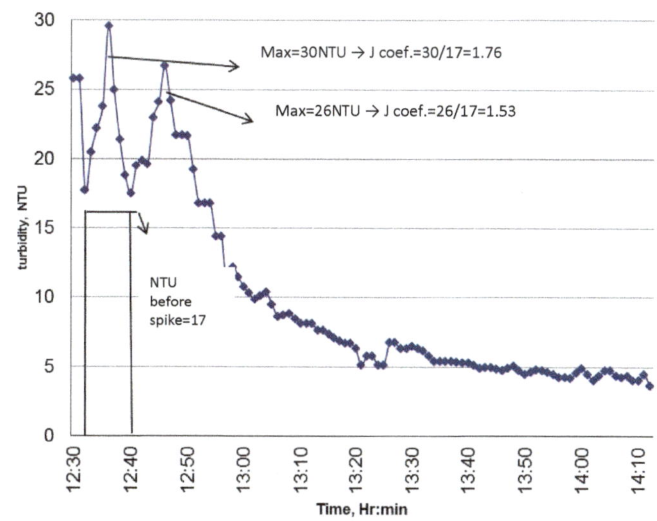

Figure 4. The turbidity pattern from UDF and schematic (graphical) representation for obtaining the algorithm for the J coefficient for a certain location.

hydraulic wall shear stress is equivalent to the minimal shear stability of deposits according to a validated hydraulic model in the conditions of the maximum consumption of water per day; the maximal shear stress to pipe walls is calculated as follows:

$$\tau = \rho g \frac{D}{4} \cdot l \tag{2}$$

where τ is boundary shear stress ($N\,m^{-2}\,m^{-1}$), ρ – density of water ($kg\,m^{-3}$), g – gravity acceleration ($m\,s^{-2}$), D – diameter of pipe (m), l = dimensionless energy gradient.

The turbidity curve predictions were calculated in a created algorithm (MS Excel). Using the PODDS model and substituting the equations therein, the output turbidity is equal to

$$\Delta C_c(t=1) = P^n_{(t_a - t_s)} \cdot 2\pi r L \tag{3a}$$

$$\Delta C_c(t \geq 2) = \underbrace{\Delta C_c(t=1)}_{\text{NTU value at previous time}} \times \underbrace{\left(1 - \frac{R_{\text{link}} \times \Delta t \times nt}{C_{\max}}\right)}_{\text{Decrease term}} \tag{3b}$$

where ΔC_c is the turbidity value at the current time at the flushing outlet. At time $t = 1$ turbidity will have the maximum possible peak value for corresponding layer strength followed by gradual decrease of maximal peak value at $t \geq 2$ and so on. Character Δt is the value of the time step in seconds, P – gradient term ($NTU\,m^{-2}$), n – power term (–), t_a – applied shear stress ($NTU\,m^{-2}$), t_s – current layer strength ($NTU\,m^{-2}$), r – radius of a pipe (m), L – length of a ling (m), nt is number of current time step, R_{link} is output peak turbidity for the corresponding link, NTU_{total} is the total accumulated mass following a flushing event.

To predict the erosion of particles and obtain the character of the turbidity curve, the following equation was used:

$$\Delta N = R \times A_s \tag{4}$$

where ΔN is the increase or decrease of turbidity (NTU), R is the rate of supply (NTU m^{-2}), A_s is the pipe surface area (m^2).

Turbidity peak for plug flow will occur at some (current) time C, after which it will gradually decrease, described by NTU decrease term. The maximum peak value was calculated from the difference of the standard maximum velocities (hourly and flushing), in terms of corresponding shear stresses regardless of the mass accumulated in pipe. The total accumulated mass was used to calculate the total turbidity mass via linear daily shear stress–turbidity relationship, NTU $m^3 = -475\,000\,\tau_{daily} + 529\,000$. Because of the total turbidity dimensions, the obtained value should be multiplied by the number of cubic meters in the pipe to obtain the total link turbidity and its maximal peak value. The magnitude rate of decrease depends on the coefficient: P and exponent n (variables in our case $P = 0.1$ and $n = 0.5$) in the rate of supply from the layer calculations:

$$R = P(\tau_{flush} - \tau_{daily})^n. \tag{5}$$

Weak layers (according to shear stress stability) can store more turbidity units than strong layers. This means that pipes with a low flow rate have a higher potential of turbidity accumulation and the same factor is referable to the fittings. By generalizing the PODDS model, it can be said that in the conditions of a constant flushing flow and a constant shear stress, the output turbidity will have only one peak without considering the local turbulence in the fittings. As it was observed from the case study, after the first turbidity peak a second or third peak may follow, which often are higher than the first one. It was calculated that time of all the next peaks mainly conform to the time it takes the plug flow to travel from the fitting of a downstream pipe. Thus, in this study we have applied an additional term to the aforementioned turbidity equation in order to account for all peaks representing the downstream fittings.

$$\text{Turbidity}\ (t \geq 2) = \text{NTU}(t=1) \times \left(1 - \frac{R \times t}{\text{NTU}_{\text{total}}}\right) + \underbrace{\text{NTU}_f \times J \times \left(1 - \frac{R \times t}{(Q \times t \times \text{NTU} \times \text{TSS} \times 10^{-6})}\right)}_{\text{Additional term describing the impact of a junction}} \tag{6}$$

where

$\text{NTU}(t = 1) = R \times A_s$
$\text{NTU}_{\text{total}} = C_{max}/(\pi r^2 L)$
$\text{NTU}_f = \text{NTU}_{\text{total}}/t$

t is time (in seconds), R – rate of supply (NTU m^{-2}), A_s – pipe surface area (m^2), r – radius of pipe (m), J – coefficient of a fitting, L – length of pipe, NTU_f denotes average output turbidity in first peak, C – the stored turbidity, calculated from layer shear strength τ' (NTU m^3).

2.6 Calculations of particle mass

The mass (kg) of particles in a fitting was calculated using a following equation:

$$M = \Sigma(Q \times t \times \text{Turb} \times \text{TSS}/10^6) \tag{7}$$

where t is time (in seconds), Q – flow rate ($l\,s^{-1}$), Turb – the turbidity data from the online photometer (NTU), TSS – the total suspended solids which was assumed as a constant (77.3 $mg\,l^{-1}$) for all the calculations (Boxall et al., 2003) where a wide range of samples obtained from different drinking water networks were analyzed.

In the Eq. (7) all the turbidity online measurements were summed up and each of the turbidity readings was converted to mass, using the above mentioned TSS constant, allowing the total mass of the flushing event to be calculated. It was assumed that the flushing plug flow velocity was equal to the flow velocity. Thus, by logging online discharge data, time from the start of the flushing event and knowing the diameters of the pipes located downstream from the flow velocity and thus past distances downstream from the pipe being flushed were calculated. As reported earlier, close similarities between online turbidity measurements peaks and downstream fitting were detected (Rubulis and Neilands, 2010).

3 Results and discussion

3.1 Field experiments

More than 67 isolated sections were flushed using UDF to collect turbidity data from washed-out particles. Both sections of CI, PVC and PE pipes were examined. The pipe diameters ranged from 100 to 300 mm. The discharge from the hydrant was maintained constant to ensure constant shear stress on particles; therefore, the possible fluctuations in hydraulic conditions could take place only due to the changes in water production/consumption ratio. In Table 3 all data corresponding only to flushed sections with T-piece are summarized, since this type of fitting has most often been represented, instead of the bends of 30°, 60°, 90°, but those data were too few to make some valid approximation trends.

The total mass of washed out particles from T-pieces in this study ranged from 0.05 to 0.29 kg. This mass was calculated by taking into account the relationship between turbidity and total suspended solids from studies of Boxall et al. (2003). The quantities of particles in our study were higher compared to the data from literature. The mass of about 0.0001–0.04 $g\,m^{-1}$ was found to be present in four different Canadian networks (Carriere et al., 2005). The quantity of particles described by the other authors was even smaller: 0.00008 $g\,m^{-1}$ (Jonas and Murphy, 2002) and

Table 3. Description of T-pieces.

Case study	$V_{\text{day max}}$ (m s^{-1})	$\tau_{\text{day max}}$ (N m^{-2})	V_{flush} (m s^{-1})	τ_{flush} (N m^{-2})	Junction type (mm)	J (–)	Mass in junct. (kg)	V/D (–)	Pipe material	Pipe length (m)
1	0.21	0.15	0.43	0.59	T-piece DN 200/100	1.87	0.24	1.05	Cast iron	457
2	0.24	0.21	1.7	7.55	T-piece DN 200/100	2.96	0.2	1.2	Cast iron	253
3	0.02	0.002	1.52	6.4	T-piece DN 200/100	8.02	0.05	0.1	Cast iron	240
4	0.39	0.22	0.37	0.45	T-piece DN 250/250	2.23	0.2	1.56	Cast iron	300
5	0.02	0.002	2.14	11.96	T-piece DN 200/100	3.4	0.11	0.1	PE	159
6	0.23	0.19	0.67	1.33	T-piece DN 150/100	1.53	0.25	1.53	Cast iron	75
7	0.27	0.24	0.39	0.47	T-piece DN 250/100	1.86	0.15	1.08	Cast iron	170
8	0.02	0.002	0.77	1.72	T-piece DN 200/100	3.94	0.18	0.1	Cast iron	295
9	0.27	0.24	1.13	3.6	T-piece DN 250/150	5.6	0.17	1.08	Cast iron	204
10.1	0.21	0.15	0.46	0.62	T-piece DN 250/200	1.4	0.29	0.84	Cast iron	745
10.2	0.21	0.15	0.46	0.62	T-piece DN 150/150	2.75	0.22	1.4	Cast iron	400
11.1	0.21	0.15	0.52	0.81	T-piece DN 250/200	1.53	0.2	0.84	Cast iron	745
11.2	0.21	0.15	0.52	0.81	T-piece DN 150/150	1.52	0.2	1.4	Cast iron	400
12	0.02	0.002	1.39	5.01	T-piece DN 150/150	2.7	0.17	0.133	Cast iron	245
13	0.04	0.004	1.5	5.97	T-piece DN 100/100	1.16	0.14	0.4	PE	360
14	0.25	0.22	0.54	0.86	T-piece DN 200/100	3.35	0.15	1.25	Cast iron	170

$0.026\ \text{g m}^{-1}$ (Delanoue and McMath, 1997). The conclusion in this study was that neither flushing (0.37–$2.14\ \text{m s}^{-1}$) nor daily maximum velocity (0.02–$0.39\ \text{m s}^{-1}$) correlated with the washout mass of the particles. It should be noted that a similar conclusion about flushing velocity (0.65–$2.3\ \text{m s}^{-1}$) and washed out mass of particles has previously been observed (Carriere et al., 2005). Analyses of the impact of daily maximum velocity on washed out mass of particles normally should be done by applying a careful hydraulic analysis using a "bottom-up" demand allocation principle, e.g. SIDMEUM (Blokker et al., 2008), rather than the "top-down" approach used in this study.

In the previous studies, the concentration of particles (or water turbidity) in distribution networks were studied depending on (1) source water quality and treatment process. Unfiltered systems have the greatest mass of solids compared to those applying filtration, with the mass decreasing based on source water and level of treatment (Ryan et al., 2008; Vreeburg et al., 2008); (2) the distance in the network, namely the increase of the particle concentration towards the mains dead ends (Gauthier et al., 2001, Matsui et al., 2007); the amount of deposits does not compulsorily increase at the end of pipe mains. If there are no suspended particles transported to the end of a pipe, which may settle, then there is also deposit formation; (3) the topography which is indirectly confirmed by the number of customer complaints about water quality in undulating terrains compared to those in flat terrains (Polychronopolous et al., 2003); (4) diameter of pipes where increased particle concentration in trunk mains, which in the network acts as sedimentation tanks, has been previously shown (Husband et al., 2010; Saldarriaga et al., 2010). (5) An incontrovertible factor that affects the concentration of particles is the maximum daily velocity, which can be related to particles by applying "bottom-up" demand modeling, has been described earlier (Blokker et al., 2008). The same opinion is postulated in the cohesive transport modeling approach called PODDS (Boxall et al., 2001; Boxall and Saul, 2005; Husband and Boxall, 2010) which predicts that the particle deposits on pipe walls are conditioned by shear stress.

This conclusion is not fully consistent with the approach postulated in the PODDS since it may indicate that the particles will not accumulate in an annular manner. In this study, the asymmetrical accumulation of particles within the fittings was found, which is consistent with the scientific studies performed in other areas of research (Fokker et al., 2003). The flow structure and, specifically, vortex pairs formed downstream of the 90° bend have been investigated in both micro-channel systems, e.g. 0.209 mm (Xiong and Chung, 2008), and pipe systems with diameters typical for drinking water mains, e.g. 150 mm (Peters and Ruppel, 2004). It should be noted that in both cases, the velocity was significantly higher than that found in drinking water networks, respectively 0.55–$8.23\ \text{m s}^{-1}$ (Xiong and Chung, 2008) and 8.3–$31.1\ \text{m s}^{-1}$ (Peters and Ruppel, 2004), and in both cases a fully developed vortex formed while circular motion (outer/inner wall vortex after fittings) started to develop already at lower velocities which correspond to the conditions in drinking water systems. Some tracer studies have reported an incomplete mixing at the pipe fittings (e.g. cross and double T-piece) in water flow, and simulations with computational fluid dynamic (CFD) models at velocities (0.35–$1.4\ \text{m s}^{-1}$) have shown flow trajectories and possible vortex zones suitable for accumulation of particles after fittings (Ho et al., 2006; Webb and van Bloemen Waanders, 2006). Guided by the concept of particle rotational

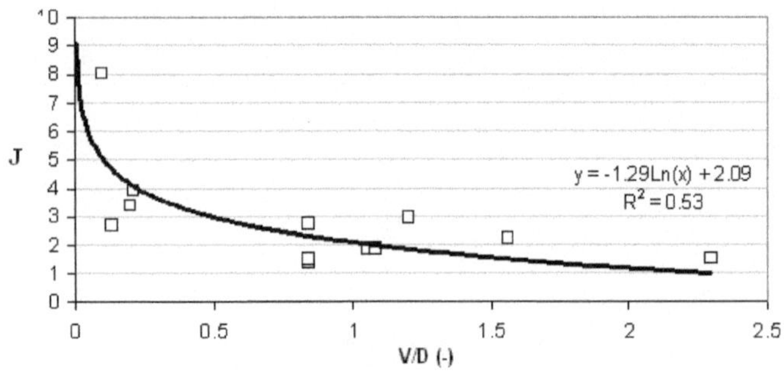

Figure 5. Turbidity potential of accumulated particles (J) in T-piece versus hydraulic conditions.

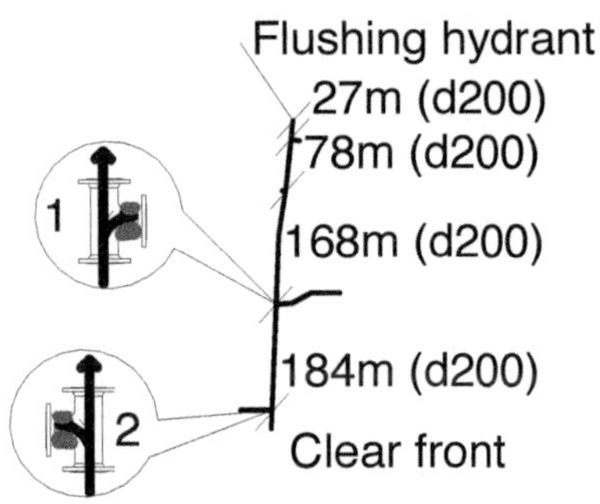

Figure 6. The section of CI pipe layout for site 1 (see Fig. 1). The locations of hydrant and fittings are indicated. The T-piece with reduction on right and left, respectively, to 100 mm is indicated under No. 1 and 2. The diameters of pipe segment in mm is shown in brackets.

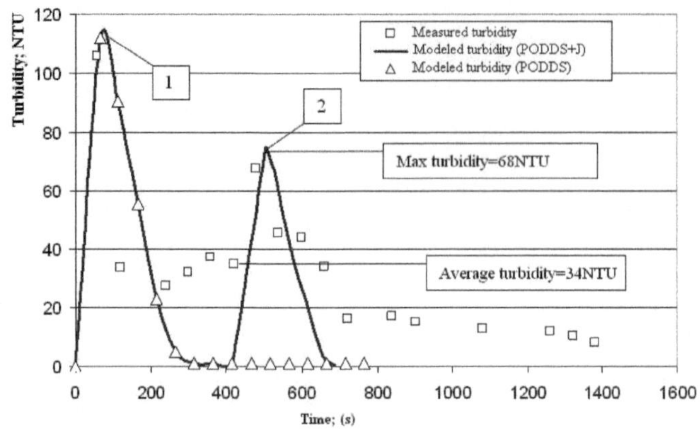

Figure 7. Turbidity curve for flushing site 1.

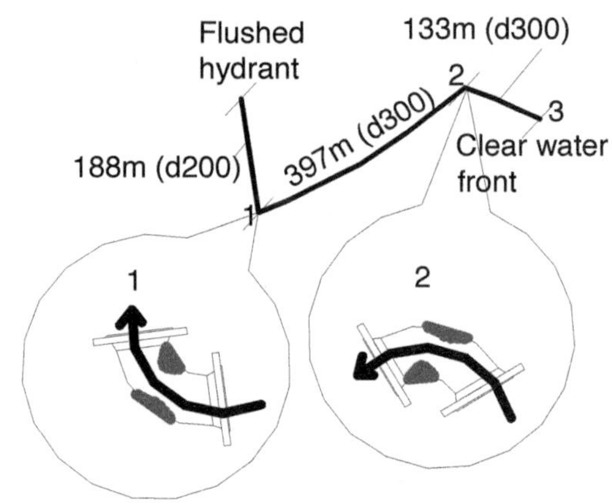

Figure 8. The section of CI pipe layout with two 90° bends. The diameters in mm are shown in brackets.

trajectories in drinking water networks generated by inertia: the Magnus and the Saffman forces (van Thienen et al., 2011b), the development of radial processes in fittings instead of gravitational settling can be expected.

To generalize the observations from field experiments and to compare results of the erosion of accumulated particles in different T-pieces, the relationship between the maximum and average turbidity value from each specific section was calculated (J coefficient), which characterized the potential of the accumulated particles. The hydraulic conditions in specific T-piece were expressed in the relationship V/D (maximal velocity per hour in typical day and diameter of inlet section of T-piece) as non-dimensional parameter.

With the increase of the maximum velocity per hour in the same fitting, the accumulated concentration of particles will decrease (Fig. 5). This relationship obtained from field experiments will supplement the knowledge presented in the PODDS model (Boxall et al., 2001).

V/D reaches "self-cleaning" proportion, which is approximately 0.35 m s^{-1} (Blokker et al., 2010); theoretically, after that J continues in a linear line with a value of 1. The same function can be obtained for other types of fittings.

3.2 Modeling of particle erosion

In order to model particle erosion from field experiments where network sections with T-pieces were flushed, an algorithm was added to the PODDS and the modeling results were compared with the original PODDS parameters. This improved modeling approach allows us to divide the modeled straight pipe sections in segments between the fittings and application of the coefficient for the potential particle accumulation (J) in every step. The first site presented here had a typical straight section of CI transport main with a nominal diameter of 200 mm and two T-pieces for connections (Fig. 6). The location of this site is shown in Fig. 1 and other parameters which describe this site are shown in Table 3 as Case Study No. 1.

Figure 7 shows the temporal turbidity response within flushing, reaching two peaks: 110 and 68 NTU. Measured particle erosion was compared with the simulations of PODDS, which replicates only the rise and exponential decay of first peak of turbidity within the first 200 s of flushing. Introduction of the coefficient J into PODDS and dividing

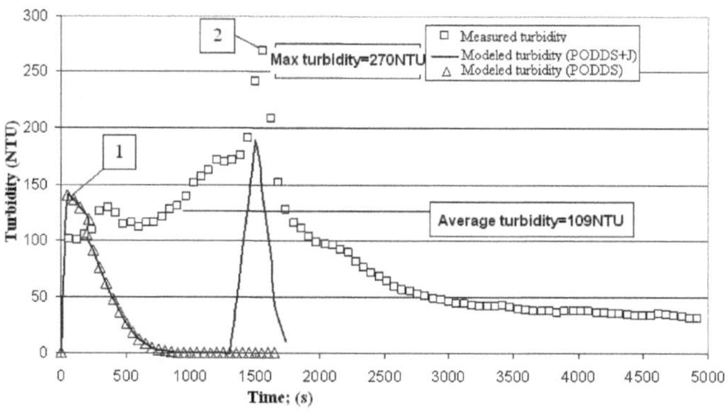

Figure 9. Turbidity curve for flushing site 2 containing two 90° bends.

the simulated section in two segments allowed us to gain two peaks which approximately repeat measured data.

The second flushing site was chosen since it had two 90° bends which are infrequently found in drinking water systems, unlike T-pieces (Fig. 8). The flushed section was 718 m long and consisted of sections of 200 and 300 mm. The velocity obtained during flushing reached 0.57 and 0.87 m s^{-1}, respectively.

The online turbidity results are shown in Fig. 9 which shows two visible turbidity peaks. The simulation results with PODDS and with PODDS supplemented with the coefficient J are also shown here. Both simulation results only partially follow the magnitude of the first peak and display a much steeper decrease in turbidity compared to the measured values. The upgraded model, having coefficient J, is in a better agreement with online data compared to the original PODDS. The repeatability of accumulation of particles in fittings was proven with UDF in one CI pipe section of 1358 m length with 200/300 mm diameter two times within a 5-month period, showing same pattern of turbidity (Rubulis and Neilands, 2010).

Due to the non-disclosure agreement, the authors of the PODDS do not present the values of model parameters (k, b, P and n). Thus, in this study these were estimated using a trial and error approach. Some authors have attempted to replicate the erosion of particles using the PODDS approach and have used the following parameter values $C_{max} = 10$, $k = -0.15$, $b = 1$, $P = 0.0000065$ and $n = 2.5$ (Aisopou et al., 2010) and $C_{max} = 5.0$, $k = -1.5$, $b = -2.5$, $P = 0.0065$, $n = -3.5$ (Naser, 2006). By using the parameter values from Aisopou et al. (2010) and Naser (2006) in our case studies, the modeled results differed significantly. These results indicate that the parameters are site specific and therefore cannot be generalized.

Various aspects need to be improved in prediction and measurement of particle accumulation/erosion: the optical tomography presented by van Thienen et al. (2011b) can be one of the instruments; RPM could be updated based on the fitting quantity in a particular pipe section. Future work in this field should concentrate upon gaining a greater understanding of the development and formation of vortex or centrifugal forces in pipe fittings of drinking water distribution networks.

4 Conclusions

Measurements of turbidity, flow during UDF and analysis of pipe geometry showed significance of accumulation of particles in pipe fittings. Turbidity potential of accumulated particles (J) was determined for T-pieces of several sizes from DN 100/100–DN 250/250 for CI and PE pipes. The existing turbidity model PODDS was updated with an additional term for pipe fittings – (J). The total mass of particles accumulated in T-pieces ranged from 0.05 to 0.29 kg. The study showed that pipe fittings (T-pieces, valves, bends etc.) act as catchment areas for particle accumulation in drinking water networks.

Appendix A

Notation

A_s	pipe surface area (m^2)
C	stored turbidity, calculated from layer shear strength
ΔC_c	changes in turbidity potential (NTU m^2 s^{-1})
D	diameter of pipe (m)
f	a dimensionless coefficient called the Darcy friction factor
g	gravity acceleration (m s^{-2})
ΔH	the head loss due to friction (mH_2O)
J	coefficient of a fitting/turbidity potential of accumulated particles
L	the length of the pipe (m)
M	mass of particles (kg)
ΔN	in/decrease of turbidity (NTU)
Q	flow rate (m^3 s^{-1} or l s^{-1})
r	radius of pipe (m)
R	rate of supply (NTU m^{-2})
t	time (s)
TSS	total suspended solids (mg l^{-1})
V	the average velocity of the fluid flow (m s^{-1})
τ	boundary shear stress (N m^{-2} m^{-1})
ρ	density of water (kg m^{-3})
π	a mathematical constant approx. equal to 3.14

Acknowledgements. This work has been supported by the European Social Fund within the project "Support for the implementation of doctoral studies at Riga Technical University".

This work has been undertaken as a part of the research project "Technology enabled universal access to safe water – TECHNEAU"

(Nr. 018320) which is supported by the European Union within the 6th Framework Programme.

There hereby follows a disclaimer stating that the authors are solely responsible for the work. It does not represent the opinion of the Community and the Community is not responsible for any use that might be made of data appearing herein.

The authors would like to acknowledge the help of Talis Juhna for his guidance throughout the method development for the measurement of turbidity in field. The authors are thankful to Sandis Dejus for assistance in calibration of the hydraulic model and the director of Adazi udens Ltd. Janis Neilands for his advice and guidance during the field campaign. Special thanks must be expressed to Edgars Grundbergs and Janis Neilands Jr. for their assistance in the flushing and tracer campaigns. Special thanks are due to Simona Larsson for reviewing the English of this paper.

References

Aisopou, A., Stoianov, I., and Graham, N.: Modelling Discoloration in WDS Caused by Hydraulic Transient Events, in: Proceedings of the Water Distribution System Analysis Conference, Tucson, Arizona, ASCE Conf. Proc., doi:10.1061/41203(425)49, 2010.

Blokker, E., Vreeburg, J., Schaap, P., and van Dijk, J.: The self-cleaning velocity in practice, in: Proceedings of the Water Distribution System Analysis Conference, Tucson, Arizona, ASCE Conf. Proc., doi:10.1061/41203(425)19, 2010.

Blokker, E. J. M., Vreeburg, J. H. G., Buchberger, S. G., and van Dijk, J. C.: Importance of demand modelling in network water quality models: a review, Drink. Water Eng. Sci., 1, 27–38, doi:10.5194/dwes-1-27-2008, 2008.

Boxall, J. B. and Saul, A. J.: Modeling discoloration in potable water distribution systems, J. Environ. Eng., 131, 716–725, 2005.

Boxall, J. B., Skipworth, P. J., and Saul, A. J.: A novel approach to describing sediment movement in distribution mains, based on measured particle characteristics, Water Softw. Syst., 1, 263–273, 2001.

Boxall, J. B., Saul, A. J., Gunstead, J. D., and Dewis, N.: Regeneration of discolouration in distribution systems, World Water and Environmental Resources Congress, Philadelphia, ASCE Conf. Proc., 434–442, doi:10.1061/40685(2003)132, 2003.

Carriere, A., Gauthier, V., Desjardins, R., and Barbeau, B.: Evaluation of Loose Deposits in Distribution Systems through Unidirectional Flushing, J. AWWA, 97, 82–92, 2005.

Delanoue, A. and McMath, S. v. M.: Effect of Pipe Materials in Biofilm Growth and Deposition Formation in Water Distribution Systems, Proc. Of the Water Quality Technology Conference, Denver, Colo., AWWA, 1997.

Fokker, S., Kingman, S., Lowandes, I., and Reynolds, A.: Characterisation of the cross sectional particles concentration distribution in horizontal dilute flow conveying – a review, Chem. Eng. Process., 43, 677–691, 2003.

Gauthier, V., Gerard, B., Portal, J.-M., Block, J.-C., and Gatel, D.: Organic matter as loose deposits in a drinking water distribution system, Water Res., 33, 1014–1026, 1999.

Gauthier, V., Barbeau, B., Millette, R., Block, J.-C., and Prévost, M.: Suspended particles in the drinking water of two distribution systems, Wa. Sci. Technol., 1, 237–245, 2001.

Ho, C. K., Orear Jr., L., Wright, J. L., and McKenna, S. A.: Contaminant Mixing at Pipe Joints: Comparison between Laboratory Flow Experiments and Computational Fluid Dynamics Models, in: Proceedings of the Water Distribution System Analysis Symposium, Cincinnati, Ohio, ASCE Conf. Proc., doi:10.1061/40941(247)153, 2006.

Husband, S. and Boxall, J. B.: Field studies of discoloration in water distribution systems: model verification and practical implications, J. Environ. Eng., 36, 86–94, 2010.

Husband, S., Williams, R., and Boxall, J. B.: Discolouration risk management for trunk mains, in: Proceedings of the Water Distribution System Analysis Conference, Tucson, Arizona, ASCE Conf. Proc., doi:10.1061/41203(425)50, 2010.

Husband, P. S., Boxall, J. B., and Saul, A. J.: Laboratory studies investigating the processes leading to discolouration in water distribution networks, Water Res., 42, 4309–4318, 2008.

Jonas, S. and Murphy, B.: Distribution System Flushing Optimization to Reduce Environment Impact, Proc. Of the Water Quality Technology Conference, Seattle, AWWA, 2002.

Matsui, Y., Yamagishi, T., Terada, Y., Matsushita, T., and Inoue, T.: Suspended particles and their characteristics in water mains: development of sampling methods, J. Water Supply Res. T., 56, 13–24, 2007.

Naser, G.: Water quality transformations in pipelines: a two-dimensional multi-component simulation model, Academic dissertation, University of Toronto, 2006.

Neilands, K., Rubulis, J., and Juhna, T.: Application of Online Monitoring, Resuspension Potential Method and Unidirectional Flushing for Control of Drinking Water Quality in a Distribution Networks, Latvia, in: Proceedings 1st Eastern European Regional Young Water Professionals Conference, Minsk, Belarus, ISBN 987-985-525-145-4, 2009.

Peters, F. and Ruppel, C.: A pressure probe for the detection of the flow direction close to walls, Case study: flow through a bend, Exp. Fluids, 36, 813–818, 2004.

Polychronopolous, M., Dudley, K., Ryan, G., and Hearn, J.: Investigation of factors contributing to dirty water events in reticulation systems and evaluation of flushing methods to remove deposited particles., Wa. Sci. Technol, 3, 295–306, 2003.

Rossman, L. A.: EPANET 2 Users Manual, EPA/600/R-00/057, US Environmental Protection Agency, Cincinnati, OH, 2000.

Rubulis, J. and Neilands, K.: Interpretation of loose deposits motion in drinking water distribution network, in: Proceedings of the Water Distribution System Analysis Conference, Tucson, Arizona, ASCE Conf. Proc., doi:10.1061/41203(425)61, 2010.

Ryan, G., Mathes, M., Haylock, G., Jayaratne, A., Wu, J., Noui-Mehidi, N., Grainger, C., and Nguyen, B. V.: Particles in Water Distribution Systems, Tech. Rep. 33, Cooperative Research Centre for Water Quality and Treatment, 2008.

Saldarriaga, J., Nieto, L., Hernández, F., and Osorio, C.: Discolored Water Analysis and Modeling for Trunk Networks: The Case of Bogotá, Colombia, in: Proceedings of the Water Distribution System Analysis Conference, Tucson, Arizona, ASCE Conf. Proc., doi:10.1061/41203(425)48, 2010.

van Thienen, P., Vreeburg, J. H., and Blokker, E. J.: Radial transport processes as a precursor to particle deposition in drinking water

distribution systems, Water Res., 45, 1807–1817, 2011a.
van Thienen, P., Floris, R., and Meijering, S.: Application of optical tomography in the study of discolouration in drinking water distribution systems, Drink. Water Eng. Sci., 4, 61–69, doi:10.5194/dwes-4-61-2011, 2011b.
Vreeburg, J. H. G.: Discolouration in drinking water systems: a particular approach, Academic dissertation, Delft University of Technology, Delft, p. 183, 2007.
Vreeburg, J. H. G., Schaap, P., and van Dijk, J. C.: Measuring Discoloration Risk: Resuspention Potential Method, IWA Leading Edge Conference, Prague, 2004.
Vreeburg, J. H. G., Schippers, D., Verberk, J. Q. J. C., and van Dijk, J. C.: Impact of particles on sediment accumulation in a drinking water distribution system, Water Res., 42, 4233–4242, 2008.
Verberk, J. Q. J. C., Hamilton, L. A., O'Halloran, K. J., van der Horst, W., and Vreeburg, J. H. G.: Analysis of particle numbers, size and composition in drinking water transportation pipelines: results of online measurements, Wa. Sci. Technol., 6, 35–43, 2006.
Webb, S. W. and van Bloemen Waaders, B. G.: High fidelity computational fluid dynamics for mixing in water distribution systems, in: Proceedings of the Water Distribution System Analysis Symposium, Cincinnati, ASCE Conf. Proc., doi:10.1061/40941(247)154, 2006.
Xiong, R. and Chung, J.: Effects of miter bend on pressure drop and flow structure in micro-fluidic channels, Int. J. Heat Mass Tran., 51, 2914–2924, 2008.

17

WaterMet^2: a tool for integrated analysis of sustainability-based performance of urban water systems

K. Behzadian[1,2], Z. Kapelan[1], G. Venkatesh[3], H. Brattebø[4], and S. Sægrov[3]

[1]Centre for Water Systems, College of Engineering, Mathematics and Physical Sciences, University of Exeter, Exeter, UK

[2]Environmental Research Centre, Amirkabir University of Technology, Tehran, Iran

[3]Department of Hydraulic and Environmental Engineering, Norwegian University of Science and Technology, Trondheim, Norway

[4]Department of Energy and Process Engineering, Norwegian University of Science and Technology, Trondheim, Norway

Correspondence to: K. Behzadian (k.behzadian-moghadam@exeter.ac.uk)

Abstract. This paper presents the "WaterMet^2" model for long-term assessment of urban water system (UWS) performance which will be used for strategic planning of the integrated UWS. WaterMet^2 quantifies the principal water-related flows and other metabolism-based fluxes in the UWS such as materials, chemicals, energy and greenhouse gas emissions. The suggested model is demonstrated through sustainability-based assessment of an integrated real-life UWS for a daily time-step over a 30-year planning horizon. The integrated UWS modelled by WaterMet^2 includes both water supply and wastewater systems. Given a rapid population growth, WaterMet^2 calculates six quantitative sustainability-based indicators of the UWS. The result of the water supply reliability (94 %) shows the need for appropriate intervention options over the planning horizon. Five intervention strategies are analysed in WaterMet^2 and their quantified performance is compared with respect to the criteria. Multi-criteria decision analysis is then used to rank the intervention strategies based on different weights from the involved stakeholders' perspectives. The results demonstrate that the best and robust strategies are those which improve the performance of both water supply and wastewater systems.

1 Introduction

One of the conventional approaches to model an urban water system (UWS) is to use a physically-based model to simulate the hydraulic behaviour of the UWS, and to identify water-quality characteristics. However, physically-based models are typically sophisticated and very detailed models which need a lot of input data. This is demanding and tedious for many case studies. Additionally, these models usually can only simulate a part of the UWS. In contrast, conceptually-based models with the ability of quantifying flow paths and contaminant loads in an UWS enable understanding of the impacts of the interaction of water within an integrated urban water management systems where potable water, storm water and wastewater need to be considered together.

Some of the instances of these conceptually-based models that have been developed in the past are AQUACYCLE (Mitchell et al., 2001), UWOT (Makropoulos et al., 2008), UVQ (Mitchell and Diaper, 2010) and CWB (Mackay and Last, 2010). These models aim to simulate the integrated water system within an urban area and estimate the contaminant loads and the volume of the water flows throughout the UWS, from source to sink. Such a simulation enables the planners to explore a wide range of conventional and emerging techniques in water supply, storm water and wastewater services to an UWS.

Despite a plethora of studies for modelling integrated urban water systems, they generally start with potable water from the points of water consumption. Therefore, potable water is modelled as an external supply and its demand is calculated as the sum of the neighbourhood demands not met by local or decentralised supply schemes (Mackay and Last, 2010). The present work strives to extend the modelling of potable water to water resources, and integrates it with other components in the water supply, sewerage and drainage subsystems. This is handled by means of a simplified and integrated approach for modelling water distribution and wastewater systems. Then, the physical metabolism of this integrated UWS is evaluated through some key performance indicators covering all sustainability-related issues (environmental, economic and social). All this, in turn, will enable the planners to assess the impact of a combination of future intervention strategies including technologies and their operation on different parts of the UWS.

Furthermore, the focus of all of the previously-developed models is mainly based on the quantification of water-related flows and their final destinations in different parts of the UWS. However, the key performance indicators employed in this paper aim to quantify both water flows and other main fluxes of sustainability-related issues such as all types of direct and indirect (embodied) energy, material flows and greenhouse gas (GHG) emissions resulted from the activities in different elements of the UWS.

Further details of the developed model are briefly outlined below. The case study is elaborated upon subsequently. Results and discussions pertaining to the case study are presented thereafter. The paper closes with the conclusions and recommendations for further research. Note that this paper is an extension of the one presented at the 12th edition of the International Conference on "Computing and Control for the Water Industry – CCWI2013" (Behzadian et al., 2014b).

2 WaterMet^2 methodology

WaterMet^2 is a conceptual, simulation-type, mass-balance-based, integrated UWS model which quantifies the metabolism-related performance of a generic UWS with focus on sustainability related issues (Behzadian et al., 2014a). WaterMet^2 is also a standalone piece of software which runs in a Windows™ screen with the capability of navigational devices to build a new UWS model. It defines mains flows and storages of UWS through four main sub-systems including water supply, water demand, wastewater and cyclic water recovery (Behzadian and Kapelan, 2012).

WaterMet^2 recognises the entire urban water system as four spatial scales including indoor, local, subcatchment and city areas. The indoor areas represent the smallest spatial scale, and include single properties (e.g. residential, industrial, commercial, public, etc.) without any surroundings (e.g. gardens or public open outdoor spaces). Water demand profiles at this scale can be defined based on either the daily average of per-capita water demand or detailed information of the water consumption in the single properties. For the latter case, water demand can be further split into six types of appliances and fittings as (1) hand basin, (2) bath and shower, (3) kitchen sink, (4) dish-washer, (5) washing machine and (6) toilet. The Local area scale represents a group of similar typical households/properties with a surrounding area. It can contain any number of indoor areas (i.e. properties) but they all must be of the same type, i.e. with identical per-capita water demand. The surrounding area is divided into pervious and impervious surfaces. The main tasks of WaterMet^2 Local area are to handle outdoor water demands, rainfall-runoff modelling and on-site water treatment options. The subcatchment area scale represents a group of neighbouring local areas. The main tasks of WaterMet^2 Subcatchment area is to represent as (1) "water consumption points" of potable water in a simplified water supply system outlined below and (2) "collection points" of a storm water runoff and sanitary sewage in a simplified sewer system outlined below. A daily time-step and a user-defined period of N years are adopted as the time interval and duration of the model simulation in WaterMet^2. All this, in turn, will enable WaterMet^2 to model a wide range of elements in UWS from residential appliances, fittings and water recycling schemes to simplified water supply and sewer systems. Details of specifications and functional processes modelled at each scale can be found at the relevant references such as Behzadian et al. (2012).

The WaterMet^2 model adopts a simplified approach for the water supply subsystem in which "source to tap" modelling is performed. The elements modelled in the water supply subsystem which are shown in Fig. 1 are:

1. three key "storage" components including raw water resources, water treatment works (WTWs) and service reservoirs;
2. three principal flow "routes" including water supply conduits, trunk mains and distribution mains;
3. Subcatchments as water consumption points.

The simulation of the water supply subsystem is carried out in two steps in which water demand is calculated upwards in the first step and water is allocated downwards in the second step. More specifically, the first step deals with the calculation of the daily water demand starting from the most downstream points (i.e. subcatchments) and aggregating in the upstream direction until it reaches most upstream points (i.e. water resources). Through this step, the calculated water demand in the components may be limited by their capacity. The second step distributes water flow in the downstream direction starting from water resources. At the most upstream point, the water release (abstraction) from each water resource is supplied provided there is enough storage in

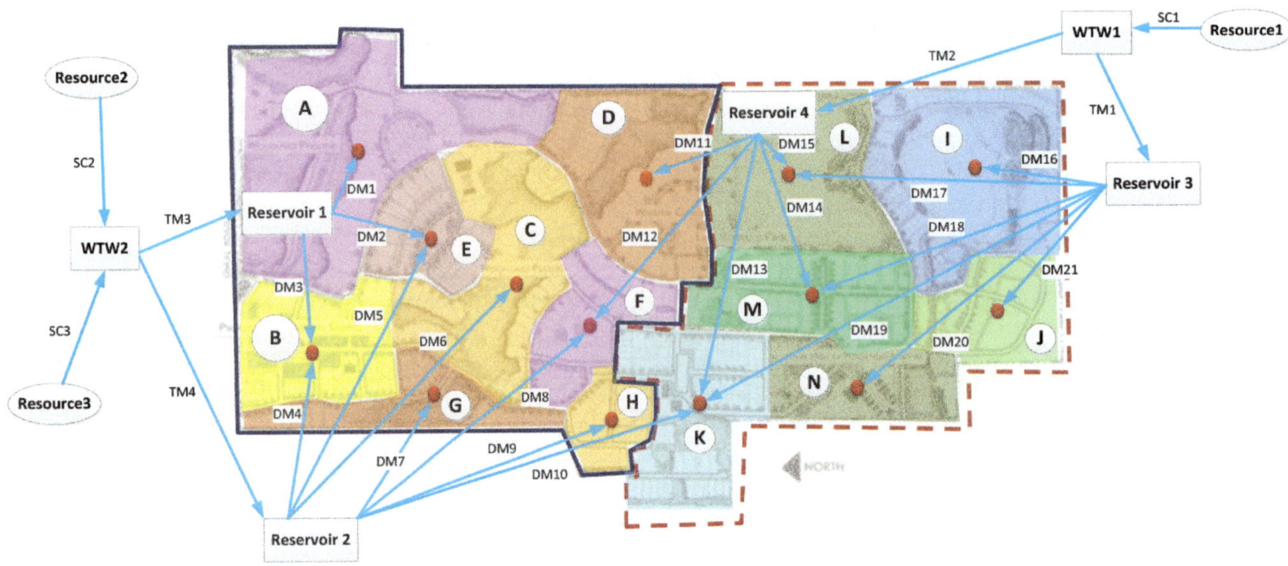

Figure 1. A schematic example of a water supply system representation in WaterMet^2; SC = water supply conduit; TM = trunk main; DM = distribution main; WTW = water treatment works.

the water resource. The water flow is allocated from multiple upstream components to multiple downstream components based on predefined allocation coefficients. The released/allocated water is transferred to the downstream elements subsequently until it reaches the Subcatchments in which water is distributed between the water demand points in the Local area(s).

Further, wastewater and storm water generated in the local areas of the Subcatchment are aggregated and represented as wastewater/storm water of the Subcatchment and start point in the simplified wastewater system represented in Fig. 2. This system comprises three key "storages":

1. Separate/combined sewer system interconnecting between Subcatchments themselves or between Subcatchment and wastewater treatment works (WWTWs);
2. WWTWs;
3. Receiving waters (only as "sink" points).

Storm water/wastewater exceeding the daily transmission/storage capacity of sewer systems overflow through combined sewer overflow (CSO) and storm tank overflow (STO) structures into receiving waters. More details of these simplified systems can be found in Behzadian et al. (2014a).

Each of the integrated UWS components used in WaterMet^2 (e.g. WTWs, trunk mains, etc.) is predefined in terms of its specific characteristics (e.g. WTWs capacity, trunk main flow capacity and specific energy used for conveying water, etc.). When a new WaterMet^2 model is being built, the required UWS components are selected and interlinked in a suitable configuration which represents the analysed UWS in the best way. A model simulation run is then performed by calculating the water flows in the UWS for a pre-specified duration and driven by the pre-specified system load (water demand and rainfall). This is done in a simplified way outlined above, based on the principles of mass balance (but by respecting relevant capacities and other characteristics of the modelled UWS components). As a result of this model simulation, principal flows/fluxes (e.g. energy, GHG emissions, chemicals and pollutants) are derived from the previously-estimated water flows by using suitable component characteristics (e.g. specific energy consumption for pumping). The detailed WaterMet^2 model outputs are then simply aggregated (spatially and/or temporally) to estimate the quantitative evaluation criteria values.

Furthermore, since WaterMet^2 is a conceptual, mass-balance-based model, it simply cannot model variations in pressure heads (such as hydraulic and physically-based models) and hence cannot calculate the results of such hydraulic simulations (such as pressure head or leakage variations). As a compromise, WaterMet^2 assumes the total leakage of the flow routes can be expressed as a percentage of water demand. This assumption can be considered as a reasonable approximation for the long-term, strategic level assessment of the UWS performance and was made by other similar models such as UVQ (Mitchell and Diaper, 2010), UWOT (Makropoulos et al., 2008) and CWB (Mackay and Last, 2010).

3 Case study

3.1 Introduction

The urban water system of the city of Oslo in Norway is used here as a reference city combined with assumptions when

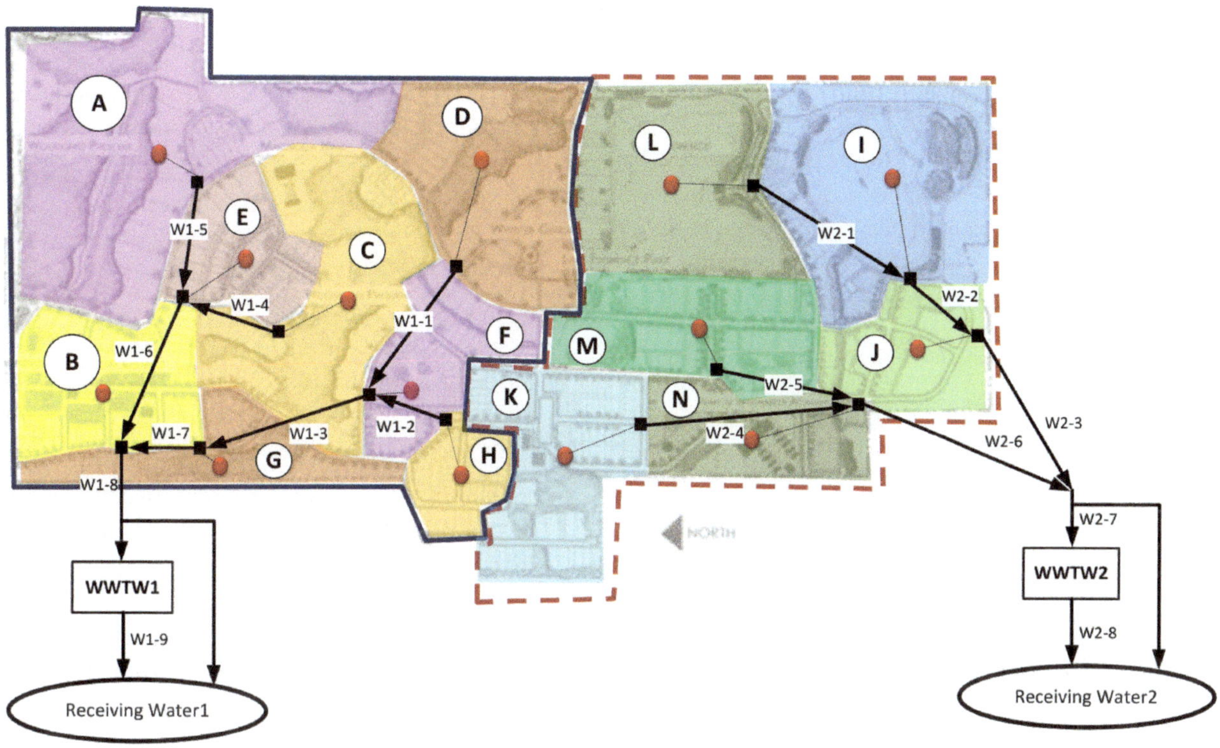

Figure 2. A schematic example of a sewer system representation in WaterMet^2; W1- & W2- = sewer system 1 and 2 respectively; WWTW = wastewater treatment works.

necessary. The existing Oslo UWS contains two main raw water resources each supplying to one of the two WTWs. The split between these raw water resources as far as freshwater supply is concerned is in the ratio 9 : 1 (VAV, 2011a). The UWS has a blend of combined and separate sewer systems and two WWTWs collecting 63 and 27 per cent of wastewater from the wastewater flow (VAV, 2006). The two water resources of the UWS are of limited capacity (60 and 13.8 million cubic metres – MCM) and inflow (287 and 12 MCM year^{-1}). The daily time series of the last 30 years inflows (1981–2010) into these water resources are selected and assumed to be the time series of inflow over the 30-year planning horizon considered in this paper.

The WaterMet^2 model is demonstrated here through sustainability-based assessment of the integrated UWS over a 30 year planning horizon. The integrated UWS modelled by WaterMet^2 includes both the water supply and the wastewater systems as described above. WaterMet^2 quantifies the sustainability-based performance of both existing UWS and new intervention strategies which will be described in the following.

3.2 Building a (replicable) UWS WaterMet^2 model

UWS of the reference city (i.e. Oslo) is modelled using a single WaterMet^2 Subcatchment with a single Local area. The total number of properties in the city is 320 000 with a population of 610 000 in 2011. The highest rate of water demand as a consequence of the highest population growth projection is assumed for future water consumption in the UWS. Water demand of the local area is split into domestic, industrial, irrigation, frost tapping and unregistered public use (VAV, 2011b). Domestic (indoor) water demand per capita is assumed to be 180 L day^{-1}. The existing leakage from the pipelines is assumed to be 22 % of total water demand. The Oslo WaterMet^2 model was calibrated for the existing flow conditions in the UWS. The calibration was carried out using historical daily measurements of water production at WTWs and wastewater treated at WWTWs. The UWS model is first calibrated for water supply systems and then for wastewater systems using a manual, trial and error approach. For the first part, two years of recorded daily water production at WTWs split into two periods using 2011 for calibration and 2012 for validation. The calibration parameters for the WSS model include (1) monthly coefficients of water demand profiles; (2) percentage contribution of daily temperature in daily variation of water demand profiles. The wastewater system model is subsequently calibrated for two years (2010–2011) of recorded daily wastewater inflows to the WWTWs, again split into two one-year periods for calibration and validation. The relevant calibration parameters are hydrologic parameters of the Subcatchment and the principal hydraulic features of the WWTWs and sewer system. Further details of the model and its calibration can be found in Behzadian and Kapelan (2012) and Behzadian et al. (2014a).

3.3 Intervention strategies

To improve the performance of the UWS – especially an increase in the reliability of water supply (will be shown in the results), some intervention strategies are suggested to be analysed using WaterMet^2 to find out how sustainability-based indicators are affected over the planning horizon. The following five intervention strategies are defined and evaluated for improvement of the performance of the UWS over a 30-year planning horizon (2011–2040):

Strategy#1: business-as-usual (BAU) strategy: as a benchmark strategy resembling "do nothing" over the planning horizon;

Strategy#2: addition of a new water resource: one new water resource and relevant WTWs are added from year-2020 to the UWS (it refers to option A2 out of the four options in the relevant report at VAV (2011a), Behzadian and Kapelan (2012).

Strategy#3: 1 % increase in annual pipeline rehabilitation rate: current annual rate of pipeline rehabilitation (i.e. 1 % of the total length of water supply pipelines) will be increased by 1 % from 2015 and the new rate will be 2 % over the rest of the planning horizon. Note that it is assumed that the current rate would cause the leakage percentage to remain constant but additional rate would proportionally decrease the leakage percentage (Venkatesh, 2012).

Strategy#4: 0.5 % increase in annual pipeline rehabilitation rate plus 10 % additional annual water meter installation: the new rate will be 1.5 % and water metering coverage of customers will annually increase 10 % of total domestic customers, both from 2015. Note that it is assumed that installing a new water meter would decrease a constant rate of 10 % for the water demand per capita (VAV, 2011b).

Strategy#5: addition of RWH and GWR systems at local level: single rainwater harvesting (RWH) and grey water recycling (GWR) systems representing all many small water treatment units across the city assuming that they are adopted by 50 % of households are added from 2015. It is assumed a tank capacity of 0.48 MCM and 39 000 m^3 for the represented RWH and GWR system, respectively (Ward et al., 2012a, b; Memon et al., 2005) which both provide water demands of toilet flushing, irrigation and industrial usages. It is assumed that RWH system collects runoff from roofs, roads and pavements and GWR system collects grey water from hand basin, shower, frost tapping. Then it is assumed that both RWH and GWR systems supply water demands for toilet flushing, irrigation and industrial usages. The electricity consumption of RWH and GWR systems is assumed to be 0.54 and 1.84 kWh m^{-3} respectively (Ward et al., 2012a; Memon et al., 2005).

3.4 Evaluation criteria

The aforementioned intervention strategies are compared by using the following multiple evaluation criteria covering different dimensions of UWS sustainability (Alegre et al., 2012):

1. *Total capital costs:* the capital investments of the intervention options are discounted in year 2011 with 3 % discount rate.
2. *Total O&M costs:* both fixed (e.g. salary) and variable (e.g. electricity per cubic metre used) costs related to different components of the UWS over the planning horizon are discounted in year 2011 with 3 % discount rate.
3. *Reliability of water supply:* ratio between total water delivered to customers and total water demand over the planning horizon.
4. *Annual average of water leakage:* leakage volume in water distribution systems over the planning horizon relative to annual pipeline rehabilitation rate.
5. *Annual average of GHG emissions:* both types of direct GHG resulted from electricity and fossil fuel; and indirect GHG resulted from embodied energy over the planning horizon.
6. *Annual average of CSOs volume:* overflows from CSO structures from both combined sewer system and WWTWs over the planning horizon.
7. *Social acceptance:* as a qualitative sustainability indicator, it examines the extent of support that an intervention strategy receives from the society in order to fulfil the requirements of water services. In other words, this criterion reflects how much water users are willing to accept a strategy. This typically depends on a number of factors such as water quality and interruption to supply issues. This indicator is rated by expert opinion between 1 and 10 with 1 being the least acceptance and 10 the highest acceptance rate.

Note that the first six criteria values are calculated using the WaterMet^2 model whilst the last criterion value is estimated using the expert judgement. The quantitative criteria are selected such that the performance of all UWS components is included. More specifically, criteria 3 and 4 are related to the performance of the water supply system components while criterion 6 is dependent on the performance of the wastewater system components. The other quantitative criteria (i.e. 1, 2, 5) are dependent on the performance of all UWS components. Also, note that the aforementioned criteria values are likely to change when UWS system modifications are introduced by means of different interventions analysed (as these modify the UWS component characteristics).

3.5 Comparison of intervention strategies

To demonstrate the WaterMet^2 model capabilities for strategic planning of the UWS, the intervention strategies will be compared with respect to either a single criterion or multiple criteria. By single criterion comparisons with respect to each of the evaluation criteria separately, advantages and shortcomings associated with the application of the intervention strategies are envisaged. The multiple criteria comparison performed here by a Compromise Programming (CP) multi-criteria decision analysis (MCDA) technique provides a ranking for the intervention strategies. The CP method originally proposed by Zeleny (1973) calculates a distance function for each strategy that is nearest with respect to an "ideal" point for which all the criteria are optimized (André and Romero, 2008).

When ranking the strategies with respect to multiple criteria, weights can be assigned to each of the criteria to indicate the relative importance of those criteria. As these weights may be a key factor for some stakeholders, and may play a significant role in the final ranking, four various perspectives each representing a viewpoint of specific involved stakeholders in the UWS, are used to specify the weights. Then, the obtained rankings from various perspectives are finally combined to specify a single and final ranking. Four main stakeholders' perspectives are analysed here as: (1) equal weight (no biased view on criteria); (2) environmentalist; (3) water company; and (4) public. The weights of the evaluation criteria associated with these perspectives are given in Table 1.

4 Results and discussion

4.1 UWS performance using WaterMet^2

The business-as-usual (BAU) strategy is first analysed in WaterMet^2 for a daily time-step over a 30-year planning horizon. Figure 3 shows the WaterMet^2 simulation for the monthly water demand and percentage of the delivered water demand in the city for 'do nothing' strategy over the planning horizon. As it can be seen, the UWS in this strategy is unable to fully supply the increasing potable water demand due to the population growth. The monthly water deficit starting slightly in the beginning years will expand rapidly over the following years with a great magnitude (less than 75 % of monthly water demand delivered in the last years as shown in Fig. 3). Thus, the water supply reliability calculated by WaterMet^2 is 94 % for the BAU (Table 2).

Figure 4 shows the contribution of three main components towards annual average of per-capita GHG emission for the BAU. WWTWs contributes the most to the total per-capita GHG emissions (91 kg CO_2-eq) owing to the considerable share of embodied energy in the chemicals used for wastewater treatment in WWTWs. This implies that intervention options aimed at decreasing the inflow of wastewater to WWTWs can be highly effective in decreasing GHG

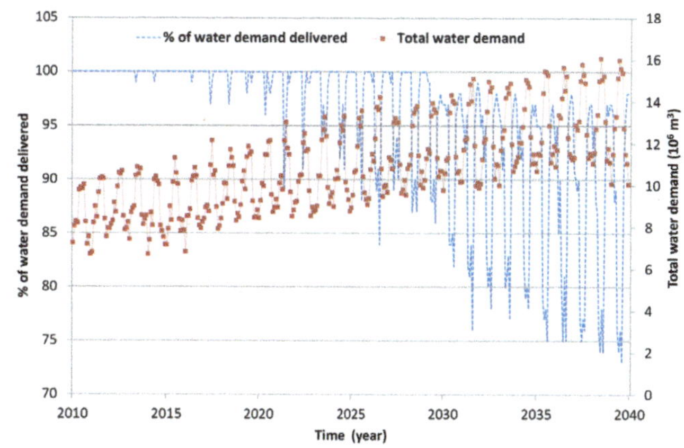

Figure 3. Water demand projection and monthly percentage of water demand delivered over the planning horizon for the BAU.

emissions. In addition, although fossil fuel is basically categorised as a source of high GHG emissions per unit volume consumption, its share here as shown in Fig. 4 is the least among all components of the UWS because of relatively negligible consumption in the UWS components (e.g. 0.002 and 0.004 L m^{-3} in WTWs and distribution systems, respectively).

The evaluation of the five intervention strategies calculated by the WaterMet^2 model is shown in Table 2. Each new intervention strategy relative to the BAU can gain some noticeable enhancement with respect to each of the criteria. In particular, the water supply reliability increases from 94 % in the BAU to at least 96 % in Strategy 3 and 100 % when new water sources are added (Strategy 2). However, an increase in reliability of water supply is achieved at a cost – huge capital investment required for building new water resources and associated WTWs in Strategy 2 or less expensive strategies (#3–5).

Strategy 5 has the highest O&M cost compared to other strategies although this strategy reduces both clean water demand and wastewater generated, and thus is expected to decrease energy cost in all relevant components. This increase can be linked to the fact that the fixed and operational costs of both RWH and GWR systems defined in Strategy 5 are far more than the reduction in O&M expenses in other UWS components. Strategy 2 is the second most-expensive O&M expenditure which is far beyond that of the BAU strategy. This is due to the provision of more clean water in water supply, and the concomitant increase in expenditure on energy in WTWs and distribution systems. New strategies 3 and 4 have a relatively negligible priority compared to the BAU with respect to O&M expenses (Table 2 and Fig. 5). This can be attributed to the fact that the reduction in O&M expenses, owing to less clean water demand (as a result of leakage reduction and water consumption) and less wastewater generation (Strategy 4), is slightly more than the increase in expenses incurred by pipeline rehabilitation.

Table 1. Weights of the criteria from different perspectives.

Criteria	Capital cost	O&M Cost	Reliability	Leakage	GHG emissions	CSOs volume	Social acceptance
Perspective							
Equal weight	1	1	1	1	1	1	1
Public	0.33	0.33	0.67	0.33	0.33	0.33	1
Environmentalist	0.5	0.5	0.5	0.5	1	1	0.5
Water company	0.2	0.2	1	0.2	0.2	0.33	0.67

Table 2. Evaluation of intervention strategies.

Criteria	Capital cost	O&M Cost	Reliability of supply	Leakage	GHG emissions	CSO volume	Social acceptance
Units	Million Euro	Million Euro $year^{-1}$	%	MCM $year^{-1}$	10^3 t $year^{-1}$	MCM $year^{-1}$	–
Objective type	Min	Min	Max	Min	Min	Min	Max
Strategy #1 (business as usual)	0	43.1	94	23	95	306	5
Strategy #2 (additional water source)	389	49.4	100	25	98	311	7
Strategy #3 (1% additional annual rehabilitation)	132	43.0	96	19	96	307	6
Strategy #4 (0.5% additional annual rehabilitation & 10 % additional annual water meter installation)	63	42.9	97	21	93	298	5
Strategy #5 (RWH and GWR systems)	270	51.0	99	19	89	209	3

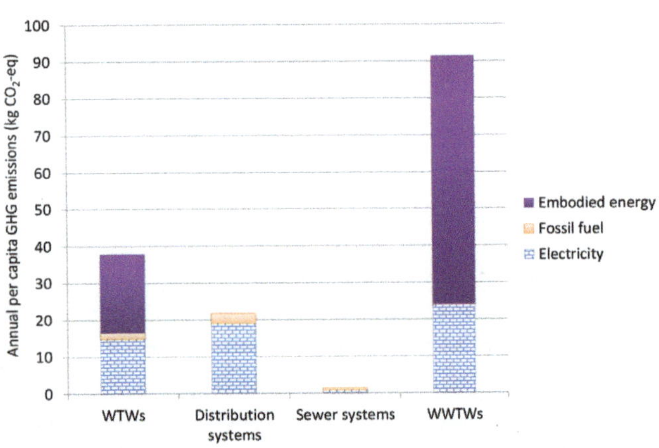

Figure 4. Contribution of three sources of GHG emissions towards annual average of per capita GHG emission in the main UWS components for the BAU.

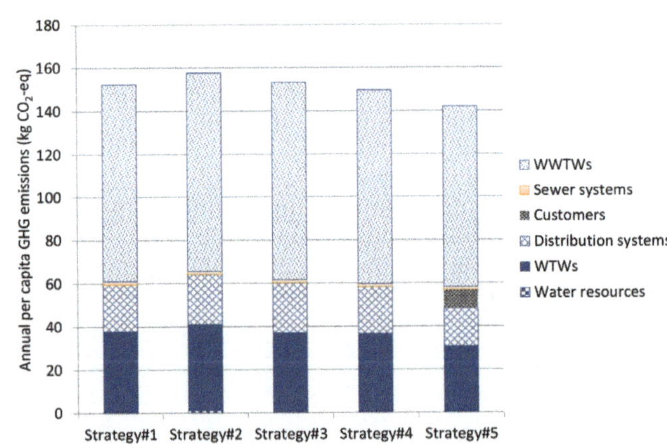

Figure 5. Annual average of per capita GHG emission for intervention strategies.

The performance of the existing UWS with respect to leakage can be improved substantially over the planning horizon by Strategies 3–5 (Table 2). This improvement can be either directly due to additional annual rehabilitation (Strategies 3–4) or indirectly due to GWR and RWH systems (Strategy 5) by reducing potable water consumption. On the other hand, Strategy 2 has more leakage than the BAU since Strategy 2 provides more potable water in the UWS and the leakage as a constant percentage of water supplied increases.

Figure 5 represents the annual average of GHG emissions in the UWS and its components for five intervention strategies. Strategy 5 compared to other strategies generates the minimum amount of annual average of per-capita GHG (142 kg CO_2-eq) although only this strategy generates GHG in the customers component due to the use of Local water recycling schemes. This can also be attributed to fact that this strategy in favour of water recycling schemes cuts down the amount of both potable water demand and wastewater generation. The overall reduction in the annual average of per-capita GHG emissions for this strategy compared to the BAU is 18.8 kg CO_2-eq including 7.18, 4.16 and 7.47 kg CO_2-eq for WTWs, distribution systems and WWTWs respectively. However, this strategy causes GHG emissions of households (end-users) to increase as much as 8.54 kg CO_2-eq per capita. On the other hand, Strategy 2 generates more GHGs than the BAU because of more potable water supplied in this strategy would cause more energy required for both water supply and wastewater sub-systems.. A thorough comparison of the annual averages of per-capita GHG emissions in Strategy 3 with those in the BAU reveals that less potable water demand as a consequence of leakage reduction can cause a negligible reduction of GHG emissions in WTWs (0.71 kg CO_2-eq) while increasing the same by 1.36 and 0.82 kg CO_2-eq for additional pipeline rehabilitation and 2 % increased wastewater generation, respectively. Thus, Strategy 3 generates slightly greater GHG emissions than the BAU (Table 2). However, Strategy 4 amends the intervention option of Strategy 3 by decreasing the magnitude of pipeline rehabilitation plus water demand consumption. Hence, Strategy 4 reduces the annual average of per-capita GHG emissions in WTWs and WWTWs by 1.30 and 1.47 kg CO_2-eq, respectively, while offsetting the emissions in distribution systems. This would lead this Strategy to generate slightly less GHG emissions than the BAU.

The performance of the UWS with respect to the "CSO volume" criterion for different strategies is influenced by generated runoff and sanitary sewage of customers in sewer systems (Table 1). More specifically, Strategy 5 with the lowest level of CSO discharge causes the overflow volume in sewer systems to significantly reduce (by 32 %). This can be attributed to both reduced runoff entering the combined sewer system courtesy runoff-collection by the RWH systems, and reduced wastewater generated owing to the fact that GWR systems reuse grey water for specific indoor and outdoor consumptions. The second lowest CSO discharge which is slightly less than the BAU is related to Strategy 4 as it can only mitigate wastewater generated by reducing water consumption as a result of water meter introduction. Strategies 2 and 3 both cause more wastewater to be generated as a result of providing more clean water supply, and thus the CSO volume increases slightly. It is also noted that although Strategy 3 can cause 17 % reduction in leakage amount and subsequently augment reliability by 2 %, it would increase the wastewater generated and subsequently magnify the CSO volume, thanks to a 2 % increase in potable water supply.

Table 2 also presents a qualitative assessment (social acceptance) of the UWS for intervention strategies quantified by expert judgement. Strategy 2 is the most socially-acceptable, as it can fulfil the requirements of the customers, in a better way. Strategies which include rehabilitation (#3 and #4) follow Strategy 2, as far as social acceptance is concerned, as in both these cases, both the incidence of breaks and the quantity of leakage is likely to be reduced.

4.2 Ranking of intervention strategies

Further analysis is carried out by ranking the intervention strategies based on both single criterion and multiple criteria approaches. Figure 6 shows the single criterion ranking of the strategies with respect to each of the seven evaluation criteria. Whereas Strategy 2 compared to other strategies is ranked number one relative to water supply reliability and "social acceptance", it is ranked the worst (fifth strategy) with respect to four criteria (i.e. capital cost, leakage, GHG emissions and CSO volume). For three criteria (i.e. leakage, GHG emissions and CSO volume) out of the four in which Strategy 2 is ranked the lowest, Strategy 5 is the highest while it is ranked the lowest relative to two criteria (i.e. O&M cost and social acceptance).

Table 3 shows the ranking of the intervention strategies based on multiple criteria. In this Table, the CP method calculates four rankings related to the four different perspectives based on the criteria weights specified in Table 1. It can be seen from Table 3 that the ranking from "environmentalist's" perspective is almost similar to those of "equal weight's" although environmentalist has a bias towards the environmental criteria (i.e. GHG and CSO). This can be due to the high influence of these strategies by the environmental criteria in "equal weighting" perspective. In addition, the close similarity of rankings between "public's" and "water company's" perspectives stem from the fact that their main concern over the UWS is more similar to each other, compared to other perspectives. More specifically, the best and the worst strategies for these two perspectives are Strategies 2 and 1, respectively, while Strategy 2 is ranked the worst from "equal weight's" and "environmentalist's" perspectives.

To incorporate these four rankings obtained from different perspectives into one final ranking, the sum of the ranks for each intervention strategy is used to calculate final ranking which is shown in the right-hand most column of Table 3. The following can be inferred from the comparison of the rankings:

1. The highest ranked strategies are complex ones (#4–5) containing two individual intervention options.

2. The ranks of these two strategies (4 & 5) and Strategy 1 are the most robust ones due to the fact that they are almost in the top and bottom strategies, respectively, from different perspectives.

3. On the other hand, Strategy 2 has the least robustness owing to the major changes of its rank in different rankings. Due to this reason, this strategy is finally ranked relatively low although it is ranked high from two perspectives.

Finally, note that the above analysis and the relevant rankings shown here should be used for illustrative purposes only in order to represent the type of the post-analysis which can done by using the WaterMet^2 model. Further analysis is

Table 3. Ranking based on the CP method.

Criteria / Strategy	Equal weight	Public	Environmentalist	Water company	Sum of rankings	Final rank
Strategy #1 (business as usual)	4	5	4	5	18	5
Strategy #2 (additional water source)	5	1	5	1	12	3
Strategy #3 (1 % additional annual rehabilitation)	3	2	3	4	12	3
Strategy #4 (0.5 % additional annual rehabilitation & 10 % additional annual water meter installation)	1	3	2	3	9	1
Strategy #5 (RWH and GWR systems)	2	4	1	2	9	1

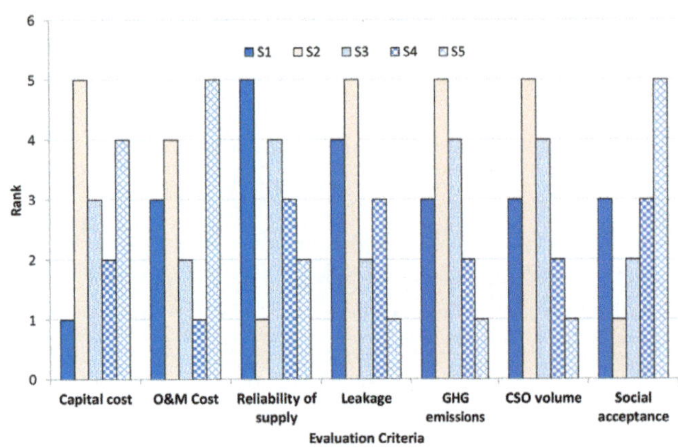

Figure 6. Ranking of the five intervention strategies (S1–S5) with respect to each of the evaluation criteria

needed to consider different scenarios and risk-based criteria in order to achieve a robust solution in the UWS.

5 Summary and conclusions

The new WaterMet^2 model was demonstrated here for strategic planning of the real-life UWS facing a problem of supply/demand balance in the future. The long-term analysis of the existing UWS by WaterMet^2 revealed the potential shortcomings especially water supply reliability. To tackle this problem, WaterMet^2 calculates the performance indicators of the UWS for five potential intervention strategies. They were then compared with respect to each of the six quantitative criteria calculated by WaterMet^2 and one qualitative criterion rated by expert's opinions. For strategic planning of the UWS, the CP method was finally used to rank the intervention strategies analysed by the WaterMet^2 model.

The results obtained demonstrate how an integrated modelling approach such as WaterMet^2 can be used to assist planners in defining the best future intervention strategy. More specifically, WaterMet^2 quantifies both water-related and metabolism-related performance which can define sustainability type indicators such as GHG emissions in UWS components. To overcome the difficulties in UWS, WaterMet^2 enables the planners to track down the detailed impact of the performance of new intervention strategies on the UWS components. WaterMet^2 as an integrated modelling tool in UWS also enables the planners to analyse the long-term impact of intervention options on both water supply and wastewater systems simultaneously. Furthermore, the complex strategies with the aim of improving the performance of both water supply and wastewater systems were ranked the highest. The rankings of these complex strategies and the BAU are reasonably robust from the perspectives of all parties due to the similar criteria prioritization.

Although the results shown here indicate some promising strategies, to obtain a real-life solution, a wide range of different intervention strategies needs to be defined and further tested and evaluated by the WaterMet^2 model for multiple future scenarios and risk type criteria. Note that the analysis conducted and the corresponding results obtained in this paper do not reflect the views of the Oslo VAV and have been used only to demonstrate possible application and functionality of the WaterMet^2 simulation model and software tool.

Acknowledgements. This work was carried out as part of the "TRansition to Urban water Services of Tomorrow" (TRUST) project. The authors wish to acknowledge the European Commission for funding TRUST project in the 7th Framework Programme under Grant Agreement No. 265122. The authors also wish to thank Oslo VAV (project partners) especially Jadranka Milina, Lars Hem and Arnhild Krogh for their collaboration.

Edited by: B. Brunone

References

Alegre, H., Cabrera Jr., E., Hein, A., and Brattebø, H.: Framework for Sustainability Assessment of UWCS and development of a self-assessment tool, TRUST report, 2012.

André, F. J. and Romero, C.: Computing compromise solutions: On the connections between compromise programming and composite programming, Appl. Math. Comput., 195, 1–10, doi:10.1016/j.amc.2007.04.064, 2008.

Behzadian, K. and Kapelan, Z.: Oslo Case Study Report, University of Exeter, Exeter, UK, 2012.

Behzadian, K., Kapelan, Z., Rozos, E., and Makropoulos, C.: WaterMet^2 Conceptual Model, University of Exeter, NTUA, 2012.

Behzadian, K., Kapelan, Z., Venkatesh, G., Brattebø, H., Sægrov, S., Rozos, E., Makropoulos C., Deliverable D33.2: Quantitative UWS performance model: WaterMet^2, TRansition to Urban water Services of Tomorrow (TRUST) project report, 2014a.

Behzadian, K., Kapelan, Z., Venkatesh, G., Brattebø, H., Sægrov, S., Rozos, E., Makropoulos, C., Ugarelli, R., Milina, J., and Hem, L.: Urban water system metabolism assessment using WaterMet^2 model, 12th International Conference on Computing and Control for the Water Industry, CCWI2013, Perugia, Italy, Procedia Engieering, Elsevier, 70, 113–122, 2014b.

Mackay, R. and Last, E.: SWITCH city water balance: a scoping model for integrated urban water management, Rev. Environ. Sci. Biotechnol., 9, 291–296, doi:10.1007/s11157-010-9225-4, 2010.

Makropoulos, C. K., Natsis, K., Liu, S., Mittas, K., and Butler, D.: Decision support for sustainable option selection in integrated urban water management, Environ. Modell. Softw., 23, 1448–1460, 2008.

Memon, F. A., Butler, D., Han, W., Liu, S., Makropoulos, C., Avery, L. M., and Pidou, M.: Economic assessment tool for greywater recycling systems, Proc. ICE-Eng. Sustain., 158, 155–161, 2005.

Mitchell, V. G. and Diaper, C.: UVQ User Manual: (urban water balance and contaminant balance analysis tool), CSIRO, 2010.

Mitchell, V. G., Mein, R. G., and McMahon, T. A.: Modelling the urban water cycle, Environ. Modell. Softw., 16, 615–629, 2001.

VAV: Water Supply Sewerage and Environment Customer Services Finances and Rates Organisation, Oslo Water and Sewerage Works, Oslo, Norway, 2006.

VAV: "Rough Analysis of Alternatives" translated from the original Norwegian version to English by G Venkatesh Oslo Water and Sewerage Works, Oslo, Norway, 2011a.

VAV: "Forecast of Water Usage in the Future in Oslo" translated from the original Norwegian version to English by G Venkatesh Oslo Water and Sewerage Works, Oslo, Norway, 2011b.

Venkatesh, G.: Cost-benefit analysis – leakage reduction by rehabilitating old water pipelines: Case study of Oslo (Norway), Urban Water J., 9, 277–286, doi:10.1080/1573062X.2012.660960, 2012.

Ward, S., Butler, D., and Memon, F. A.: Benchmarking energy consumption and CO_2 emissions from rainwater-harvesting systems: an improved method by proxy, Water Environ. J., 26, 184–190, doi:10.1111/j.1747-6593.2011.00279.x, 2012a.

Ward, S., Memon, F. A., and Butler, D.: Performance of a large building rainwater harvesting system, Water Res., 46, 5127–5134, doi:10.1016/j.watres.2012.06.043, 2012b.

Zeleny, M.: Compromise Programming. In: Multiple Criteria Decision Making, University of South Carolina Press, Columbia, South Carolina, 1973.

18

Shower heat exchanger: reuse of energy from heated drinking water for CO_2 reduction

Z. Deng[1,2], S. Mol[1], and J. P. van der Hoek[1,2]

[1]Waternet, P.O. Box 94370, 1090 GJ Amsterdam, the Netherlands
[2]Delft University of Technology, Department of Water Management, Stevinweg 1, 2628 CN Delft, the Netherlands

Correspondence to: Z. Deng (zdeng2014@outlook.com)

Abstract. The heating of drinking water in households contributes significantly to the emission of greenhouse gases. As a water utility aiming to operate at a climate neutral level by 2020, Waternet needs to reduce its CO_2 emission by 53 kton yr^{-1}. To contribute to this ambition, a pilot project was carried out in Uilenstede, Amstelveen, the Netherlands, to recover the shower heat energy with a shower heat exchanger from Dutch Solar Systems. An experimental setup was built in the Waternet laboratory to evaluate the claimed efficiencies. The energy recovery efficiency observed in the lab was 61–64 % under winter conditions and 57–62 % under summer conditions, while the energy recovery efficiency observed in Uilenstede was 57 % in December 2014. Based on the observations, 4 % of the total energy consumption of households in Amsterdam (electricity and gas) can be recovered with a shower heat exchanger installed in all households in Amsterdam, which also means a 54 kton year^{-1} CO_2 emission reduction can be achieved.

1 Introduction

In the Netherlands, domestic drinking water consumption is 118.9 L per capita per day (Van Thiel, 2014). Drinking water used for showering, bathing, washing dishes by machine, and washing clothes by machine, is heated and contributes to 59 % of domestic drinking water consumption. Drinking water is also warmed by room temperature during non-consumption periods (i.e., stagnant water in pipes inside the building or in toilets). A substantial amount of thermal energy is added to drinking water after the water has been used. According to Hofman et al. (2011), this heated drinking water exits the house at an average temperature of 27 °C, and it contributes to 40 % of the total heat loss of a modern house (through water by wastewater discharge or through the air by ventilation), which is equivalent to 450 kg CO_2 yr^{-1} (van der Hoek, 2012a).

Waternet, the water utility of Amsterdam and its surrounding area, has the ambition to operate climate neutral by 2020 (Van der Hoek, 2012a). This ambition is driven by the policy targets of the City of Amsterdam, which has aspired to be a climate neutral municipal organization since 2015 with respect to municipal services, buildings, and activities. For the whole city of Amsterdam, a 40 % reduction in 2025 and a 75 % reduction in 2040 in greenhouse gas emissions, compared to the 1990 emissions (City of Amsterdam, 2009), should be achieved. For Waternet, a climate neutral operation necessitates a reduction of greenhouse gas emissions of 53 kton CO_2 yr^{-1} (Van der Hoek, 2012b). For the City of Amsterdam, a 75 % reduction in greenhouse gas emissions implies a reduction of 3100 kton CO_2 yr^{-1} (City of Amsterdam, 2009).

The importance of reducing greenhouse gas emissions is even more stressed when the IPCC's (Intergovernmental Panel on Climate Change) Fifth Assessment Report is taken into account (IPCC, 2013). One of the conclusions is that continued emissions of greenhouse gases will cause further warming and changes in all components of the climate system. Limiting climate change will require substantial and sustained reductions of greenhouse gas emissions.

For Waternet, first calculations revealed that a $148\,\mathrm{kton\,yr^{-1}}$ greenhouse gas emission reduction can be reached through energy recovery from the water cycle in and around Amsterdam, of which $72\,\mathrm{kton\,yr^{-1}}$ is the heat from wastewater (van der Hoek, 2012a). Thermal energy recovery from heated drinking water is a promising way to reach climate neutrality by 2020.

Taking into account the three main components for heat recovery (a heat source, a heat exchanger, and a consumption point), suitable conditions have to be found to optimize its feasibility. Compared to the mixed heated water from households, shower water seems more attractive due to the fact that it has a high volume (about $50\,\mathrm{L\,day^{-1}}$) and a high temperature (about 35 °C). Furthermore, the consumption and recovery are simultaneous in time and location; thus, no storage system is required, and no extra losses take place during a short distance of heat delivery.

A shower heat exchanger is specially designed for recovering thermal energy from used shower water. It can be installed under a shower tray to transfer the heat from shower water to cold drinking water. As the drinking water has been preheated, total energy consumption to heat the water can be reduced.

In 2010, Liu et al. (2010) proposed a solar heat pump system to provide hot water for large-scale public shower facilities, from both thermal energy in used shower water and solar energy (Liu et al., 2010). Although this system has not yet been implemented, it has been estimated to be practically applicable. The system costs less than EUR 4000, and it would consume 88.9 % less electricity than the original electric boiler (which has an annual electricity consumption of 500 361 kWh). Moreover, its CO_2 emission would be only $76.9\,\mathrm{ton\,yr^{-1}}$. Another study done by Wong et al. (2010) also showed the high energy saving potential of a shower heat exchanger (Wong et al., 2010). They tested the efficiency of a horizontal shower heat exchanger, and estimated that an annual energy saving of 4–15 % can be achieved in a 40 floor (20 apartments per floor) high-rise residential building.

There are many commercial shower heat exchanger types on the market, but not many studies have been done to validate their recovery efficiency, or estimate their potential in energy saving and CO_2 reduction.

The company Dutch Solar Systems (DSS) claims their shower heat exchanger has an energy recovery efficiency of 47 % (horizontal version) to 62 % (vertical version), based on a given flow rate (Dutch Solar Systems, 2015). This means that about half of the heat in the shower water can be recovered to reduce energy (i.e., electricity, gas) consumption.

To validate the energy recovery efficiency of the DSS shower heat exchanger in practice, a pilot project was constructed in Campus Uilenstede, a housing estate for students in Amstelveen, in September 2014. An experimental setup was built in the Waternet laboratory with the same configuration as in the student apartments (vertical shower heat exchanger), in order to further validate the recovery efficiency of the shower heat exchanger under different conditions.

The energy recovery efficiency was studied regarding four main factors:

- flow rate
- duration of the shower
- time interval between two showers
- shower temperature and incoming water temperature.

The annual energy saving potential, and the economic payback time of the shower heat exchangers were calculated, and compared with the DSS documents and former estimations.

In addition, the contribution of shower heat exchangers to the greenhouse gas emission reduction target of Amsterdam was calculated, assuming that all households will be equipped with a shower heat exchanger.

2 Materials and methods

2.1 Installations and configurations

Considering the cost and efficiency, the DSS shower heat exchangers were chosen and installed in the Uilenstede pilot project, and therefore this specific shower heat exchanger was tested.

2.1.1 Project Uilenstede

In Uilenstede Amstelveen, 100 shower heat exchangers were installed in single-student apartments; 10 apartments were monitored: two reference apartments without a shower heat exchanger, two apartments with the horizontal version, and six with the vertical version. The vertical shower heat exchanger (62 % recovery efficiency) was preferred for the pilot project, but it could not be installed on the ground floor. Therefore, two horizontal exchangers were installed. These rooms were monitored with two flow meters (Kamstrup Multical®62) and two temperature sensors (SIEMENS QAD2012). The locations of these rooms are shown in Fig. 1. The distance between the heater and the thermostatic shower valve is 1.5 m; the distance between the shower drain and the shower heat exchanger is 0.3–0.5 m. The configuration of the setup in each room is illustrated in Fig. 2.

In Fig. 2, the purple line represents the cold water flow, which partially goes to the taps (in bathroom and kitchen) and partially goes to the shower heat exchanger and heater. Its flow and temperature are measured by the two sensors – *F total* and *T cold*. The blue line represents the preheated water, which feeds both the thermostatic valve and the heater. Two sensors, *T preheated* and *F shower*, are measuring its temperature and flow, respectively. The red line represents the water heated by the heater and leads to all hot water-consuming points.

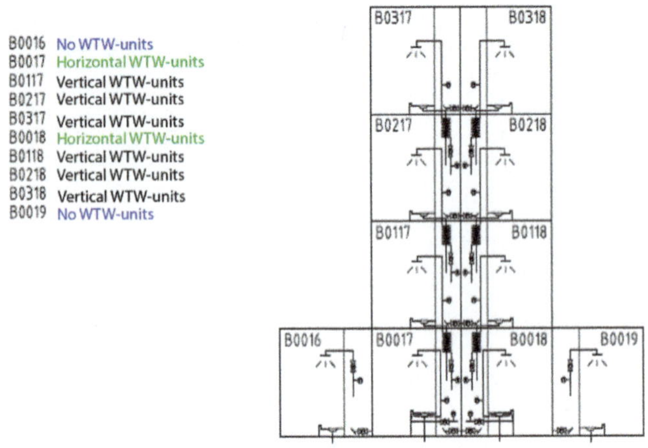

Figure 1. Student apartments monitored.

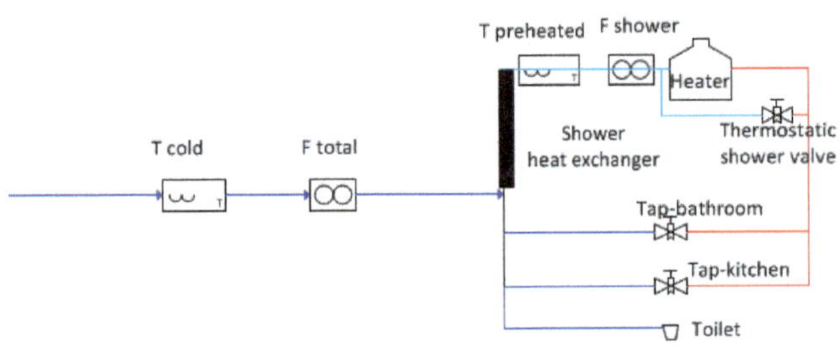

Figure 2. Project setup diagram (F is flow meter, T is temperature sensor).

These four sensors start to collect data whenever there is water consumption in the room; thus, all showers taken were recorded and stored by a SEMAPHORE T-BOX (an open station for remote control, equipped with data logger, alarm transmitter, and server), and then the recorded data were transported to the Waternet database via the Internet. The temperature of the shower water near the drain was measured manually. In Uilenstede, the conditions (flow rate, temperature) of showers taken by the occupants (students) were not controlled, but only monitored from September to December 2014. The records were used to calculate the practical energy recovery efficiency of the shower heat exchanger. The temperature of the shower water in the student's rooms was 34.5–37.5 °C, the cold water temperature was 12.5–14.5 °C (was 20 °C in a few days in September), the pre-heated water temperature was 26.0–28.0 °C, and the flow rate was 5.8–6.4 L min^{-1}.

2.1.2 Laboratory

Due to the lack of data from the project site, a laboratory setup was built to mimic the performance of the shower heat exchanger in Uilenstede. In a more controlled environment, the efficiency of the shower heat exchanger could be evaluated, and, furthermore, the relevant factors could be investigated.

Figure 3. Experimental setup of the laboratory configuration.

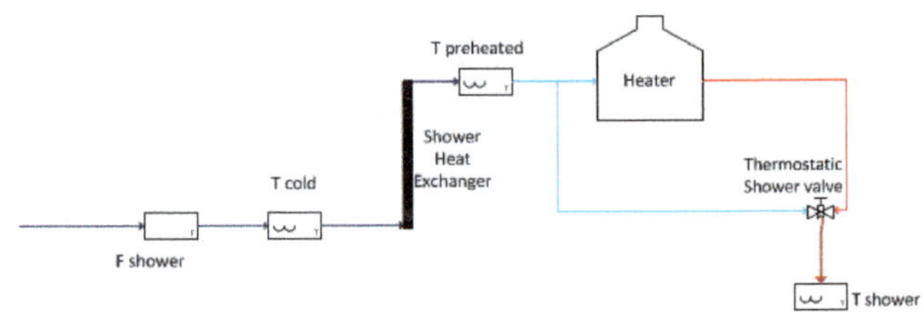

Figure 4. Flow diagram of the laboratory configuration (F is flow meter, T is temperature sensor).

The configuration of the system includes three temperature sensors (SIEMENS QAD2012) and one flow meter (Kamstrup Multical®62). Figure 3 shows the experimental setup in the laboratory while Fig. 4 shows the flow diagram. The shower flow diagram of the laboratory configuration is the same as that of Uilenstede. The black line represents the cold water, which goes to the shower heat exchanger. The blue line represents the preheated water that goes to both the heater and the thermostatic valve. The red line is the heated water from the heater, and the dark red line indicates the used shower water that goes to the drain (temperature measured by T shower).

2.1.3 Experiments

In the laboratory, two experiments were carried out. The experimental conditions are summarized in Tables 1 and 2.

In experiment 1, a high shower temperature (38 °C) and two flow rates (5.4 and 6.5 L min^{-1}, respectively) were applied to simulate the efficiency of the shower heat exchanger under winter conditions. There were six shower turns in one test, i.e., shower turn 1, 2, 3, 4, 5, and 6. Each shower turn lasted for 30 min. The time interval between each shower turn increased from 10 to 20, 30, 60, and then to 120 min. Six tests (36 shower turns in total, 18 for winter conditions and 18 for summer conditions) were conducted to take average energy recovery efficiency for each flow rate.

Table 1. Summary of experiments.

	EXP1	EXP2
Interval between showers (min)	10, 20, 30, 60, 120	15
Number of tests	6	2
Shower durations (min)	30	30
Flow rates ($\mathrm{L\,min^{-1}}$)	5.4, 6.5	5.2, 6.8
Shower temperature (°C)*	38	33
Cold water temperature (°C)	10	20

* Shower temperature is the temperature measured near the drain.

Table 2. Shower schedule per test.

EXP 1		EXP 2	
Time	Showers	Time	Showers
9:00–9:30	1	9:00–9:30	1
9:40–10:10	2		
10:30–11:00	3	9:45–10:15	2
11:30–12:00	4		
13:00–13:30	5	10:30–11:00	3
15:30–16:00	6		

In experiment 2, a lower shower temperature (33 °C) and a higher incoming water temperature (20 °C) were applied. These two temperatures were applied to create a smaller temperature difference between the cold water and shower water, in order to simulate the efficiency under summer conditions. Two flow rates of 5.2 and 6.8 $\mathrm{L\,min^{-1}}$ were compared in this experiment. Two tests were performed with three shower turns each, to get average energy recovery efficiency for each flow rate; 15 min time intervals were applied between each shower.

The shower turn durations, temperatures, and flow rates, which were applied in the two experiments, were determined based on the monitoring results and manual measurements in Uilenstede. In the experiments, room temperature and humidity were not registered.

2.1.4 Analysis methods

The energy and efficiency calculations are based on a standard method (NEN 7120+C2:2012, 2012).

$$Q_{\text{recovered}} = \Sigma\{q_{\text{cold}} \times \rho(T_{\text{cold}}) \times \left[h\left(T_{\text{preheated}}\right) - h\left(T_{\text{cold}}\right)\right] \times \mathrm{d}t\}, \quad (1)$$

$$Q_{\text{waste}} = \Sigma\{q_{\text{shower}} \times \rho(T_{\text{shower}}) \times [h(T_{\text{shower}}) - h(T_{\text{cold}})] \times \mathrm{d}t\}, \quad (2)$$

$$\eta_{\text{recover}} = \frac{Q_{\text{recovered}}}{Q_{\text{waste}}}, \quad (3)$$

where Q_{waste} is the total energy in used shower water in kilojoule, $Q_{\text{recovered}}$ is the energy recovered by the shower heat exchanger in kilojoule, q_{cold} is the drinking water flow rate through the shower heat exchanger in [$\mathrm{m^3\,s^{-1}}$], and q_{shower} is the shower water flow rate (should be the same as q_{cold} in our laboratory case) through the shower heat exchanger in [$\mathrm{m^3\,s^{-1}}$]. $\rho(T)$ and $h(T)$ are the specific density and enthalpy of the water, as functions of the temperature according to

$$\rho(T) = 999.9649 + 0.0264672 \times T - 0.0061549 \times T^2 + 1.775 \times 10^{-5} \times T^3 \text{ in } [\mathrm{kJ\,kg^{-1}}], \quad (4)$$

$$h(T) = 0.167853 + 4.18587 \times T - 0.000146789 \times T^2 + 9.38153 \times 10^{-7} \times T^3 + 8.36764 \times 10^{-9} \times T^4 \text{ in } [\mathrm{kJ\,kg^{-1}}], \quad (5)$$

η_{recover} is the energy recovery efficiency [%].

2.2 Greenhouse gas emissions

With the energy saved per shower calculated by Eq. (1), greenhouse gas emissions have been calculated with the factors and other parameters (from Waternet) from Table 3.

2.3 Payback period

The energy saved per shower was calculated by the Eq. (1), both in terms of electricity and natural gas. Based on this calculation and the other parameters presented in Table 4, the payback period calculation can be expressed as in Eq. (6).

$$\text{Payback period} = \frac{\text{Cost}_{\text{exhchanger}} + \text{Cost}_{\text{labor}}}{\text{Energy saving} \times \text{Energy price}} \quad (6)$$

3 Results and discussion

3.1 Energy recovery efficiency

With different types of heaters and fluctuations in drinking water temperature, there might be some minor variations in the time needed to stabilize the system. In general, it takes about 90 s to reach 90 % (summer) to 99 % (winter) of the final shower water temperature and preheated water temperature. In the Dutch Standard Method (NEN 7120+C2:2012), the calculation of energy recovery efficiency starts after the system becomes stable. But in this way, the energy saved during the warm-up period is excluded. The data in this study (Table 5) were collected from the beginning of the shower turn; therefore, the whole shower turn period was included. This approach describes the performance of the shower heat exchanger in a more realistic manner.

3.1.1 Impact of flow rates, shower turn durations, water temperature differences, and shower turn intervals

In winter conditions (Fig. 5a), the average energy recovery efficiency of a 5.4 $\mathrm{L\,min^{-1}}$ flow rate is in the range

Table 3. Conversion factors and parameters used for CO_2-equivalents calculation.

Parameter	Value	Unit
CO_2-eq conversion factor 1	1.63×10^{-10}	kton CO_2-eq kJ^{-1} electricity
CO_2-eq conversion factor 2	1.8×10^{-6}	kton CO_2-eq Nm^{-3} gas
Population in Amsterdam	800 000	–
Number of apartments in Amsterdam	412 000	–
Natural gas consumption per household	1600	Nm^3
Electricity consumption per household	1800	kWh
Nm^3 gas conversion to kWh	8.76×10^{-3}	kWh Nm^{-3}

Table 4. Assumptions and information for payback period calculations.

Assumption				
			Annual energy saving*	
Type of house	Occupant	Showers per day	Electricity ($kWh\,yr^{-1}$)	Natural gas ($Nm^3\,yr^{-1}$)
Single-student apartments	1	1	316.1	36
Normal household in Amsterdam	2	2	632.2	72
Normal household in the Netherlands	4	4	1264.4	144
Information				
	Shower heat exchanger ($EUR\,unit^{-1}$)	Labor ($EUR\,unit^{-1}$)	Electricity ($EUR\,kWh^{-1}$)	Natural gas ($EUR\,m^{-3}$)
Cost	390	100	0.23	0.55
Reference	DSS	Uilenstede		

* Calculated by Eq. (1).

of 64–64.5 %. When showering with a higher flow rate ($6.5\,L\,min^{-1}$), the average energy recovery efficiency was around 61.5–62 %. The recovery efficiency under the flow rate of $5.4\,L\,min^{-1}$ was 2.5–3.0 % higher than the flow rate of $6.5\,L\,min^{-1}$. The efficiency gradually increased with the shower turn durations, but only within 0.5 %, which means that the efficiency of the shower heat exchanger was roughly stable against shower duration in winter. Taking a longer shower turn does not result in higher energy recovery efficiency.

In summer conditions (Fig. 5b), the cold water temperature in practice can exceed 20 °C; thus, the temperature difference between shower water and cold water was only 13 °C (in winter conditions it could be 28 °C). In this situation, a flow rate of $5.2\,L\,min^{-1}$, which was used for the first three showers, resulted in an energy recovery efficiency of 61–62 %, and a flow rate of $6.8\,L\,min^{-1}$, which was used for the last three showers, resulted in an energy recovery efficiency of 57–58 %. In all, 4 % higher recovery efficiency was found when showering with the lower flow rate of $5.2\,L\,min^{-1}$. This was similar to the findings in winter conditions.

Considering the shower turn duration, at flow rate of $5.2\,L\,min^{-1}$, the average efficiency for 8 min shower turns was 61.0 %, and 62.4 % for 30 min showers; thus, a 1.4 % increase was achieved. In winter conditions, the increase was limited to 0.5 %. When showering at $6.8\,L\,min^{-1}$, the same phenomenon was observed: an extension from 8 min shower turns to 30 min shower turns in summer conditions resulted in an efficiency increase of 1 %, while in winter an increase of only 0.5 % was observed. In summer conditions, the energy recovery efficiency increased with shower turn durations more significantly than in winter conditions.

Temperature differences between cold drinking water and shower water were smaller in summer conditions, which resulted in a 2–3 % lower (overall) efficiency.

Figure 6 shows six consecutive shower turns with increasing time intervals between the shower turns (10, 20, 30, 60, and 120 min), both for a flow rate of 5.4 and $6.5\,L\,min^{-1}$. The time intervals between the showers affected the efficiencies. For instance, for 8 min shower turns, the recovery efficiency of shower turn 2 (which was taken 10 min after shower turn 1) was 0.5 and 1.0 % higher than shower turn 6 (which was taken 120 min after shower turn 5), namely, by taking two showers with a shorter time interval, more energy could be saved. But this effect was significant only for shorter shower

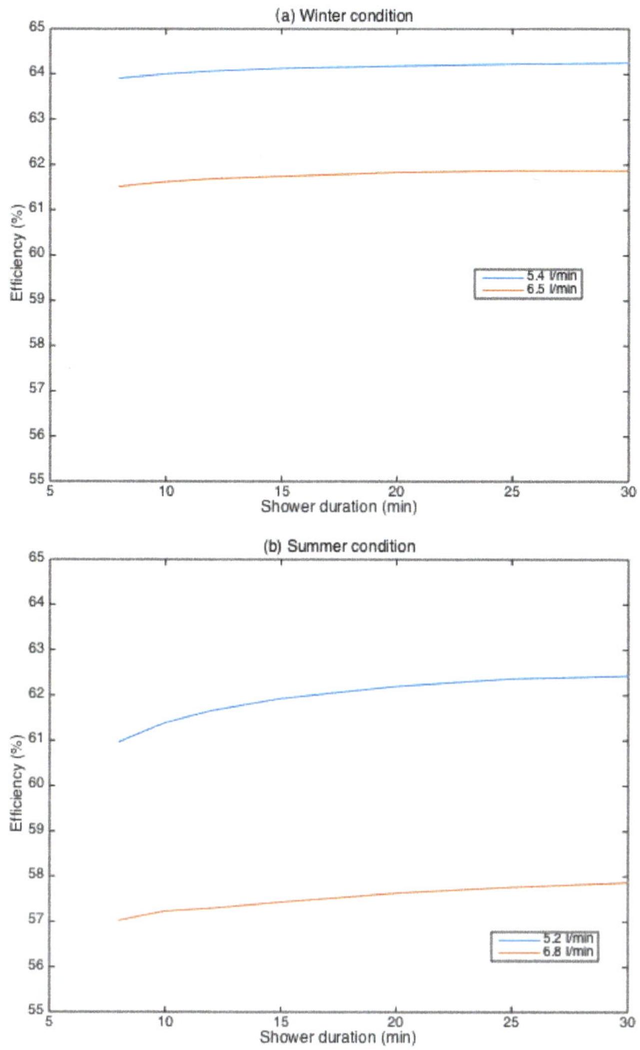

Figure 5. Energy recovery efficiencies versus impact parameters.

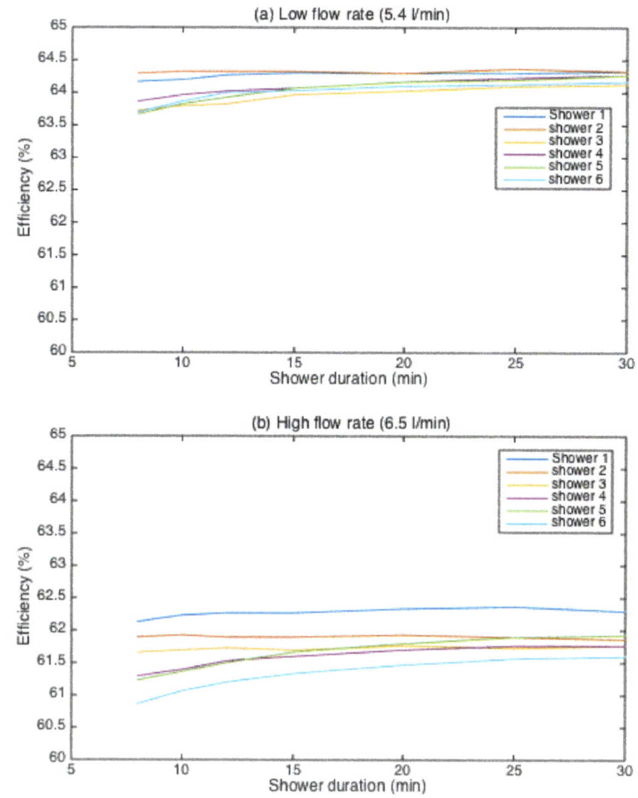

Figure 6. Energy recovery efficiency of each shower (average of three tests) in EXP 1.

turns (< 15 min), and it was negligible for long shower turns exceeding 20 min.

3.1.2 Project Uilenstede

Due to technical problems with the monitoring system, only data from one room (B0218, vertical WTW-unit) were valid in this phase. Records of four showers were found in Room B0218 (December, 2014), and an average energy recovery efficiency of 57 % was calculated.

In this student house, the shower energy recovery efficiency was 4 % lower than the laboratory winter conditions. The reasons for this might be (1) a smaller temperature difference or (2) a more fluctuated flow rate.

First, the cold water temperature inside the building was higher than the average drinking water temperature: it was 14.5 °C in September and 12.5 °C in December, while the average water temperature in the laboratory was 11 °C . Therefore, taking into account the shower water temperature between 34.5 and 37.5 °C, the temperature difference can be 2–6 °C smaller.

Second, the water flow on a higher floor (i.e., second and third) tends to fluctuate more than on the ground floor. This was observed in both the monitoring results and manual measurements in the student's apartments.

The first cause was proven to be valid by the results of experiment 2, but, unfortunately, due to the difficulty in controlling the flow rate in the laboratory, the second cause has not been tested. Additionally, the room temperature and humidity might have also had an impact on recovery efficiency; these data have still not been collected.

3.2 Energy savings and CO_2 reduction

The average electricity and gas consumption in households is about 1800 kWh yr^{-1} and 1600 Nm^3 yr^{-1} in Amsterdam (in 2012), and the total energy consumption equals 1770 kton CO_2 emission (Table 3).

Assuming people take a 10 min showers each day with a water saving shower valve (about 5 L min^{-1}), 0.4 kWh (in summer) and 1.1 kWh (in winter) per shower turn can be saved with a shower heat exchanger (calculated by Eq. 1). Under a maximum scenario, with 412 000 apartments and a

Table 5. Comparison of energy efficiencies.

	DSS*	Lab				Uilenstede
		Winter		Summer		
Shower temperature (°C)	40	38		33		34.5
Cold water temperature (°C)	10	9–10		19–20		12.5
Flow rate ($L\,min^{-1}$)	5.8	5.4	6.5	5.2	6.8	6.4
Efficiency (%)	62.7	64	61	62	58	57

* Dutch Solar Systems (2015)

Table 6. Estimation of energy recovery and CO_2 emission reduction.

Annual energy saved by shower heat exchanger	kWh	260 000 000
Electricity consumed in households	kWh	740 000 000
Gas consumed in households	kWh	5 800 000 000
Annual total consumption	kWh	6 540 000 000
Saving compared to electricity	%	35.0
Saving compared to gas	%	4.5
Saving compared to total energy	%	4.0
CO_2 reduction	kton	54

population density of 2 per dwelling in Amsterdam, the energy that can be recovered is around 900 000 $kWh\,day^{-1}$ in winter and 300 000 $kWh\,day^{-1}$ in summer. The energy recovered in 1 year is approximate 260 million kWh (122 days are counted as summer with average drinking water temperature above 20 °C; 243 days as winter). This is 4.0 % of the total households electricity and gas consumption, which equals 6540 million kWh (Table 6).

In the Netherlands, shower water is mainly heated by gas. With a shower heat exchanger in every house in Amsterdam, a 4.5 % reduction in gas consumption can be achieved per year, which is equivalent to saving about 30 million $Nm^3\,yr^{-1}$ gases, or a reduction of 54 kton CO_2. With regard to the reduction requirement of Waternet of 53 kton yr^{-1} (van der Hoek, 2012b), this would be a significant achievement.

It is difficult to forecast to what extend this maximum scenario will be realized, as it also depends on incentives and thus the policy of the city of Amsterdam. In addition, on the long term the efficiency of shower heat exchangers may decrease due to fouling and corrosion. Hence, the estimation might be too optimistic. However, it shows the benefits of installing shower heat exchangers and it stresses the importance of promoting the installation of the shower heat exchangers.

3.3 Payback period

A shower heat exchanger costs EUR 390 and the installation was about EUR 100 in Uilenstede, which brings the total costs to about EUR 500. With an average natural gas price of EUR 0.55 Nm^{-3}, and the annual gases saving per capita (about 36 Nm^3), the payback period for single-student apartments is around 13 years.

The payback period becomes shorter when the number of occupants increases. For example, in a four-person residence house, the payback period could be less than 4 years. Although the installation cost might be underestimated (no pipelines have to be changed in the new student apartments), the annual saving (EUR 152 yr^{-1}) was quite close to the DSS estimation (EUR 126 yr^{-1} for normal Dutch households with four people). Concerning the apartments using electricity for water heating (electricity price EUR 0.23 kWh^{-1}), the payback period can be as short as 2 years. If taking into account a higher installation fee, the estimation of payback period is close to the estimation (4–25 years) found by Mol (2013).

The aging of the material and the fouling of the inner pipes could lead to a deterioration of the recovery efficiency, which might also increase the payback period. This effect was not examined in these short-term observations, but it will be studied in the next stage long-term observations in the Uilenstede project site.

4 Conclusions

The energy recovery efficiency observed in this study (57–64 % observed in the lab and 57 % observed in Uilenstede) is quite close to the claimed efficiency (56–62.7 % for vertical shower heat exchanger). The performance of the shower heat exchanger is relatively stable for different shower turn durations, shower turn intervals, and seasonal impacts, while the flow rate of the shower was shown to have a more significant influence: a lower flow rate resulted in a higher energy recovery efficiency. Therefore, combining shower heat exchangers with water saving shower valves is recommended.

With a shower heat exchanger, the energy recovered by 412 000 households in Amsterdam is about 260 million $kWh\,yr^{-1}$, which equals a reduced greenhouse gas emission of 54 kton $CO_2\,yr^{-1}$. The potential of shower heat exchangers is promising, and once implemented they could provide a large contribution to the CO_2 reduction target of Waternet in 2020.

Based on the costs of the Uilenstede project, the average payback period in single-student apartments that uses gas for heating shower water is about 13 years, and about 7 years when electricity is used. The installation cost might be higher for older apartments, but it can be compensated by having more occupants using the same bathroom.

References

City of Amsterdam: Amsterdam: a Different Energy – 2040 Energy Strategy, Report City of Amsterdam, Klimaatbureau, Amsterdam, the Netherlands, 2009.

Dutch Solar Systems: DSS shower drain water heat recovery unit, Enschede, the Netherlands, available at: http://www.dutchsolarsystems.com/douche_wtw/downloads/4029152-folder%20DSS-douchepijp-ENGELS-LR.pdf, last access: 3 March, 2015.

Hofman, J., Hofman-Caris, R., Nederlof, M., Frins, J., and Van Loosdrecht, M.: Water and energy as inseparable twins for sustainable solutions, Water Sci. Technol., 63, 88–92, 2011.

IPCC: Summary for policymakers, in: Climate Change 2013: The Physical Science Basis. Contribution of Working Group I to the Fifth Assessment Report of the Intergovernmental Panel on Climate Change, edited by: Stocker, T. F., Qin, D., Plattner, G. K., Tignor, M., Allen, S. K., Boschung, J., Nauels, A., Xia, Y., Bex, V., and Midgley, P. M., Cambridge University Press, Cambridge, UK and New York, NY, USA, 4–29, http://www.ipcc.ch/pdf/assessment-report/ar5/wg1/WG1AR5_SPM_FINAL.pdf, 2013.

Liu, L., Fu, L., and Jiang, Y.: Application of an exhaust heat recovery system for domestic hot water, Energy, 35, 1476–1481, 2010.

Mol, S.: Evaluation of shower heat exchanger as a means for CO_2 reduction in the water cycle, Report Waternet, Amsterdam, the Netherlands, 2013.

NEN 7120+C2:2012: Energieprestatie van gebouwen – Bepalingsmethode (Dutch Standard "Energy performance of buildings"), the Netherlands Standardisation Institute, Delft, the Netherlands, 2012.

van der Hoek, J. P.: Climate change mitigation by recovery of energy from the water cycle: a new challenge for water management, Water Sci. Technol., 65, 135–141, 2012a.

Van der Hoek, J. P.: Towards a climate neutral water cycle, Journal of Water and Climate Change, 3, 163–170, 2012b.

Van Thiel, L.: Watergebruik Thuis 2013 (Water Use at Home 2013), Report G5707 TNS Nipo/Vewin, 17 February 2014, Amsterdam, the Netherlands, 2014.

Wong, L. T., Mui, K. W., and Guan, Y.: Shower water heat recovery in high-rise residential buildings of Hong Kong, Appl. Energy, 87, 703–709, 2010.

19

Finite volume simulation of unsteady water pipe flow

J. Fernández-Pato and P. García-Navarro

LIFTEC (CSIC) – University of Zaragoza, C/María de Luna 5, 50018 Zaragoza, Spain

Correspondence to: J. Fernández-Pato (jfpato@unizar.es)

Abstract. The most commonly used hydraulic network models used in the drinking water community exclusively consider fully filled pipes. However, water flow numerical simulation in urban pipe systems may require to model transitions between surface flow and pressurized flow in steady and transient situations. The governing equations for both flow types are different and this must be taken into account in order to get a complete numerical model for solving dynamically transients. In this work, a numerical simulation tool is developed, capable of simulating pipe networks mainly unpressurized, with isolated points of pressurization. For this purpose, the mathematical model is reformulated by means of the Preissmann slot method. This technique provides a reasonable estimation of the water pressure in cases of pressurization. The numerical model is based on the first order Roe's scheme, in the frame of finite volume methods. The novelty of the method is that it is adapted to abrupt transient situations, with subcritical and supercritical flows. The validation has been done by means of several cases with analytic solutions or empirical laboratory data. It has also been applied to some more complex and realistic cases, like junctions or pipe networks.

1 Introduction

The numerical simulation of water pipe networks is characterized by the difficulty of simulating both transient and steady states. Most of the time, the water flow is unpressurized but, under exceptional conditions, the limited storage capacity of the pipes could cause a momentary pressure peak if the water level raises quickly. Under these conditions, a pressurized flow should be taken into account, implying a change in the mathematical model in order to solve the problem accurately.

The most commonly used hydraulic network models used in the drinking water community exclusively consider fully filled pipes. However, a complete pipe system model should be able to solve steady and transient flows under pressurized and unpressurized situations and the transition between both flows (mixed flows). Many of the models developed to study the propagation of hydraulic transients solve both system of equations, e.g. Gray (1954), Wiggert and Sundquist (1977). For this purpose, numerical schemes derived from the Method of Characteristics (MOC) are employed. This method transforms the continuity and momentum partial differential equations into a set of ordinary differential equations, much easier to solve. MOC schemes can handle complex boundary conditions. However, the main issue with these kind of methods is the need for interpolation in some cases, so that they are not strictly conservative. This results in a diffusion of the wave front that implies a wrong arrival time of the waves to the boundaries. Hence, a new efficient numerical scheme is justified in order to solve accurately the wave front propagation.

Two main points of view have been considered by the researchers to study mixed flow problems along the last decades. The first one consists of solving separately pressurized and shallow flows, treating the transitions between both flow types as internal boundary conditions. These models are capable of simulating sub-atmospheric pressures in the pipe, but the need for solving both mathematical models implies a high level of complexity. Other authors, e.g. García-Navarro et al. (1994), León et al. (2009), Kerger et al. (2010) developed simulations applying the shallow water equations in a slim slot over the pipe (Preissmann slot method). An estimation of the pipe pressure can be obtained from the water level in the slot. The great advantage of this model is the use of a single equation system for solving the complete problem. This technique has been widely used to simulate local transitions between pressurized and shallow flows but some stability problems have been reported (Trajkovic et al., 1999) when simulating cases with abrupt transitions. The reason is a big difference between the wave speeds (from $\sim 10\,\mathrm{m\,s^{-1}}$ in shallow flow to $\sim 1000\,\mathrm{m\,s^{-1}}$ in pressurized flow). In order to solve this issue, the slot width could be increased, but this implies a loss of accuracy in the results, due to the no-mass/momentum conservation.

This work is focused on the development of a simulation model able to solve pipe system networks working mainly under free surface flow situations but exceptionally pressurized. Hence, it is based on the second option previously mentioned and it is aimed to generalize the shallow water equations by means of the Preissmann slot method. The explicit first order Roe scheme has been chosen as numerical method for all the simulations.

Implicit methods or parallel computation have been dismissed, as the calculation times are not long. Fluxes and source terms have been treated by means of an upwind scheme (García-Navarro and Vázquez-Cendón, 2000). All the symbols and units used are described in Table 1.

This paper is an extension of the one presented at the 12th edition of the International Conference on "Computing and Control for the Water Industry – CCWI2013" (Fernández-Pato and García-Navarro, 2014).

2 Mathematical model

2.1 Shallow water equations

The unsteady open channel water flow are usually modelled by means of the 1D shallow water or St. Venant equations. These equations represent mass and momentum conservation along the main direction of the flow and become a good description for the behaviour of the most of the pipe-flow-kind problems (Kundu et al., 2012; Abbott, 1979). The conservative form of this systems of equations is presented in Eqs. (1) and (2):

$$\frac{\partial A}{\partial t}+\frac{\partial Q}{\partial x}=0 \tag{1}$$

$$\frac{\partial Q}{\partial t}+\frac{\partial}{\partial x}\left(\frac{Q^2}{A}+gI_1\right)=gI_2+gA\left(S_0-S_f\right) \tag{2}$$

where A is the wetted cross section, Q is the discharge, g is the acceleration due to gravity, I_1 represents the hydrostatic pressure force term and I_2 accounts for the pressure forces due to channel/pipe width changes:

$$I_1=\int_0^{h(x,t)}(h-\eta)\,b(x,\eta)\,\mathrm{d}\eta \tag{3}$$

Table 1. List of symbols and units.

Symbol	Meaning	Units
A	Wetted cross-section	$\mathrm{m^2}$
A_f	Full pipe cross-section	$\mathrm{m^2}$
b_s	Preissmann slot width	m
b	Pipe/channel width	m
c	Wave speed	$\mathrm{m\,s^{-1}}$
c_{WH}	Water hammer speed	$\mathrm{m\,s^{-1}}$
D	Diameter	m
E	Pipe elastic modulus	Pa
e	Pipe thickness	m
Fr	Froude number	–
g	Accelaration due to gravity	$\mathrm{m\,s^{-2}}$
H	Elevation of the hydraulic grade line	m
h	Free surface depth	m
I_1	Hydrostatic pressure force term	$\mathrm{m^3}$
I_2	Pressure force term due to channel/pipe width changes	$\mathrm{m^2}$
K	Fluid elastic modulus	Pa
n	Manning's roughness coefficient	$\mathrm{s\,m^{-1/3}}$
P	Wetted perimeter	m
Q	Water discharge	$\mathrm{m^3\,s^{-1}}$
R	Hydraulic radius	m
S_0	Bed slope	–
S_f	Energy grade line slope	–
t	Time	s
u	Flow speed	$\mathrm{m\,s^{-1}}$
λ	Eigenvalue	$\mathrm{m\,s^{-1}}$
$\mathbf{e}$	Eigenvector	$\mathrm{m\,s^{-1}}$
v	Cross-section averaged flow velocity	$\mathrm{m\,s^{-1}}$
θ	Angle between pipe and longitudinal coordinate	rad
x	Longitudinal coordinate	m
z	Elevation	m
η	Vertical coordinate	m
ρ	Fluid density	$\mathrm{kg\,m^3}$
τ_0	Boundary shear	Pa

$$I_2=\int_0^{h(x,t)}(h-\eta)\frac{\partial b(x,\eta)}{\partial x}\mathrm{d}\eta \tag{4}$$

$$b=\frac{\partial A(x,\eta)}{\partial \eta} \tag{5}$$

in which η and $h(x,t)$ represent vertical coordinate and the free surface depth, respectively. The remaining terms, S_0 and S_f, represent the bed slope and the energy grade line (defined in terms of the Manning roughness coefficient), respectively:

$$S_0=-\frac{\partial z}{\partial x},\qquad S_f=\frac{Q\,|Q|\,n^2}{A^2R^{4/3}} \tag{6}$$

where $R=A/P$, P being the wetted perimeter. The coordinate system used for the formulation of the shallow water equations is shown in Fig. 1. A vectorial form of the Eqs. (1) and (2) is widely used:

$$\frac{\partial \mathbf{U}}{\partial t}+\frac{\partial \mathbf{F}}{\partial x}=\mathbf{R} \tag{7}$$

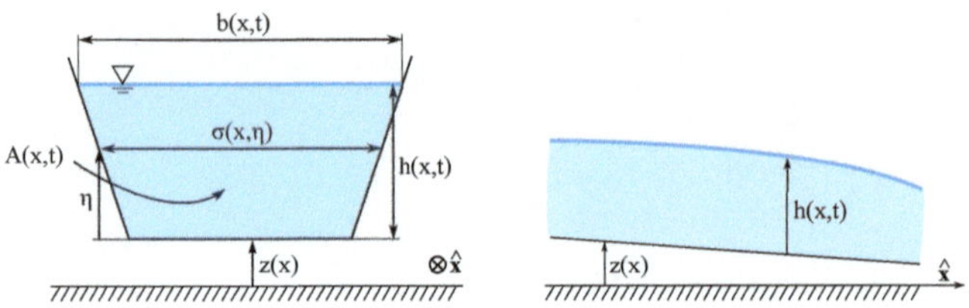

Figure 1. Coordinate system for shallow water equations.

where

$$\mathbf{U} = (A, Q)^T \tag{8}$$

$$\mathbf{F} = \left(Q, Q^2/A + gI_1\right)^T \tag{9}$$

$$\mathbf{R} = \left(0, gI_2 + gA\left(S_0 - S_f\right)\right)^T \tag{10}$$

In those cases in which $\mathbf{F} = \mathbf{F}(\mathbf{U})$, when $I_2 = 0$, it is possible to rewrite the conservative system by means of the Jacobian matrix of the system Eq. (11):

$$\frac{\partial \mathbf{U}}{\partial t} + \mathbf{J}\frac{\partial \mathbf{U}}{\partial \mathbf{x}} = \mathbf{R}, \qquad \mathbf{J} = \frac{\partial \mathbf{F}}{\partial \mathbf{U}} = \begin{pmatrix} 0 & 1 \\ c^2 - u^2 & 2u \end{pmatrix} \tag{11}$$

where c is the wave speed (analogous to the speed of sound in gases), defined as follows:

$$c = \sqrt{g\frac{\partial I_1}{\partial A}} \tag{12}$$

The set of real eigenvalues ($\lambda^{1,2}$) and eigenvectors ($\mathbf{e}^{1,2}$) of the system matrix can be used for diagonalizing it. These magnitudes represent the speed of propagation of the information:

$$\lambda^{1,2} = u \pm c, \qquad \mathbf{e}^{1,2} = (1, u \pm c)^T \tag{13}$$

It is very common to characterize the flow type by means of the Froude number $Fr = u/c$ (analogous to the Mach number in gases). It allows the classification of the flux into three main regimes: subcritical $Fr < 1$, supercritical $Fr > 1$ and critical $Fr = 1$.

2.2 Water-hammer equations

The instant response for changes in a pipe flow is closely related to the elastic compresibility of both the fluid and the pipe wall material. Unsteady flow in pipes is commonly described by the cross-section integrated mass and momentum equations (Chaudhry and Mays, 1994):

$$\frac{\partial H}{\partial t} + v\frac{\partial H}{\partial x} + v\sin\theta + \frac{c_{\mathrm{WH}}^2}{g}\frac{\partial v}{\partial x} = 0 \tag{14}$$

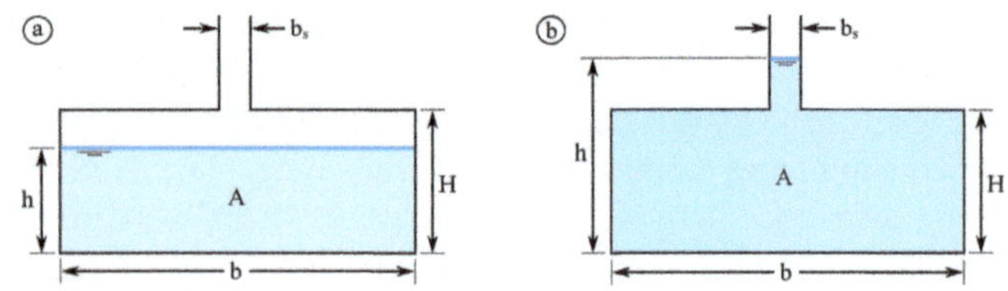

Figure 2. (**a**) Pure shallow flow (**b**) Pressurized pipe.

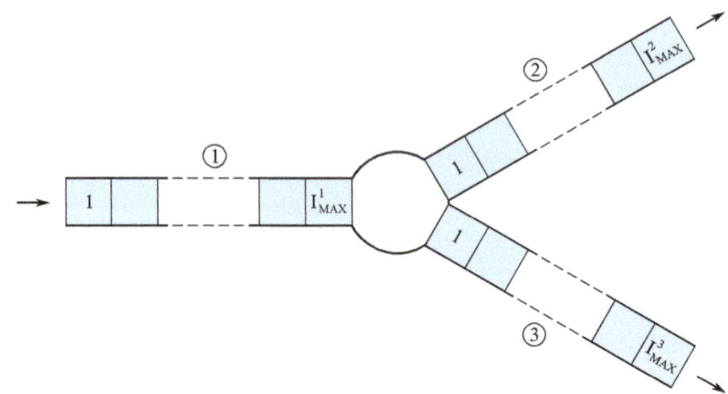

Figure 3. Example of pipe junction.

$$\frac{\partial v}{\partial t} + v\frac{\partial v}{\partial x} + g\frac{\partial H}{\partial x} + \frac{4\tau_0}{\rho D} = 0 \tag{15}$$

in which $H(x,t)$ = elevation of the hydraulic grade line, $v(x,t)$ = local cross-section averaged flow velocity, $\theta(x,t)$ = angle between the pipe and the horizontal level, D = diameter, ρ = fluid density, τ_0 is the boundary shear, typically estimated by means of a Manning or Darcy-Weissbach friction model. The magnitude c_{WH} accounts for the elastic waves speed in the pipe:

$$c_{\mathrm{WH}} = \sqrt{\frac{K/\rho}{1 + \frac{DK}{eE}}} \tag{16}$$

being e the pipe thickness and E, K the elastic modulus of the pipe material and fluid, respectively. By neglecting convective terms, it is possible to reach a linear hyperbolic equation system:

$$\frac{\partial H}{\partial t} + \frac{c_{\mathrm{WH}}^2}{g}\frac{\partial v}{\partial x} = 0 \tag{17}$$

$$\frac{\partial v}{\partial t} + g\frac{\partial H}{\partial x} + \frac{4\tau_0}{\rho D} = 0 \tag{18}$$

These equations conform a hyperbolic system, analogous to the shallow water equations, considering the pressure and the velocity as the conserved variables.

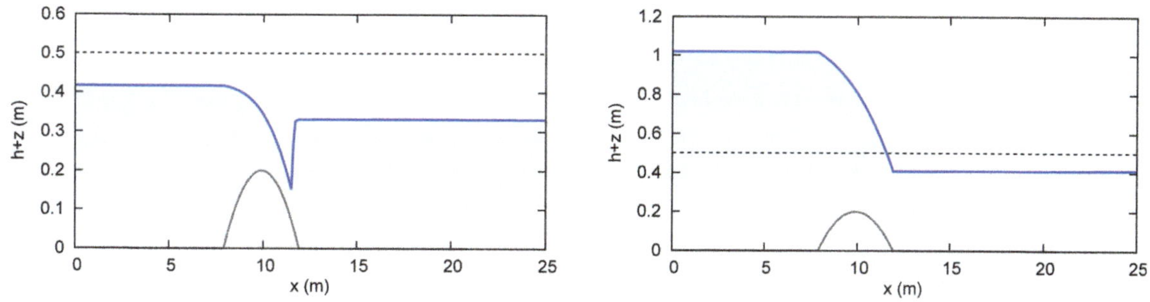

Figure 4. Steady state over a bump. Initial state (dashed line). Bed level (grey). Numerical solution (blue): **(a)** Test #1; **(b)** Test #2.

Table 2. Boundary conditions.

Flow regime and boundary	Number of physical BC to impose
Upstream subcritical flow	1
Upstream supercritical flow	2
Downstream subcritical flow	1
Downstream supercritical flow	0

Table 3. Steady state over a bump.

Test	Upstream $Q\left(\mathrm{m}^3\,\mathrm{s}^{-1}\right)$	Downstream $h\,(m)$
#1	0.18	0.33
#2	1.53	0.66 (sub)

3 Preissmann slot model

The Preissmann slot approach consist in assuming that the top of the pipe/closed channel is connected to a fictitious narrow slot, which is open to the atmosphere. This fact allows to consider a free surface flow situation, even when the pipe is pressurized, by including the slot in the shallow water equations (see Fig. 2). The slot width b_s is ideally chosen equaling the speed of gravity waves in the slot to the water hammer wavespeed. Therefore, the water level in the slot corresponds to the pressure head level. This model is based on the previously remarked similarity between the wave equations which describe free surface and pressurized flows. The water hammer flow comes from the capacity of the pipe system to change the area and fluid density, so forcing the equivalence between both models requires that the slot stores as much fluid as the pipe would by means of a change in area and fluid density. The pressure term and the wavespeed for $A \leq bH$ are:

$$h = \frac{A}{b} \tag{19}$$

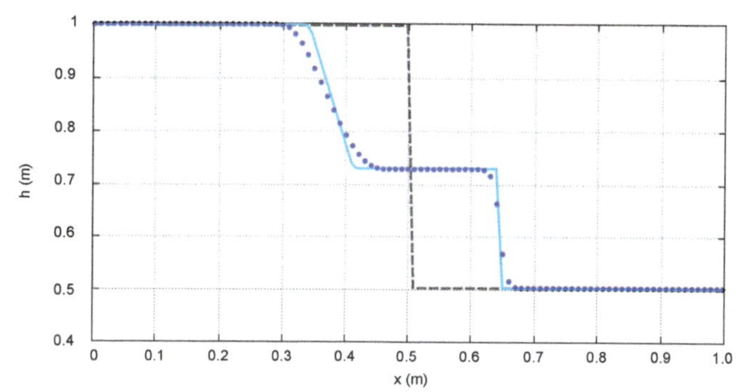

Figure 5. Dam-break test case. Initial state (dashed line). Numerical results (blue dots). Analytical solution (cyan line).

$$I_1 = \frac{A^2}{2b} \tag{20}$$

$$c = \sqrt{g\frac{\partial I_1}{\partial A}} = \sqrt{g\frac{A}{b}} \tag{21}$$

and for $A > bH$:

$$h = H + \frac{A - bH}{b_s} \tag{22}$$

$$I_1 = bH\left(\frac{A - bH}{b_s} + \frac{H}{2}\right) + \frac{(A - bH)^2}{2b_s} \tag{23}$$

$$c = \sqrt{g\frac{\partial I_1}{\partial A}} = \sqrt{g\frac{A}{b_s}} \tag{24}$$

The ideal choice for the slot width results in:

$$c_{\mathrm{WH}} = c \Rightarrow b_s = g\frac{A_f}{c_{\mathrm{WH}}^2} \tag{25}$$

in which A_f stands for the full pipe cross-section.

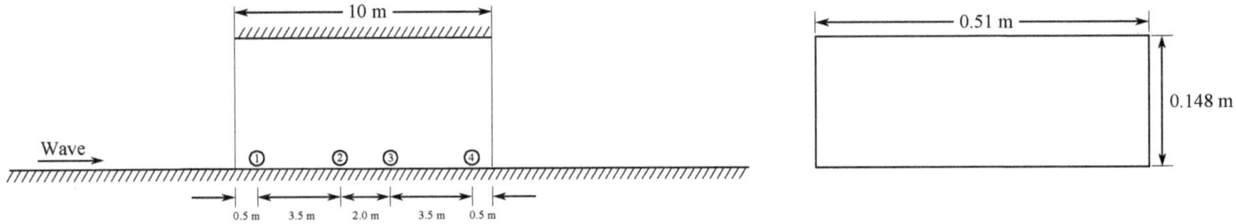

Figure 6. Wiggert experimental setup.

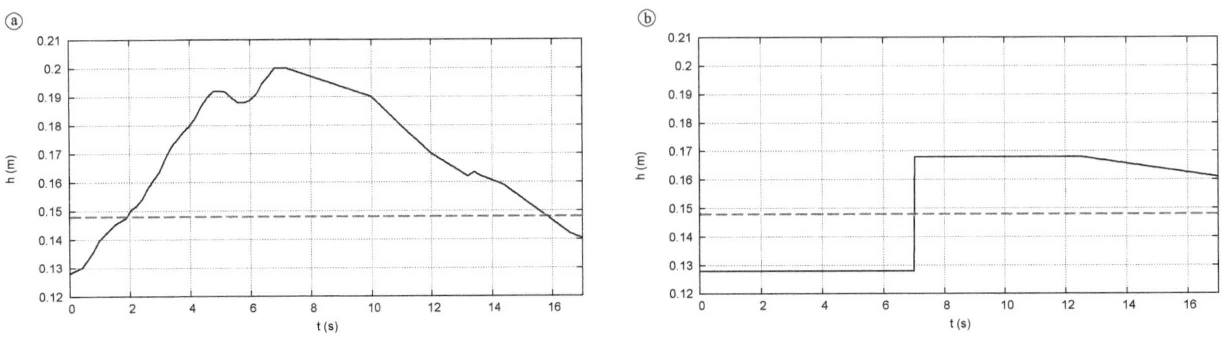

Figure 7. Upstream (**a**) and downstream (**b**) pressured head level.

4 Finite volume numerical model

4.1 Explicit first order Roe scheme

Roe scheme is based on a local linearization of the conserved variables and fluxes:

$$\delta \mathbf{F} = \tilde{\mathbf{J}} \delta \mathbf{U} \tag{26}$$

It is necessary to build an approximate Jacobian matrix $\tilde{\mathbf{J}}$ whose eigenvalues and eigenvectors satisfy:

$$\delta \mathbf{U}_{i-1/2} = \mathbf{U}_{i+1} - \mathbf{U}_i = \sum_{k=1}^{2} \left(\tilde{\alpha}_k \tilde{\mathbf{e}}_k\right)_{i+1/2} \tag{27}$$

$$\delta \mathbf{F}_{i+1/2} = \mathbf{F}_{i+1} - \mathbf{F}_i = \tilde{\mathbf{J}}_{i+1/2} \delta \mathbf{U}_{i+1/2} = \sum_{k=1}^{2} \left(\tilde{\lambda}_k \tilde{\alpha}_k \tilde{\mathbf{e}}_k\right)_{i+1/2} \tag{28}$$

The wave strenght coefficients $\tilde{\alpha}_k$ represent the variable variation coordinates in the Jacobian matrix basis. The eigenvalues and eigenvectors are expressed in terms of the average flux velocity and wave speed:

$$\tilde{\lambda}_k = (\tilde{u} \pm \tilde{c}), \qquad \tilde{\mathbf{e}}_k = (\tilde{u} \pm \tilde{c})^T \tag{29}$$

where

$$\tilde{u}_{i+1/2} = \frac{Q_{i+1}\sqrt{A_i} + Q_i\sqrt{A_{i+1}}}{\sqrt{A_i A_{i+1}}\left(\sqrt{A_{i+1}} + \sqrt{A_i}\right)} \tag{30}$$

$$\tilde{c}_{i+1/2} = \sqrt{\frac{g}{2}\left[\left(\frac{A}{b}\right)_i + \left(\frac{A}{b}\right)_{i+1}\right]} \tag{31}$$

Following (García-Navarro and Vázquez-Cendón (2000)), the source terms of the equation system (Eq. 7) are also discretizated using an upwind scheme:

$$\left(\tilde{\mathbf{R}}\delta x\right)_{i+1/2} = \left(\sum_{k+} \tilde{\beta}_k \tilde{\mathbf{e}}_k\right)_{i+1/2} + \left(\sum_{k-} \tilde{\beta}_k \tilde{\mathbf{e}}_k\right)_{i+1/2} \tag{32}$$

in which the source streghts $\tilde{\beta}_k$ take the role of $\tilde{\alpha}_k$ coefficients for the fluxes. Then, the complete discretization of the system becomes:

$$\mathbf{U}_i^{n+1} = \mathbf{U}_i^n - \frac{\delta t}{\delta x}\left[\left(\sum_{k+}\left(\tilde{\lambda}_k \tilde{\alpha}_k - \tilde{\beta}_k\right)\tilde{\mathbf{e}}_k\right)_{i-1/2} + \left(\sum_{k-}\left(\tilde{\lambda}_k \tilde{\alpha}_k - \tilde{\beta}_k\right)\tilde{\mathbf{e}}_k\right)_{i+1/2}\right] \tag{33}$$

4.2 Boundary conditions and stability conditions

In order to solve a numerical problem, it is necessary to characterize the domain limits by imposing some physical boundary conditions (BC). The number of boundary conditions depends on the flow regime (subcritical or supercritical). Hence, there are four posibilities for a 1-D numerical problem (see Table 2).

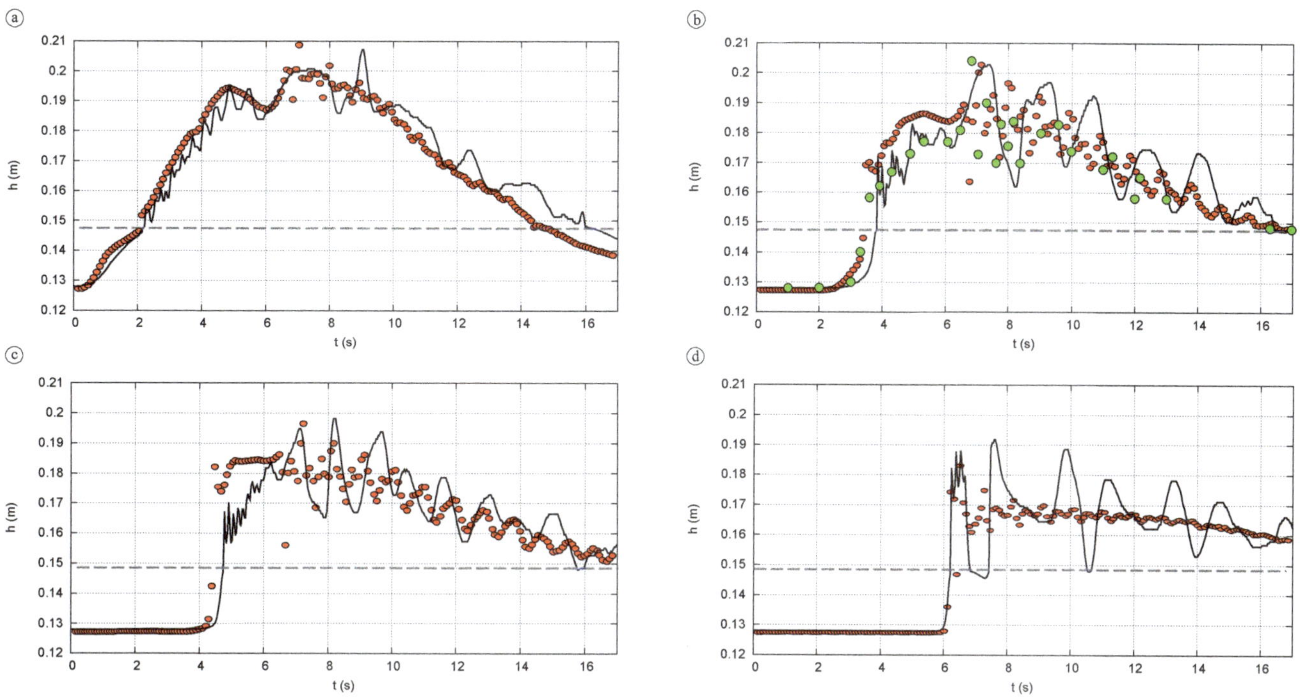

Figure 8. Comparision between numerical results (blue), experimental data (green dots) and numerical results from Kerger et al. (2010) (red dots) for the four gauges of the Wiggert test case setup.

In the cases that consider pipe junctions, additional internal boundary conditions are necessary in order to model this feature. Figure 3 shows an example of a 3 pipe junction.

The water level equality condition is imposed at the junction for all the pipes:

$$h_1 = h_2 = \ldots = h_N \tag{34}$$

Discharge continuity condition is formulated depending on the flow regime. In the cases presented, water flows in a subcritical regime:

$$\sum_{i=1}^{N} Q_i = 0 \tag{35}$$

Storage wells are another common junction type. For these cases, the boundary conditions are modified as follows:

$$h_1 = h_2 = \cdots = h_N = H_{\text{well}}, \tag{36}$$

$$\sum_{i=1}^{N} Q_i = A_{\text{well}} \frac{\mathrm{d}H_{\text{well}}}{\mathrm{d}t}$$

where A_{well} and H_{well} are the well top area and depth, respectively.

The presented method uses an explicit time discretization and it requires a control on the time step in order to avoid instabilities. Hence the Courant-Friedrichs-Lewy (CFL) condition is applied on every branch (j) of the network:

$$\Delta t^j_{\max} = \frac{\Delta x^j}{\max\left(\left|u^j\right| + c^j\right)} \tag{37}$$

Then, the minimum of all the computed time steps is chosen:

$$\Delta t_{\max} = \min\left\{\Delta t^j_{\max}\right\}, \qquad \text{CFL} = \frac{\Delta t}{\Delta t_{\max}} \leq 1 \tag{38}$$

5 Test cases

The model is applied to several analytic or experimental test cases in order to evaluate its accuracy.

5.1 Steady state over a bump

Following Murillo and García-Navarro (2012) a frictionless rectangular 25 m × 1 m prismatic channel is considered. The variable bed level is given by:

$$z\,(8 \leq x \leq 12) = 0.2 - 0.05(x - 10)^2 \tag{39}$$

and the initial conditions for the water depth and discharge:

$$h(x,0) = 0.5 - z(x), \qquad u(x,0) = 0 \tag{40}$$

The different boundary conditions for the test cases simulated are summarized in Table 3.

Figure 4a shows the generation of hydraulic jump that connects both subcritical and supercritical regimes. In the second test case, shown in Fig. 4b the connection is made without any shock wave and it can be mathematically proved (and numerically checked) that this transition takes place in the highest part of the bump.

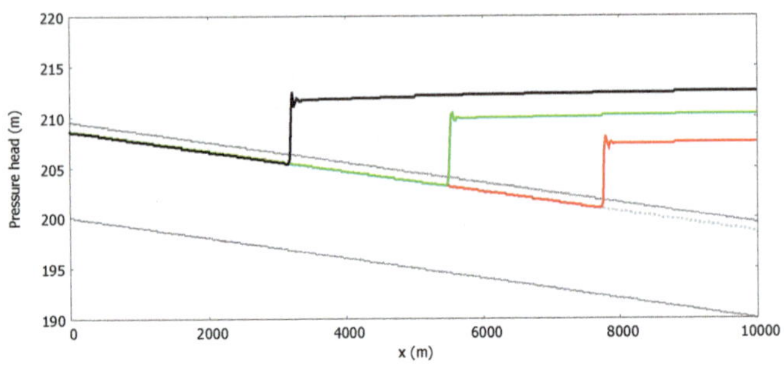

Figure 9. Shockwave propagation in transient mixed flow at $t = 100$ s (red), $t = 200$ s (green) and $t = 300$ s (blue).

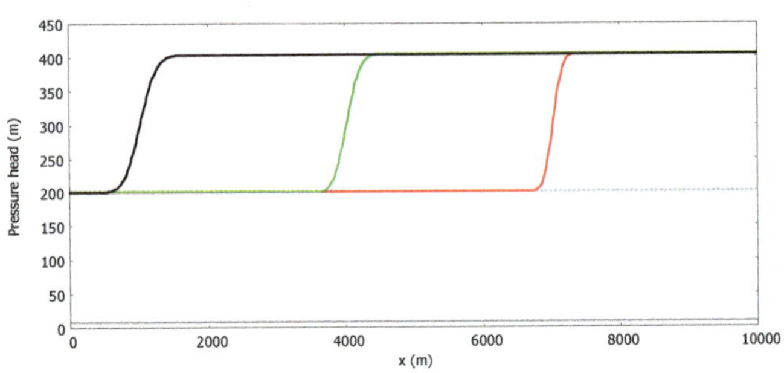

Figure 10. Waterhammer simulation. Pressured head at $t = 3$ s (red), $t = 6$ s (green) and $t = 9$ s (blue).

5.2 Dam-break

Dam-break is a classical example of non-linear flow with shocks. It has been widely used to test the accuracy and conservation of the numerical scheme, by comparing with its analytical solution. In the case presented, an initial discontinuity of 1 m : 0.5 m ratio with no friction was considered. Figure 5 shows the good agreement between numerical and analytical solution at the given time.

5.3 Wiggert test case

The experimental case designed by Wiggert (1972) and widely numerically reproduced (e.g. Kerger et al., 2010; Bourdarias and Gerbi, 2007) consists in a horizontal 30 m long and 0.51 m wide flume. A 10 m roof is placed in the middle section, setting up a closed rectangular pipe 0.148 m in height (see Fig. 6). A 0.01 $\mathrm{s\,m^{-1/3}}$ Manning roughness coefficient is assumed. As initial conditions, a water level of 0.128 m and zero discharge are considered. Then a wave coming from the left causes the pressurization of the pipe. The imposed downstream boundary conditions are the same values measured by Wiggert (1972) (Fig. 7).

Figure 8 shows the numerical results for the four gauges. The experimental comparison is done only for the second one, due to data availability issues. An overall good agreement is observed with only small numerical oscillations presented.

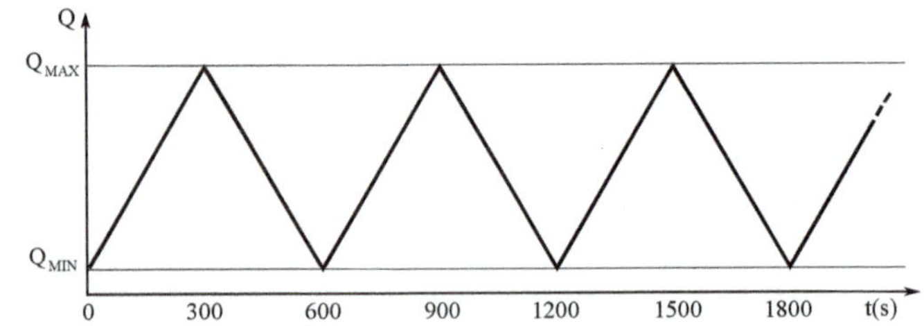

Figure 11. Triangular inlet hydrograph in the pipe 1.

5.4 Transient mixed flow

The aim of the next test case is to prove the model in the simulation of large-scale strong transients. It considers a uniform slope (0.1 %) rectangular pipe connected to a downstream valve (León, 2006; León et al., 2009). The lenght, width and height are 10 km, 10 m and 9.5 m, respectively, and the chosen Manning roughness is 0.015.

As initial condition, a uniform water depth (8.57 m) and discharge ($Q = 240\,\mathrm{m^3\,s^{-1}}$) are assumed. From this state, the downstream valve is closed, generating a shock wave moving upstream and pressurizing the pipe. Figure 9 shows the numerical results obtained for the pressure head at three different times. For this simulation, a 500 cells mesh and a timestep given by CFL = 0.5 have beed used. It is clearly observed that the pipe pressure raises gradually with the upwards shockwave movement.

5.5 Pressurized flow transients

A full-pressurized (León, 2006; León et al., 2009) 10 km flat, frictionless pipe of rectangular section ($10\,\mathrm{m} \times 7.853\,\mathrm{m}$) is connected upstream to a water reservoir that guarantees a constant pressure head of 200 m. A closure valve is placed downstream. When the valve is suddenly closed, a waterhammer wave is generated moving upwards. Figure 9 shows the numerical results with CFL = 0.8 and a 500 cells mesh.

This case simulates a completely pressurized pipe, so it is necessary to be very precise in the slot width selection in order to ensure the numerical solution accuracy. A waterhammer speed of $1000\,\mathrm{m\,s^{-1}}$ and a initial flow speed of $2.0\,\mathrm{m\,s^{-1}}$ are assumed. Following (Eq. 25) condition:

$$c_{\mathrm{WH}} = c \Rightarrow b_s = g\frac{A_f}{c_{\mathrm{WH}}^2} = 0.77\,\mathrm{mm} \qquad (41)$$

6 Pipe networks

6.1 Transient flow in a 3-pipe junction

A case proposed in Wixcey (1990) is presented first. A 5 km long prismatic main channel (1) dividing into two secondary branches (2 and 3) also 5 km long with the same 1 m wide rectangular geometry have been considered, as presented in

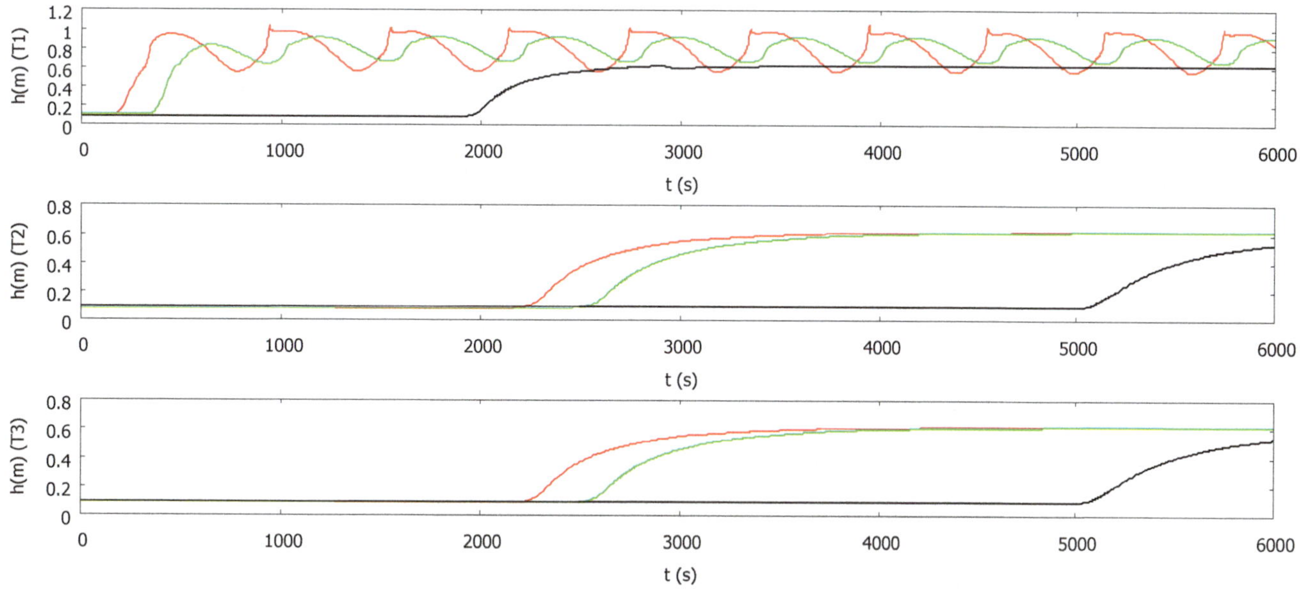

Figure 12. Pressurized transient in a pipe junction. Pressured head vs time $x = 500$ m (red), $x = 1000$ m (green) and $x = 5000$ m (blue).

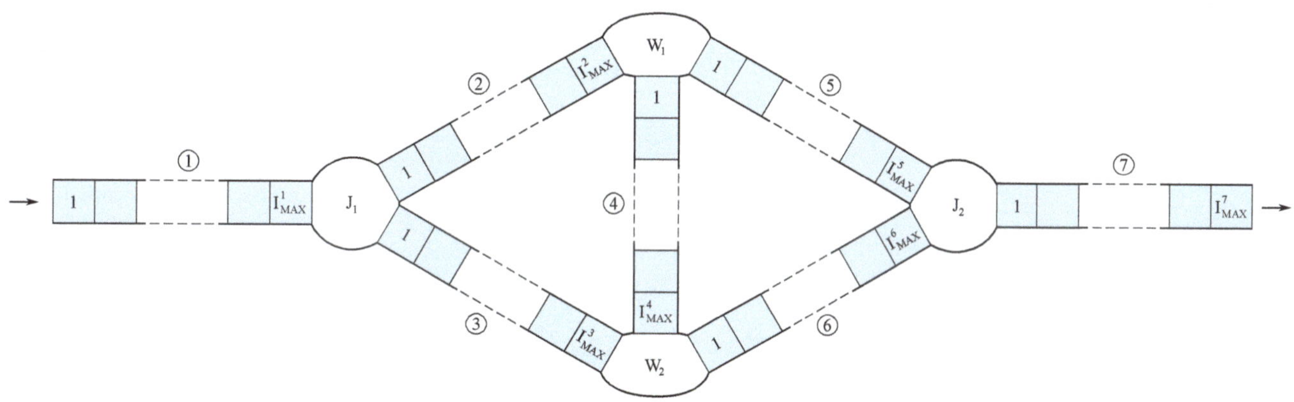

Figure 13. (Scheme of the looped pipe network.

Fig. 3. Manning roughness coefficient for all the pipes is $0.01\ \mathrm{s\,m^{-1/3}}$. The main branch has a slope of 0.002 and the secondary ones 0.001.

First, a steady state has been calculated from the initial conditions:

$$Q_1(i,0) = 0.1\,\mathrm{m^3\,s^{-1}}, \qquad (42)$$

$$Q_2(i,0) = Q_3(i,0) = 0.05\,\mathrm{m^3\,s^{-1}}$$

$$h_1(i,0) = h_2(i,0) = h_3(i,0) = 0.2\,\mathrm{m} \qquad (43)$$

using as boundary condition the inlet discharge:

$$Q_1(1,t) = 0.1\,\mathrm{m^3\,s^{-1}} \qquad (44)$$

and free outflow boundary condition at the outlet of pipes 2 and 3.

Using this steady state as initial condition for the transient calculation, a triangular function of peak discharge $Q_{\mathrm{MAX}} = 3.2\,\mathrm{m^3\,s^{-1}}$ and a period of 600 s (Fig. 11) is imposed at the beginning of the pipe 1. Figure 12 shows the numerical results using a 100 cells mesh and a CFL $= 0.9$ condition.

6.2 Transient flow in a 7-pipe looped network

In this section a simple looped pipe network (see Fig. 13) is presented Wixcey (1990). Each pipe is closed, squared, 1 m wide and 100 m long. The bed slopes are $S_{0,1} = 0.002$, $S_{0,2} = S_{0,3} = 0.001$, $S_{0,4} = 0.0$, $S_{0,5} = S_{0,6} = 0.001$, $S_{0,7} = 0.002$ and a $0.01\ \mathrm{s\,m^{-1/3}}$ Manning friction coefficient is considered. A first calculation provided the steady state from the following initial conditions:

$$Q_1(i,0) = Q_7(i,0) = 0.1\,\mathrm{m^3\,s^{-1}} \qquad (45)$$

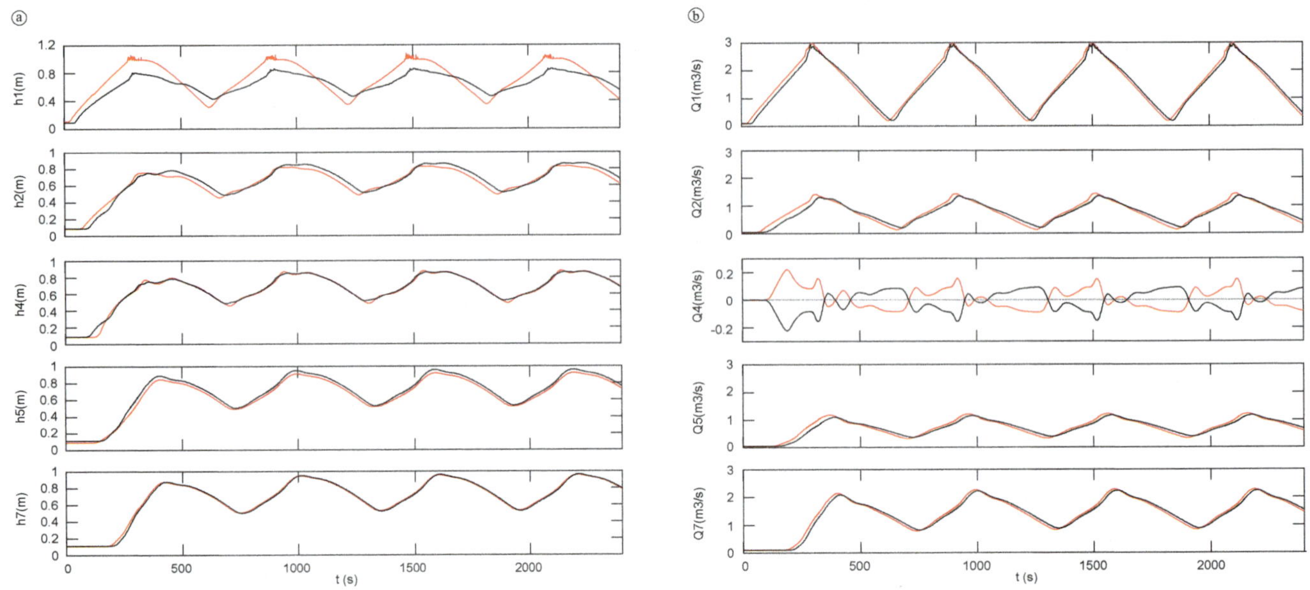

Figure 14. Time histories at grid points $i = N/2$ (red) and $i = N$ (blue) for punctually pressurized flow: **(a)** Water depth; **(b)** Discharge. The discharge for pipe 4 is recorded at locations $i = 1$ (red), $i = N/2$ (grey) and $i = N$ (blue).

$$Q_2(i,0) = Q_3(i,0) = Q_5(i,0) = Q_6(i,0) = 0.05\,\mathrm{m}^3\,\mathrm{s}^{-1} \tag{46}$$

$$Q_4(i,0) = 0 \tag{47}$$

$$h_j(i,0) = 0.2\,\mathrm{m}, \qquad (j = 1,\ldots,7) \tag{48}$$

and the upstream boundary condition:

$$Q_1(1,t) = 0.1\,\mathrm{m}^3\,\mathrm{s}^{-1} \tag{49}$$

Free outflow downwards boundary condition has been imposed.

Junctions J_1 and J_2 are treated as normal confluences (Eq. 34) and a storage well of 5 m^2 top surface is assumed in junctions W_1 and W_2 (Eq. 36). The previously computed steady state is now used as an initial condition for a second calculation. A triangular function of peak discharge Q_{MAX} and a period of 600 s (Fig. 11) is imposed at the beginning of the pipe 1. The minimum discharge is 0.1 m^3 s^{-1}.

Two different situations were studied. In the first one the peak discharge is fixed to $Q_{\mathrm{MAX}} = 2.0\,\mathrm{m}^3\,\mathrm{s}^{-1}$ and the flow remains unpressurized all over the pipe system. In the second case, the maximum discharge is increased to $Q_{\mathrm{MAX}} = 3.0\,\mathrm{m}^3\,\mathrm{s}^{-1}$, so pressurization occurs in some points of pipe 1. Figure 14 shows the results for the pressurized case. The results in pipes 3 and 6 have been omitted because of symmetry reasons. The discharge at the center of the pipe 4 is constantly zero and equal in magnitude but opposite in sign in the symmetric points, which verify the simmetric character of the pipe network.

7 Conclusions

The present work leads to the conclusion of the reasonably good applicability of the Preissmann slot model for an estimation of the pressure values in unsteady situations between both shallow and pressurized flows. In this second case, a small deviation of the ideal slot width induces a mass and momentum conservation error. On the other hand, it is clear that wider slots increases the stability of the numerical model. Hence, in benefit of the results accuracy, it is remarkably important to find an agreement between both facts.

The model takes advantages of the similarity of both equation systems (shallow water and pressurized flow) and provides a simple way to simulate occasionally pressurized pipes, treating the system as an open channel in the Preissmann slot. This results in an easier implementation of the model because it avoids the managing of two separate systems of equations (shallow water and water hammer) in order to model separately the pressurized and the free surface flows.

The use of an explicit scheme implies a limitation on the computational time step, which increases the calculation time, specially in the in the pressurized cases, due to the small width of the slot.

Finally, the authors want to remark the possibility of adapting the method to more realistic systems, like pipe networks which can be punctually pressurized. Some additional internal boundary conditions are necessary in these cases, in order to represent pipe junctions and storage wells.

References

Abbott, M. B.: Computational Hydraulics. Elements of the Theory of the Free Surface flows, Pitman Publishing Ltd., London, 324pp., 1979.

Bourdarias, C. and Gerbi, S.: A finite volume scheme for a model coupling free surface and pressurised flows in pipes, J. Computat. Appl. Mathemat., 209, 109–131, 2007.

Chaudhry, M. H. and Mays, L. W. (Eds.): Computer modeling of free-surfaces and pressurized flows, Kluwer Academic Publishers, Boston, 741 pp., 1994.

Fernández-Pato, J. and García-Navarro, P.: A Pipe Network Simulation Model with Dynamic Transition between Free Surface and Pressurized Flow, Proc., 12th Int. Conf. on "Computing and Control for the Water Industry – CCWI2013", Perugia, Procedia Engineering, Elsevier, 70, 641–650, 2014.

García-Navarro, P. and Vázquez-Cendón, M. E.: On numerical treatment of the source terms in the shallow water equations, Comput. Fluids, 29, 951–979, 2000.

García-Navarro, P., Alcrudo, F., and Priestley, A.: An implicit method for water flow modelling in channels and pipes, J. Hydraul. Res., 32, 721–742, 1994.

Gray, C. A. M.: Analysis of Water Hammer by Characteristics, Trans. Am. Soc. Civ. Engin., 119, 1176–1189, 1954.

Kerger, F., Archambeau, P., Erpicum, S., Dewals, B. J., and Pirotton, M.: An exact Riemann solver and a Godunov scheme for simulating highly transient mixed flows, J. Computat. Appl. Mathemat., 235, 2030–2040, 2010.

Kundu, P. K., Cohen, I. M., and Dowling, D. R.: Fluid Mechanics, Academic Press, 891 pp., 2012.

León, A. S.: Improved Modeling of Unsteady Free Surface, Pressurized and Mixed Flows in Storm-sewer Systems (Dissertation), 2006.

León, A. S., Ghidaoui, M. S., Schmidt, A. R., and García, M. H.: Application of Godunov-type schemes to transient mixed flows, J. Hydraul. Res., 47, 147–156, 2009.

Murillo, J. and García-Navarro, P.: Augmented versions of the HLL and HLLC Riemann solvers including source terms in one and two dimensions for shallow flow applications, J. Computat. Phys., 231, 6861–6906, 2012.

Trajkovic, B., Ivetic, M., Calomino, F., and D'Ippolito, A.: Investigation of transition from free surface to pressurized flow in a circular pipe, Water Sci. Technol., 39, 105–112, 1999.

Wiggert, D. C.: Transient flow in free-surface, pressurized systems, Journal of the Hydraulics Division, Proc. Am. Soc. Civ. Engin., 98, 11–27, 1972.

Wiggert, D. C. and Sundquist, M. J.: Fixed-grid Characteristics for Pipeline Transients, Journal of the Hydraulics Division, Proc. Am. Soc. Civ. Engin., 103, 1403–1416, 1977.

Wixcey, J. R.: An investigation of algorithms for open channel flow calculations, Numerical Analysis Internal Report, 21, Departament of Mathematics, University of Reading, 1990.

20

Functioning conditions of the Casale pumping station in Mantova, Italy

C. Capponi[1], M. Ferrante[1], M. Pedroni[2], B. Brunone[1], S. Meniconi[1], M. Zaghini[2], and F. Leoni[2]

[1]Dipartimento di Ingegneria Civile ed Ambientale, The University of Perugia, via G. Duranti 93, 06125 Perugia, Italy

[2]TeaAcque Srl, Via Taliercio 3, 46100 Mantova, Italy

Correspondence to: C. Capponi (caterina.capponi@studenti.unipg.it)

Abstract. Pump speed, input power, pressure and discharge data are analyzed for evaluating functioning conditions of Casale pumping station in Mantova, Italy, managed by TeaAcque. A model based on the affinity laws is used to simulate the behavior of the installed variable speed pumps. Quadratic and cubic polynomial curves are used to fit the pump data given by the affinity laws. Such curves allow predicting the efficiency when the functioning conditions change. The relationship between the rotation speed and the efficiency is also derived.

1 Introduction

Variable speed pumps (VSPs) are widely used in water distribution systems because they can change the functioning conditions more efficiently than fixed speed pumps so as to meet the network requirements. When the system functioning conditions require a change in pump flow and head, the pump speed and characteristic curve vary consequently. In order to simulate the VSP behavior, taking into account the variation of both the system demand and the pump speed, a model based on the affinity laws can be used (Simpson and Marchi, 2013). Such a model allows to simulate the VSP curve by means of three dimensionless quantities, depending on the pump speed. This work is based on the research project involving the University of Perugia and TeaAcque, the manager of the water supply and distribution system in Mantova. Within the project, data acquired at pumping stations are analyzed to improve their efficiency; in this paper attention is focused on the Casale pumping station. The affinity laws are used to model measured data and fit the pump curves. In particular, a quadratic polynomial curve is used to approximate data in terms of dimensionless head and flow, according to Ulanicki et al. (2008). Hydraulic power and efficiency data are fitted by a cubic polynomial curve. Such curves allow predicting the efficiency when the functioning conditions change. The relationship between the rotation speed and the efficiency is also determined. This paper is an extension of the paper by Capponi et al. (2014), presented at the 12th edition of the International Conference on "Computing and Control for the Water Industry – CCWI2013", where a preliminary analysis of the behavior of the Casale pumping station is reported.

2 Collected data

The set of data from 22–29 May 2013, provided by TeaAcque, concerns one of the four VSPs installed in the station (Fig. 1): tank level, h_L, pressure downstream of the pump, p_d, pump flow, Q (measured by means of an electromagnetic flowmeter), and pump speed, N, are acquired at 1/60 Hz, while the input power, P_{act}, is acquired at 1/900 Hz.

With reference to Fig. 2 scheme, where $z_m = 2.20$ m, since local head losses are negligible, the pump head, H, is given by:

$$H = \frac{p_d}{\gamma} + z_m - h_L \tag{1}$$

with γ = water specific weight. To have a general picture of the pumping station functioning conditions, in Fig. 3 the rate of occurrence of H and Q is shown, i.e. the number of data for specific ranges of values.

Figure 1. Casale pumping station: the four pumps in parallel.

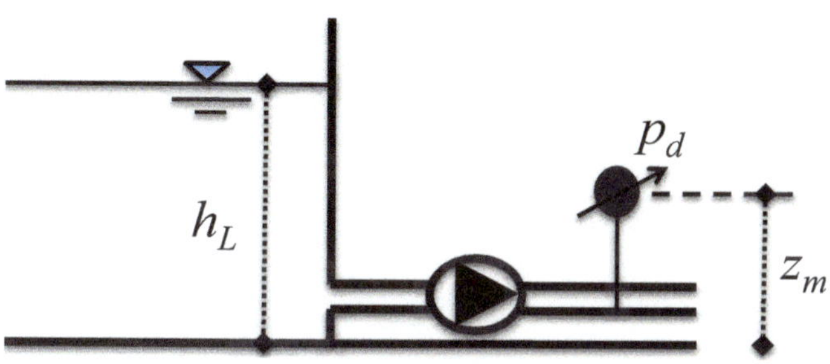

Figure 2. Scheme of the pump setup installation.

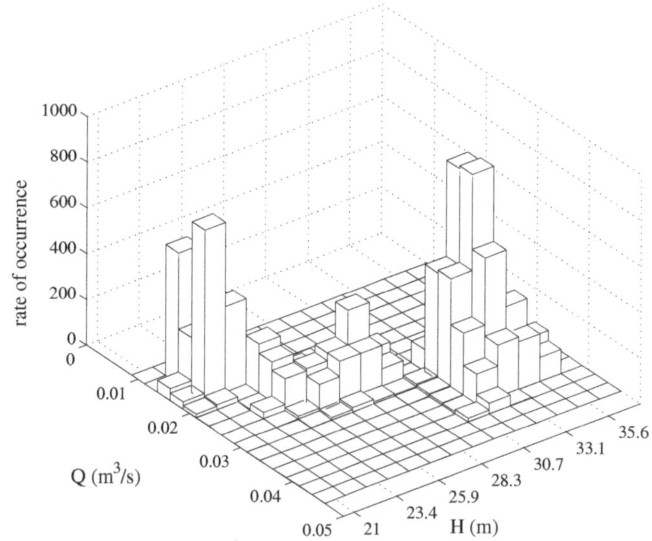

Figure 3. Histogram with the rate of occurrence of H and Q.

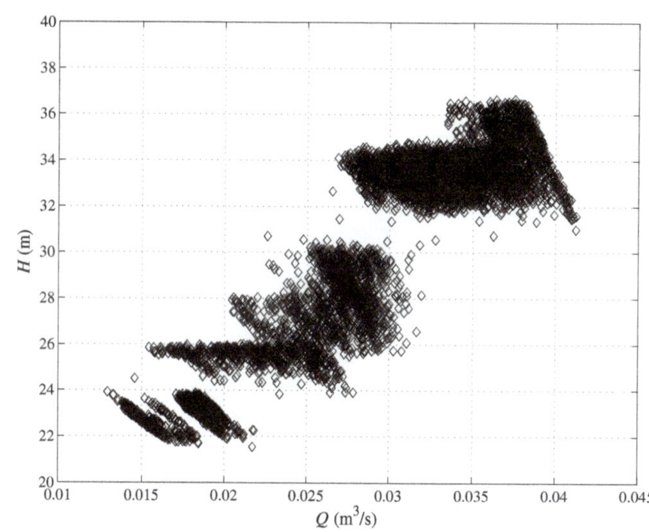

Figure 4. Pump head, H, vs. pump flow, Q.

Such a histogram points out two zones with the highest rate of occurrence corresponding to pairs of values (Q, H) that represent the most frequent functioning conditions. Such zones merit the attention of managers since any intervention on them gives rise to possible more effective functioning conditions.

In Fig. 4 data are reported in the Q-H plane. Since such data correspond to different pump speeds, they do not lie on a single curve. Hence, to characterize the behavior of the considered pumping station, the rotation speed has to be taken into account.

3 The model: affinity laws

The proposed model is based on the following dimensionless quantities defined by the affinity laws:

$$C_\mathrm{q} = \frac{Q}{N\phi^3} \tag{2}$$

$$C_\mathrm{h} = \frac{gH}{N^2\phi^2} \tag{3}$$

$$C_\mathrm{p} = \frac{P}{\rho N^3\phi^5} \tag{4}$$

where ρ = water density, g = acceleration of gravity, and ϕ = impeller diameter. Dimensionless quantities, C_q, C_h and C_p defined by Eqs. (2–4), with $\phi = 0.204$ m, are more effective in modeling pump curves and in characterizing the behavior of the pumping station, with respect to the dimensional quantities Q, H, and P (Simpson and Marchi, 2013). As shown in Fig. 5, by considering C_h and C_q, measured data lie on a single curve; specifically the quadratic polynomial proposed by Ulanicki et al. (2008) fits the experimental data with a good agreement ($R^2 = 0.9967$):

$$C_\mathrm{h} = -485.3915 {C_\mathrm{q}}^2 + 27.2944 C_\mathrm{q} + 4.8914 \tag{5}$$

As a useful reference, in Fig. 5 the pump characteristic curve, supplied by the manufacturer, is also plotted, as well as the relevant band of confidence (±10 %). Differences between

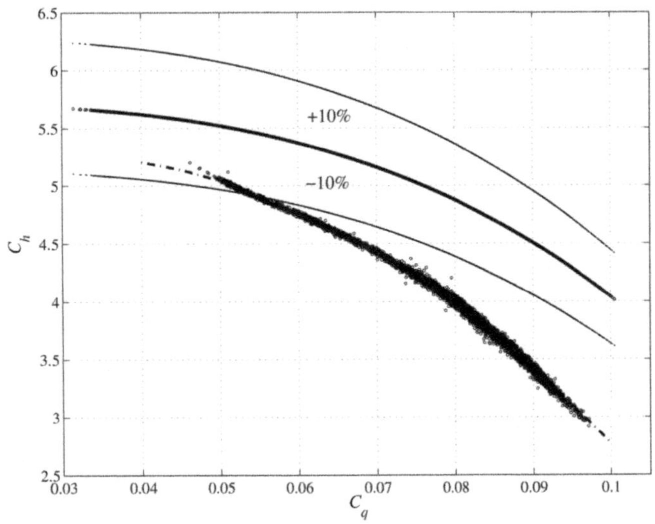

Figure 5. Experimental data, pump characteristic curve (solid line) and the fitting (dashed line) made by Eq. (5) plotted in the C_q-C_h plane (thinner solid lines define the band of confidence given by the manufacturer).

field and manufacturer's data are mainly due to the fact that Eq. (5) curve takes into account more factors related to the system actual functioning conditions.

4 Efficiency evaluation

By assuming in Eq. (4) $P = P_{th}$ – with $P_{th} = \gamma\ QH$ being the theoretical hydraulic power – and P_{act} – with P_{act} = measured hydraulic power – the theoretical value, $C_{p,\ th}$, and the actual one, $C_{p,\ act}$, of C_p can be obtained, respectively. In Fig. 6, $C_{p,\ th}$ and $C_{p,\ act}$ are compared: the 45° dashed line corresponds to ideal conditions and the distance of the data from it measures the energy dissipation. Precisely from Fig. 6 it can be observed that the $C_{p,\ act}$ values are about twice the $C_{p,\ th}$ ones. If $C_{p,\ th}$ data are plotted against C_q data, the points have a shape that reminds a third degree curve (Fig. 7). In fact, according to Ulanicki et al. (2008), experimental data can be fitted by means of a cubic polynomial curve:

$$C_{p,\ th} = -1.124C_q^{\ 3} + 0.1612C_q^{\ 2} - 4.224 \times 10^{-3}C_q + 2.014 \times 10^{-4} \tag{6}$$

Also in this case, the agreement between data and fitting is quite good ($R^2 = 0.9874$).

The efficiency, η, is a dimensionless parameter that allows evaluating the performance of the pumping station:

$$\eta = \frac{P_{th}}{P_{act}} \tag{7}$$

Values obtained by means of Eq. (7) are reported in the η-C_q plane in Fig. 8. Firstly, it can be noted that the η values are rarely above 0.6. It must be said that such an efficiency is a

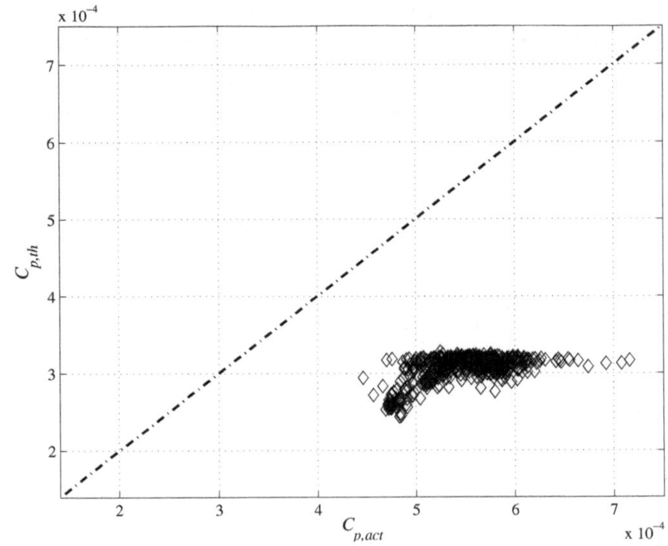

Figure 6. Dimensionless streamflow power, $C_{p,\ th}$ vs. dimensionless input power, $C_{p,\ act}$. The dashed line is the 45° line.

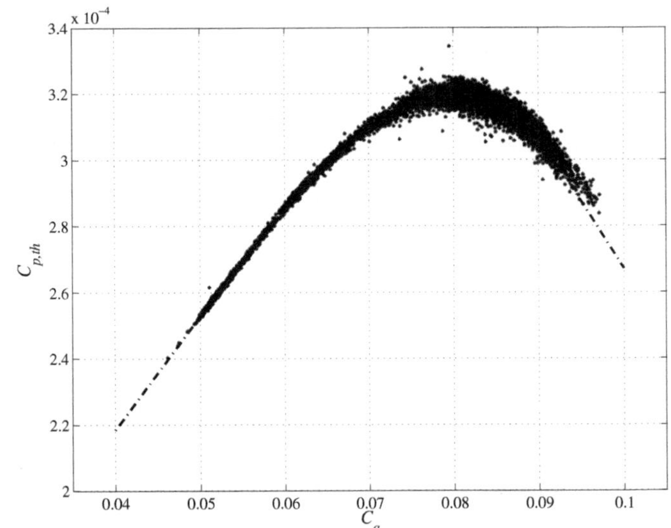

Figure 7. Dimensionless streamflow power, $C_{p,\ th}$, vs. dimensionless flow, C_q: experimental data and fitting.

"wire to water" efficiency. In fact it takes into account many factors and then such quite low values are not unexpected. Secondly, although data spreading is higher than in the previous graphs, a fitting can be made by a cubic polynomial curve:

$$\eta = -903.3C_q^{\ 3} + 87.44C_q^{\ 2} + 1.625C_q + 0.3437 \tag{8}$$

with $R^2 = 0.772$. For the sake of clarity, both fittings of C_h (Fig. 5) and η (Fig. 8) in terms of C_q are plotted in Fig. 9. These two curves allow to characterize the Casale pumping station in its actual functioning conditions.

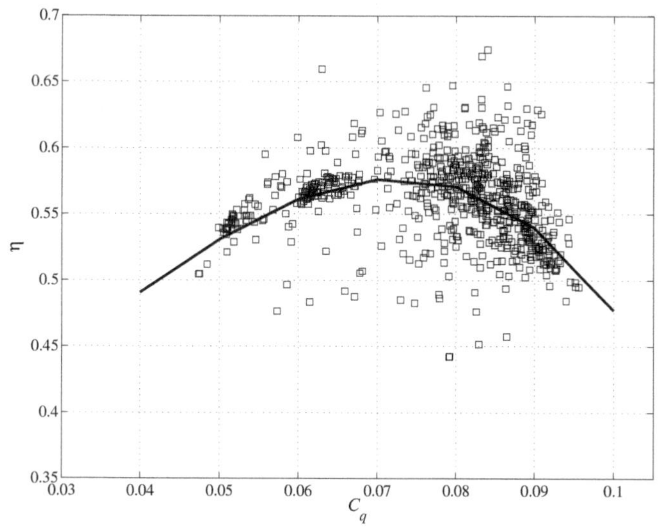

Figure 8. Efficiency values in terms of dimensionless flow, C_q: experimental data and fitting.

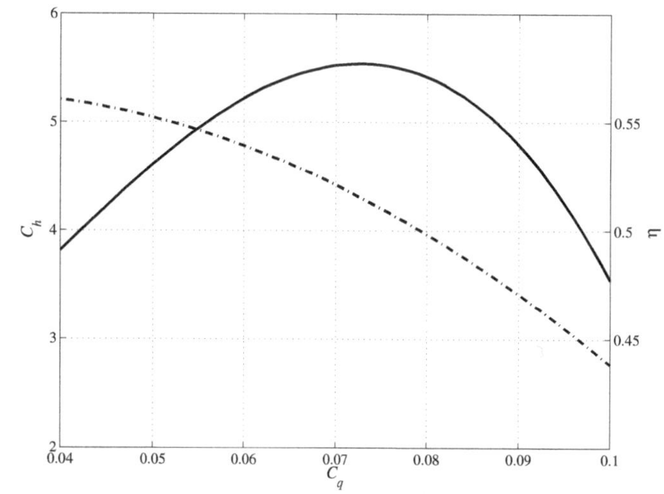

Figure 9. Fittings of data of Figs. 5 (dashed line) and 8 (continuous line) plotted on the same plane.

When the system functioning conditions require a change in pump flow and head, the pump speed and characteristic curve are varied consequently by means of an inverter drive – also referred to as "variable frequency drive" – installed in the system. In particular, the pump speed is adjusted so that the pump works at a chosen pump head value within a certain time slot: this value is usually called "set point". So, for each hour of the day, a set point is determined. Within each interval defined by the set points, the pump speed, N, increases with the flow, according to the trend shown in Fig. 10. On this figure also the efficiency values against flow are plotted. It can be observed that for both the lowest and highest values of N and Q, η values decrease. By means of this analysis it is possible to predict the behavior of the pumping station in terms of efficiency, for the given set points.

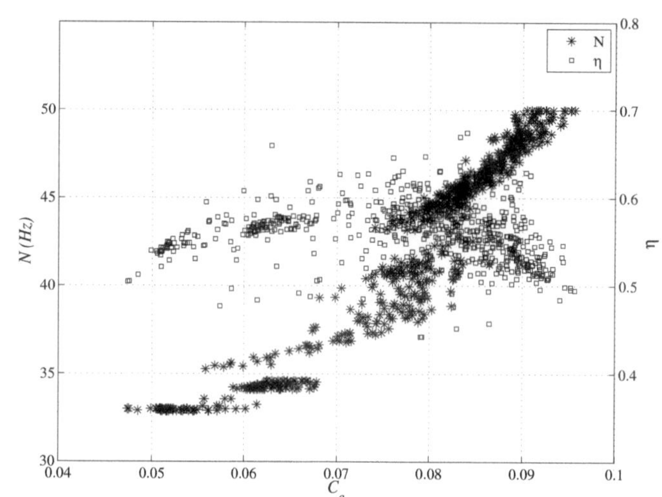

Figure 10. Pump speed, N, and efficiency, η, values against dimensionless flow, C_q.

5 Conclusions

Measurements of pump speed and input power, besides pressure and flow data, can be useful for management purposes. Pump speed data allow to scale hydraulic pump characteristics, i.e. pump head and flow, and to model the pump behavior, in terms of pump characteristic curves, according to affinity laws. In this paper a set of measurements executed at the Casale pumping station is considered, for the duration of a week. Experimental data are firstly analyzed in terms of pump head and flow, evaluating the rate of occurrence, in order to find the main functioning conditions of the pumping station. Secondly, data are analyzed using the three dimensionless quantities introduced by affinity laws. Data in the dimensionless plane C_q-C_h can be fitted with a quadratic polynomial with a quite good agreement. The dimensionless streamflow power, $C_{p,\,th}$, can be expressed in terms of C_q by a cubic polynomial, with a quite good agreement. The streamflow power is also compared with the measured input power: the ratio between these two quantities allows evaluating the "wire to water" efficiency, η, of the pumping station. The η experimental data show a high spreading and the fitting is characterized by a worse agreement if compared with the other fittings. Trends of pump speed and efficiency against pump flow, if plotted on the same figure, show how the efficiency changes with the variation in the rotation speed, which is related to the setting of the inverter drive. Combining this analysis with a larger set of data will allow to improve the pumping station functioning conditions and then to reduce the costs of the system management. The knowledge of the network behavior could be combined with the results shown in this paper in order to predict the response of the network to any change in the pumping station functioning conditions.

Acknowledgements. This research has been supported by the Italian Ministry of Education, University and Research (MIUR) under the Projects of Relevant National Interest "Advanced analysis tools for the management of water losses in urban aqueducts", "Tools and procedures for an advanced and sustainable management of water distribution systems" and Fondazione Cassa Risparmio Perugia under the project "Hydraulic characterization of innovative pipe materials (no. 2013.0050.021)".

References

Capponi, C., Ferrante, M., Pedroni, M., Brunone, B., Meniconi, S., Zaghini, M., and Leoni, F.: Real data analysis and efficiency of the TEA Mantova Casale (Italy) variable-speed pumping station, 12th Int. Conference on "Computing and Control for the Water Industry – CCWI2013", Perugia, Procedia Engineering, Elsevier, 70, 248–255, 2014.

Simpson, A. R. and Marchi, A.: Evaluating the approximation of the affinity laws and improving the efficiency estimate for variable speed pumps, J. Hydraul. Engin., 139, 1314–1317, 2013.

Ulanicki, B., Kahler, J., and Coulbeck, B.: Modeling the efficiency and power characteristics of a pump group, J. Water Resour. Plann. Manage., 134, 88–93, 2008.

21

Water Expert: a conceptualized framework for development of a rule-based decision support system for distribution system decontamination

J. L. Gutenson[1], A. N. S. Ernest[1], J. R. Fattic[2,a], L. E. Ormsbee[3], A. A. Oubeidillah[1], and X. Zhang[1]

[1]The University of Alabama, Tuscaloosa, Alabama
[2]Western Kentucky University, Bowling Green, Kentucky
[3]University of Kentucky, Lexington, Kentucky
[a]now at: Akebono Brake Corporation, Glasgow, Kentucky

Correspondence to: J. L. Gutenson (jlgutenson@eng.ua.edu)

Abstract. Significant drinking water contamination events pose a serious threat to public and environmental health. Water utilities often must make timely, critical decisions without evaluating all facets of the incident. The data needed to enact informed decisions are inevitably dispersant and disparate, originating from policy, science, and heuristic contributors. *Water Expert* is a functioning hybrid decision support system (DSS) and expert system framework that emphasizes the meshing of parallel data structures in order to expedite and optimize the decision pathway. Delivered as a thin-client application through the user's web browser, *Water Expert*'s extensive knowledgebase is a product of inter-university collaboration that methodically pieced together system decontamination procedures. Decontamination procedures are investigated through consultation with subject matter experts, literature review, and prototyping with stakeholders. This paper discusses the development of *Water Expert*, analyzing the development process underlying the DSS and the system's existing architecture specifications. *Water Expert* constitutes the first system to employ a combination of deterministic and heuristic models which provide decontamination solutions for water distribution systems. Results indicate that the decision making process following a contamination event is a multi-disciplinary effort. This contortion of multiple inputs and objectives limit the ability of the decision maker to find optimum solutions without technological intervention.

1 Introduction

Decontamination decisions made following contamination events for water distribution networks (WDNs) are a complex entanglement of empirical practices. Solutions are selected through heuristic solution processes. Oftentimes in such events, a lack of synthesized information does not afford the decision maker with the ability to rationally determine the optimum remediation and recovery strategy. The data needed in the decision making progression can be supplied through field observations, mathematical models, regulatory requirements, and organizational policy. These data vary both spatially and conceptually, as science and policy from the local, state, and federal scales are pooled to produce conclusions. Coalescing these diverse sources of data into usable information presents a challenge. Thus, to make a timely decision, the managers of WDNs are unlikely to take into account all characteristics of the emergency situation. In this regard, web-delivered information systems, driven by artificial intelligence (AI) such as expert systems, provide a means by which to supply the decision maker with recommendations. The expert system compiles divergent data sources within an integrated decision support system platform. The *Water Expert* system framework, a hybrid DSS with expert system capabilities, provides such a means. Developed by a consortium of universities, this application provides integrated decision making tools that are driven by all of the identified data inputs and presented to the user through his or her local web browser. Integrated with KYPipe's Network Decontamination Model (NDM), data can be acquired from third party applications, uploaded into *Water Expert* via the internet, and used to drive *Water Expert* recommendations and actions.

Water Expert provides recommendations for decontamination of WDNs using regulatory requirements, manuals of practice, academic and trade association journals, and industry-established procedures. This knowledge was gathered from a series of tabletop, technology demonstration, and technology deployment workshops hosted by the investigating consortium. This system can be dynamically augmented to include other knowledge domains such as local and state regulations as needed. Further, additional third-party applications can similarly be linked into *Water Expert*, either replacing or augmenting its standard toolset.

This paper describes the groundwork for the *Water Expert* concept and the decision making process undertaken during contamination events. Primarily, the current methodology of decision making during WDN emergency operations, how *Water Expert* augments this process, and the technical and contextual composition of *Water Expert* are described. A case study with a small WDN will demonstrate the system's current capacity. Finally, a discussion of the future vision for the *Water Expert* platform will follow. The core of this research addresses how computational and heuristic tools can be coalesced to address contamination incidents in WDNs. In addition, the authors elicit insight from the process by which utilities typically decontaminate their networks.

2 Literature review

Decontamination procedures are critical to successful water system recovery following a contamination event. They have repeatedly been highlighted as a significant issue for the water sector by the numerous working groups and councils that have strategized and formulated the most appropriate courses of action (Water Sector Coordinating Council, 2007; Water Sector Decontamination Workshop, 2008). In addition, governmental agencies such as the US Environmental Protection Agency (USEPA), and trade associations such as the American Water Works Association (AWWA) and the Water Research Foundation (WRF), develop and maintain guidance documents (USEPA, 2012; Water Research Foundation, 2009), case studies (USEPA, 2008; Murray et al., 2010), and tools (Sandia National Laboratories, 2010; USEPA, 2013a, 2014b, c) that are intended to enhance water infrastructure resiliency to contamination events.

Of particular interest to those in the United States, the USEPA has published a series of guides, collectively known as the Response Protocol Toolbox (USEPA, 2003a, c, d, e, f; 2004b, c, d), for addressing security and protection of the drinking water and wastewater sectors. The USEPA actively maintains the Threat Ensemble Vulnerability Assessment (TEVA) program (Murray et al., 2004). However, synthesized information for water systems to make informed decisions regarding decontamination of compromised drinking water systems is lacking. This disconnect exists because disparate sources of information cloud the resolution process. No preexisting software application provides utility managers and planners with recovery guidance after a contamination event. Therefore, a significant disconnect remains between the numerical modeling applications, such as EPANET2, and taking the appropriate course of action during contamination episodes. Appropriate decontamination procedures are further contorted by the actions incorporated at the individual utility level, through site-specific Vulnerability Assessments (USEPA, 2002a, c), Emergency Response Plans (ERPs) (USEPA, 2003b, 2004a), and Sanitary Surveys (USEPA, 1995, 2001).

Vulnerability assessments include a series of analyses: (1) a review of pipes and constructed conveyances; (2) physical barriers; (3) water collection, pretreatment, treatment, storage and distribution facilities; (4) electronic, computer or other automated systems which are utilized by the public water system; (5) the use, storage, or handling of various chemicals; and the operation and maintenance of the system. ERPs include plans, procedures, and identification of equipment that can be implemented or utilized in the event of a terrorist or other intentional attack on the public water system. The ERP also includes actions, procedures, and identification of equipment which can prevent or significantly lessen the impact of terrorist attacks or other intentional actions on public health and safety and the supply of drinking water provided to communities and individuals. Sanitary surveys review a water system's source water (identifying sources of contamination using results of source water assessments where available), facilities, equipment, operation, maintenance and monitoring compliance in order to evaluate the adequacy of the system, its sources and operations, and the distribution of safe drinking water. The findings from these efforts at the local level are paired with information produced by government and trade associations and provided to output decision models for contamination events.

Artificial intelligence (AI), in the form of computer algorithms, have successfully been applied to bridge the gap between data acquisition and heuristic decision making. AI has been used extensively in the industrial sector (Fonseca et al., 2003; Delen and Pratt, 2006; Moynihan, 2004; Moynihan et al., 2006) and business community (Ahn et al., 2000; Nemati et al., 2002; Bahrammirzaee, 2010; Shen et al., 2011) to turn data into knowledge. AI application has also migrated into additional disciplines, including the water industry, where it has been demonstrated as an aid to modeling water quality (Panda et al., 2004; Purkait et al., 2008; Kisi et al., 2013) estimating water quantity (Nourani et al., 2012), predicting wastewater treatment performance (Rene et al., 2008), and optimizing distribution system design (Suribabu and Neelakantan, 2006). Thus, AI demonstrates tremendous, cross-disciplinary functionality in decision support.

Expert systems are a form of AI designed to mimic the decision making faculties of human experts. As AI inherently mimics human intelligence, expert systems replicate the intelligence of the human expert (Engelmore and Feigenbaum, 1993). Expert systems are commonly composed of three modules: a knowledgebase, inference engine, and user interface. The fact base represents a composition of knowledge broken down into its most fundamental form. The rule base describes the interaction between different facts. The fact base and rule base form if-then conditional relationships and compose the knowledgebase. The knowledgebase consists of knowledge acquired from models, sensors, the literature, experts, and the user. The inference engine drives the interaction between the user, rule base, and fact base. The user interface provides the user a means of interaction with the system (Feinsten and Hadden, 1989). Interaction is typically accomplished through the system asking the user a series of questions. Thus, the interaction resembles a conversation between a user and a human expert.

Expert systems have been used to improve distribution system performance and provide recommendations for sensor placement in water distribution networks. Sandeep and Rakesh (2011) demonstrate that expert systems are an effective means of improving the performance of aging distribution systems through the linkage of heuristic and mechanistic models. In particular, Sandeep and Rakesh have shown that the C Language Integrated Production System (CLIPS) is an effective expert system shell in distribution system contexts. Chang et al. (2012a, b) successfully used rule-based decision support for guidance in optimum sensor placement across large and small scale distribution systems. This research indicated that the functional capacity is adaptable to both large and small systems. These examples demonstrate scalable rule based, expert system capability in drinking water decision support in management, maintenance, and emergency scenarios. In this optic, The University of Alabama, Western Kentucky University, The University of Kentucky, the University of Missouri, and the University of Louisville teamed together and developed an analysis and decision support system named *Water Expert*. *Water Expert* includes an expert system shell and provides guidance to water sector owners and operators for the decontamination of water systems after a significant chemical or biological agent propagates within their distribution system. This tool differs from those developed by the USEPA in that it unites computational tools with heuristic decision making in order to provide the user with recommendations on appropriate decontamination strategies.

Water Expert provides guidance to local, state, and regional water system owners and operators during their initial response and subsequent decision making regarding contaminated water systems. The system provides guidance by incorporating information acquired through literature reviews, tabletop exercises, and technology demonstration and deployment workshops conducted by each participating university into a knowledgebase. *Water Expert* includes both graphical tools and user-assisted menus for helping water system personnel select the appropriate course of action relative to the nature and extent of contamination. This tool uses the Network Decontamination Model (NDM), a modular extension to the hydraulic modeling software KYPipe. The NDM identifies what valves need to be closed to isolate a particular pipe. It calculates the associated volume of water that may have to be remediated and identifies hydrants that are connected to the contaminated section. The NDM also determines whether closing different valves will causes pressure changes within the system (KYPipe and University of Kentucky, 2012; KYPipe, 2014).

Water Expert then generates a local fact base derived from the user and compares the fact base to the knowledgebase using a pattern matching, inference engine. Interaction with this system occurs through an online Content Management System (CMS), Drupal 6 (Drupal, 2008), embedded with the expert system shell, (CLIPS, 2013). *Water Expert* generates warnings regarding the public and environmental health effects of a contaminant. It also informs the user if contaminant concentration exceeds both drinking water and source water regulatory concentrations. Additionally, the system offers recommendations for the most effective treatment technologies, including detailed procedures and recommendations and provides fact sheets on common contaminants.

The tool has the ability to integrate third party applications, including EPANET2 and Geographic Information System (GIS) data. It also provides the adaptive extensibility needed to refine the system to meet the needs of individual utilities.

With this framework, the meshing of heuristic decision making and numerical measurement is accomplished by gaining perspective on how decontamination decisions are currently made. The following sections detail how the knowledge domains in *Water Expert* were developed and intertwined with the computational, regulatory, trade association, and academic knowledge domains. This accomplishment represents the first instance of combining heuristic knowledge with computational tools to provide assistance to WDNs during contamination events. The design of the system encompasses the numerous disciplines that guide final decisions. Thus, the decision making process was fully investigated when developing *Water Expert*. The process of designing *Water Expert* looked to examine three specific research questions: (1) what is the current decision process undertaken by water utilities during contamination events? (2) Can a computer system design encompass the decision needs of contaminated distribution systems? (3) Can we combine heuristic decision making with existing distribution system computational models?

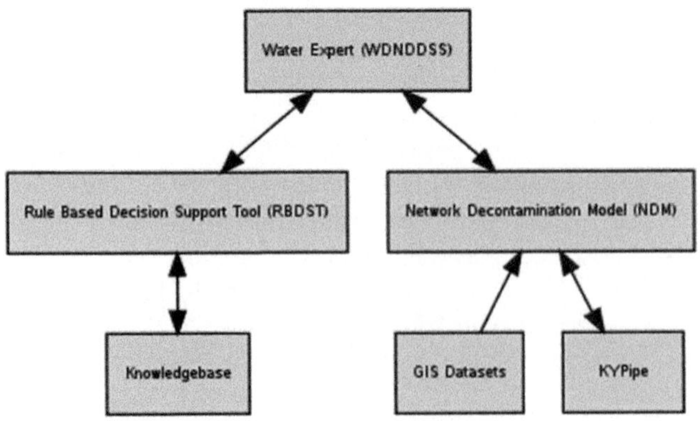

Figure 1. Flow diagram of the *Water Expert* system.

3 Methodology

Water Expert is a product coalesced from research conducted at six water research and development centers at five participating universities; research by the participating entities eventually amalgamated and comprised the architecture of WDNDDSS. Figure 1 illustrates the flow of the system.

The Water Resources Research Center and the Community Policy Analysis Center at the University of Missouri conducted a literature review that elicited a large portion of the Knowledgebase of Guidance Rules depicted in Fig. 1. The CIR at the University of Louisville conducted a series of tabletop exercises that examined the decision-making practices of key agencies involved in decontamination, further feeding the Knowledgebase. The KWRRI at The University of Kentucky and KYPipe partnered to develop the Network Decontamination Model (NDM), where GIS data and KYPipe were integrated to provide visualization and analysis tools to feed data into *Water Expert*.

These projects then coalesced within the *Water Expert* system framework by the EI at The University of Alabama and CWRS at Western Kentucky University. This meshing provides an integrated, novel platform with which utilities make timely and informed decisions on how to best decontaminate their distribution system. CWRS and EI hosted a series of technology demonstration workshops in which utility administrators and regional planners examined the final product. Three participating utilities then hosted a subsequent round of technology deployment workshops in order to display the final product. Each research product provides intelligence regarding the current decision making process during WDN contamination events. Thus, the following text provides insight into the current decision paradigm. Further, *Water Expert* addresses how to effectively combine heuristic decision making and mathematical model outputs to assist WDN decontamination processes. To determine the validity of such software, researchers queried academic literature, consulted water utilities, and developed a computational tool. The following sections detail how each of these objectives were accomplished.

3.1 Literature review and state of the art decontamination report

The Water Resources Research Center and the Community Policy Analysis Center at the University of Missouri conducted an extensive literature review of government and research databases. This review examined the current methodology and technology available for decontaminating drinking water systems. Examination produced a robust knowledge domain for validating decontamination strategies.

With the inception of the Bioterrorism Act of 2002 and the Homeland Security Presidential Directive 7 (HSPD 7), the USEPA took charge in facilitating resilient water infrastructure (US Department of Homeland Security, 2003; USEPA, 2002b). The USEPA then produced a series of guidance documents and tools for utilities to utilize in order to improve their contamination resiliency. These guidance documents form the girth of fused information currently available for contamination prevention and recovery. These products included detailed systematic methodologies for response and recovery during contamination events (USEPA, 2003a, c, d, e, f, 2004b, c, d), guidelines for preparing Emergency Response Plans (USEPA, 2003b, 2004a), procedures for preparing vulnerability assessments (USEPA, 2002a, c), and a database of contaminants that pose significant threat to water systems (USEPA, 2013a). In particular, Module 6 of the USEPA Response Protocol Toolbox provided a consolidated table of treatment technologies and their effectiveness in treating broad classifications of contaminant species. Table 1 presents a variation (USEPA, 2004c).

Review of publicly available research databases, such as *Academic Search Premier* and *Scopus*, provided additional detailed knowledge concerning the decontamination process. Researchers reviewed and condensed relevant literature item into a concise summary. This search found nearly 900 pieces of relevant literature with information on technologies and methodologies pertinent to the decontamination of a water system.

Overall, results indicated that the course a utility takes towards decontamination will likely depend upon factors including contaminant characteristics, distribution system characteristics, volume of effected water, extent of the contamination, and resources available for response. *Water Expert*'s design considers these decision making characteristics.

3.2 Tabletop exercise

The CIR at the University of Louisville hosted a workshop involving utilities, emergency responders, and other stakeholders in order to gain perspective on the relevant decision points in making an informed decontamination decision. These sessions follow the tabletop exercise format developed

Table 1. Technologies used to decontaminate water and their relative effectiveness, adapted from USEPA Module 6: Response and Recovery Guide (USEPA, 2004c).

Technology	Inorganic Chemicals	Microbes	Radionuclides	Non-volatile Organic Chemicals	Volatile Organic Chemicals
Activated Alumina	More Effective	Not Effective	More Effective	Insufficient Data	Insufficient Data
Activated Carbon	Less Effective	Insufficient Data	Less Effective	More Effective	More Effective
Air Stripping	Not Effective	Not Effective	Not Effective	Not Effective	Less Effective
Chloramination	Insufficient Data	Less Effective	Not Effective	Insufficient Data	Insufficient Data
Chlorination	Less Effective	More Effective	Not Effective	Insufficient Data	Less Effective
Chlorine Dioxide	Less Effective	More Effective	Not Effective	Insufficient Data	Less Effective
Coagulation/Filtration	Less Effective	More Effective	Not Effective	Less Effective	Not Effective
Direct Filtration	Insufficient Data	More Effective	Insufficient Data	Insufficient Data	Insufficient Data
Ion Exchange	More Effective	Not Effective	More Effective	Not Effective	Not Effective
Microfiltration/Ultrafiltration	Not Effective	Insufficient Data	More Effective	Insufficient Data	Not Effective
Ozonation	Less Effective	More Effective	Not Effective	Less Effective	Less Effective
Reverse Osmosis/Ultrafiltration	Not Effective	More Effective	More Effective	More Effective	More Effective
Ultraviolet Disinfection	Not Effective	More Effective	Not Effective	Insufficient Data	Insufficient Data
Advanced Oxidation	Less Effective	More Effective	Not Effective	More Effective	More Effective

by the Homeland Security Exercise and Evaluation Program (HSEEP) (US Department of Homeland Security, 2013) and focused explicitly on recovery of the distribution network. A centralized mind-mapping theme organized the thoughts created during the exercise. A mind-mapping expert effectively transfered the interconnectivity and centralized themes of the discussions (Margulies and Maal, 2002). As the decision making activities during the recovery process are nonlinear, this mind-mapping format translated well as an approach to organizing the thoughts generated during the tabletop exercises (Stephens and Hermus, 2007).

Participants engaged amongst themselves and with CIR researchers within an exercise that examined the decision making process which would occur during a simulated contamination event. Participants began the exercise at a point in the contamination narrative in which the contaminant ceased movement through the system and the situation was stable. From this point, they then generated a set of questions that must be answered in order to effectively make the correct decision. These questions became the critical decision points elicited from the exercise. Table 2 is a summary table of these decision points categorized by general areas of concern. Figure 2 presents the groups mind map.

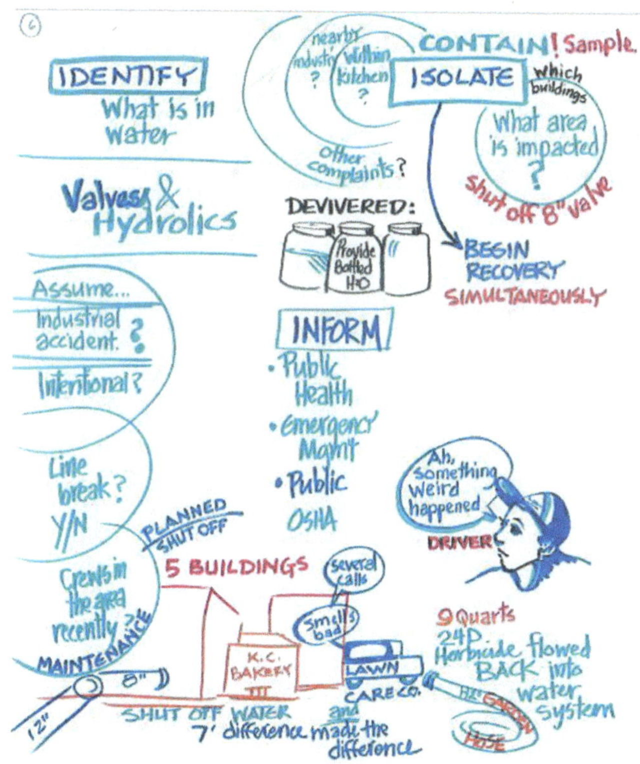

Figure 2. Mind map of hypothetical contamination scenario in which tabletop exercise participants determined key events and participant questions.

3.3 Network Decontamination Model (NDM)

The NDM, a module to KYPipe and product of cooperation with the KWRRI at the University of Kentucky, addresses the spatial issues related to potential contamination events. KYPipe is a widely used commercial, water distribution modeling software. The NDM is a graphical utility that can generate a schematic of a utility's distribution system using preexisting geographic information system (GIS) data. The NDM analyzes the extent of contamination within the distribution network.

The Kentucky Infrastructure Authority's (KIA's) preexisting online portal of water infrastructure data, the Water Resources Information System (WRIS) (Kentucky Infrastructure Authority, 2013), prototypes the NDM and GIS interoperability. The WRIS database includes GIS information on pipelines, water tanks, water treatment plants, water meters,

Table 2. Summarized decision points critical to response following a contamination event.

General Area of Concern	Decision Points
Contaminant	(1) What is the contaminant? (2) Is the contaminant isolated to single site? Or impacting multiple sites? (3) Is the contaminant biological? Is it contagious? (4) What if the contamination event occurs during inclement (freezing temperatures)? (5) Is this contaminant a vehicle for another contaminant? (6) Is the type of contaminant known or unknown? (7) What are the environmental and public health risks with the associated contaminant?
Laboratory Support	(1) What laboratory support will be needed and how will this be coordinated? (2) Will local laboratory have the capacity to meet testing demands? (3) If additional facilities are needed, how will they be selected? Overseen? (4) Where can processes and procedures for coordination of laboratory analytical support be found? (5) What EPA laboratory network member laboratories may be available to provide analytical support and how can these laboratories be identified?
Recovery	(1) When does response end and recovery begin? (2) What facilities and/or activities are most critical and require immediate attention at the conclusion of response activities? (3) What tools are available to assist in identification of contaminants, water-treatment methods and infrastructure decontamination? (4) What procedures are in place to expedite the recovery process and return to normal business operations? (5) When is the drinking water considered safe and how will the public be convinced that it is safe? (6) What is the recommended sampling protocol? (7) How will the utility ensure that the backflow does not re-contaminate the system? (8) What are the economic impacts of long-term outages? (9) If the contamination was intentional, how does that change the response and responsibilities? (10) At what point would you have a debriefing to identify lessons learned?
Critical Customers	(1) Have the vulnerable customers been identified? Do you have appropriate contact information for them? (2) How will impacted critical care customers be handled during long-term outages? Evacuation? Shelter in place? (3) What if a school is impacted? Alternative schools? (4) What are the short-term and long-term health effects of the contaminant on humans?
Infrastructure	(1) Do we know the pipeline material affected by the contaminant? (2) What is the permeability of the pipeline? (3) What is the age of the affected pipeline? (4) Is there information on the contaminant's short-term and long-term impact on pipeline materials, valves and gaskets? (5) Is there information on the contaminant's interactions with permeable materials in the distribution system? (6) Are the locations of the existing pipelines and valves known? (7) Is there is a hydraulic model of the distribution system? (8) Can the impacted area be isolated quickly in an interconnected system? (9) What can remote-sensing systems provide? (10) Where does the utility's responsibility to clean impacted infrastructure end? At the meter? (11) What is the contaminant's impact on residential infrastructure and water-using appliances? Copper pipes? Water heaters? Ice makers? (12) Whose responsibility is it to replace or fix impacted residential infrastructure and appliances?
Communications	(1) How will communications be handled, including to internal personnel, to the public, to the business community and to elected and other government officials? (2) Which local government officials and offices need to be informed? (3) Who is responsible for internal, public and other communication activities during the recovery from this incident? (4) Who is responsible for communicating with the local critical customers (medical facilities, nursing homes, dialysis facilities, public school system) impacted by the event? (5) What state and local regulators (Division of Water, Emergency Management, Public Health, OSHA, and EPA) need to be informed? (6) What procedures have been previously developed for use during this type of emergency? (7) How would internal communication procedures change if normal communication links are disrupted? (8) How would it be determined if an advisory or public notification (such as boil water or do not use) needs to be issued for this incident? How would this occur without power and if normal communication media such as TV and radio outlets are unavailable? What alternative resources could be available to issue an advisory? (9) Where can you find information for release to the media regarding potential water contaminants? (10) How will the utility restore public confidence in the product provided? (11) How would the utility handle communications if the contamination incident was an accident versus intentional? (12) How does a utility discern between an intentional and accidental contamination event?

Table 2. Continued.

General Area of Concern	Decision Points
Alternative Water Sources	(1) Does the impacted utility have the capacity to supply potable water after the contamination event? If not, what are the alternative sources? (2) How will the alternative water sources be provided during long-term outages? (3) Does the utility have recommended delivery methods? Central distribution location? Temporary lines? (4) Who will pay for providing alternative water sources during long-term outages? Does this change if the contamination event was intentional? (5) How will the impacted utility recoup the associated costs? (6) How will critical care customers be supplied during long-term outages? (7) At what point can treatment facilities be established? (8) How will fire protection services be addressed in the contaminated area?
Coordination	(1) What agencies or groups will the utility coordinate with concerning recovery and remediation? (2) What are the expectations for their support? (3) What specific coordination procedures have been developed to ensure successful coordination?
Business Continuity	(1) During emergency operations including recovery and remediation, what outside contract services are available to provide services? (2) What priority will your contractors give you if other business and city operations are competing for the same equipment or services? (3) Who is responsible for covering the cost of impacted equipment (ice makers, laboratory equipment and manufacturing equipment)? (4) How does the utility assist small businesses?
Logistics	(1) What procedures have been developed to address logistical support (e.g., food, shelter and equipment for responders) during this incident? (2) What procedures and/or provisions are in place to support personnel with special needs (e.g., lost or damaged personal property, injured or killed family/friends and psychological impact from devastation) due to the incident? (3) If service is out for a prolonged period, how is alternative water going to be provided to impacted customers?
Local and State Coordination	(1) How do current financial systems that track reimbursable expenses coordinate with local and state requirements?
Federal Coordination	(1) What type of federal support do you expect to receive related to your long-term recovery operations? (2) Many types of federal aid are tied to restoring infrastructure to pre-event status – what are your options regarding incorporating enhancements or planned improvements as part of this restoration? (3) If the conditions in this scenario were to escalate and federal assistance was needed, what are the expectations for support and which agency would be contacted? Who will coordinate this effort at your utility?
Mutual Aid/Assistance (MAA)	(1) If your utility does not have enough personnel during the incident, what other options are available and have arrangements been made in advance? (2) How are you going to manage the demands for long-term recovery, and should these needs be addressed in your MAA agreement?
Law Enforcement	(1) What inspection or surveillance programs are in place to detect any physical security breaches in utility appurtenances, such as water storage facilities and the distribution system? Does your utility have a "neighborhood watch" or "water watchers" program in place within the served community?
Finance and Administration	(1) What procedures have been developed to address incident-related expenses? (2) How are financial and other incident records maintained? How will the lost revenue of impacted business be covered? What about the cost of lost products (food spoilage, contaminated products)? (4) Once a recovery plan is established, what is the financial impact on customers and the utility? (5) Will the utility's actions or inactions lead to legal liability?

and pump stations. A direct linkage to this repository exists within the NDM and allows the user to seamlessly download and generate a model for his or her distribution system. Figure 3 is an example NDM network model.

Functionally, the NDM provides the user with the ability to examine the spatial distribution of valves and hydrants, important for either isolating the contaminant or for ex-situ remediation of the network. Users can alter contaminant spread through opening and closing the valves and hydrants. Additionally, the use of icons and backdrop maps visualize the locations of critical customers, such as hospitals and schools (Fig. 4).

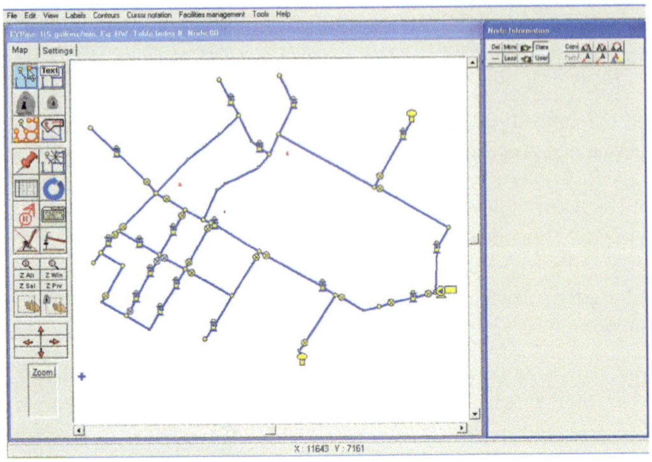

Figure 3. Example network in the Network Decontamination Model (NDM). The NDM is an extension to KYPipe, a commercially available, hydraulic modelling software.

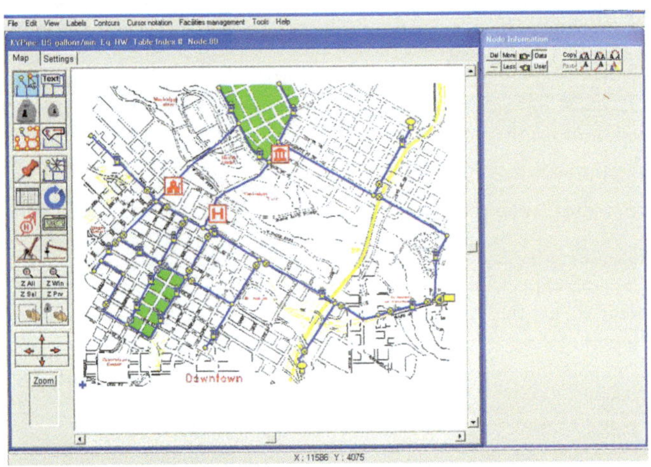

Figure 4. NDM with additional icons depicting the location of critical users and a backdrop map.

Once the NDM conceptualizes a full contamination event, the model calculates the volume of contaminated water, summarizes the valves and hydrants in the contamination area, identifies the age and material of pipe in the contamination area, and identifies the hydrant with the lowest elevation, important if the system is to be flushed. These outputs are then reported to the user, where he or she can either use the data independently or export it for use within *Water Expert*. Figure 5 is an example of hypothetical contamination event, created using the NDM (KYPipe and University of Kentucky, 2012).

The NDM constitutes the first instantiation of a mathematical model plugin developed for *Water Expert*. The outputs of an NDM run are exported into *Water Expert* to refine the knowledgebase and provide additional refinement of *Water Expert* recommendations.

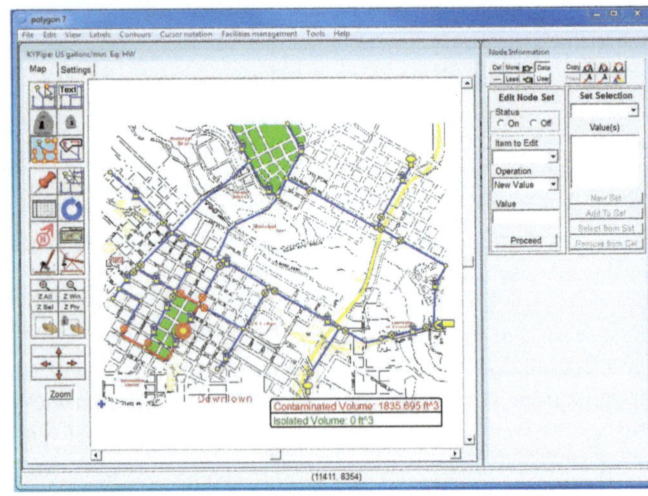

Figure 5. Example contamination event within the NDM.

3.4 Rules Based Decision Support Tool (RBDST)

The knowledgebase in *Water Expert* is a product of the guiding principles of knowledge engineering. According to McCorduck and Feigenbaum (1983), this philosophy takes the knowledge associated with complex problem-solving abilities, normally requiring an advanced level of human expertise, and converts this knowledge into a computerized format. The process of performing this conversion includes four steps: (1) information is gathered using literature reviews, workshops, and seminars; (2) information is extracted from these sources; (3) information is compounded into a taxonomical structure that comprises a knowledgebase; and (4) the knowledgebase is integrated into computerized modelling or simulation environment for functionality and additional knowledge extraction. The CWRS at Western Kentucky University and the EI at The University of Alabama developed the *Water Expert* knowledgebase. *Water Expert* unites data gathered during the initial literature review, state of the art decontamination report, tabletop exercises, and through discussions with an external advisory board of industry, regulatory, and trade association representatives. A MySQL relational database management system (RDBMS) holds the data (MySQL, 2014) and the Drupal 6 CMS brings the data to the user (Drupal, 2008). The CLIPS inference engine drives the rules and facts that comprise the knowledgebase (CLIPS, 2013).

3.5 Technology deployment and implementation workshops

Four deployment workshops evaluated the prototype version of *Water Expert*. Western Kentucky University, the University of Kentucky, and KYPipe hosted these workshops. Participants included water utility personnel, emergency preparedness representatives, public health officials, regional planners, and regulators. Training materials introduced participants to the software's functionality and guided him or her through usage of *Water Expert* and the NDM. Participants filled out two separate evaluation forms for each piece of software to solicit feedback on the presentation, the presentation materials, and the features of *Water Expert*.

Subsequent to the technology demonstration workshops, project personnel used the feedback gathered therein to refine the software and ultimately to develop a revised version of the prototype *Water Expert*. This refined product was then launched at three utilities who aided the research team in conceptualizing plausible contamination scenarios for formal validation. These three hypothetical contamination scenarios provided further demonstration and evaluation of the software's effectiveness.

4 Results

Water Expert includes both graphical tools and menus that allow the user to select the appropriate course of action relative to the nature and extent of contamination. Through a preferred web browser, users can access *Water Expert*. The primary target users are water utility managers or operators. Researchers assume that *Water Expert* will be useful to academics, regulators, public health officials, emergency preparedness representatives, and local and regional planners. The *Water Expert* code is separate from the core CMS installation and is a PHP modular extension to Drupal 6.

The following sections summarize the functionality of the system. Figure 6 is a system flow chart for the final software product that summarizes its functionality. The decision points provided in Table 2 constitute the basis of this system flowchart. Each of these portals are entrances where the user interacts with the *Water Expert*. They exist on the Drupal 6 CMS, the central hub of interaction with *Water Expert*. User input gathered at these portals become "local facts" that compare against the system-wide "global facts", derived from preceding subprojects. The CLIPS inference engine drives this comparison. All reports and recommendations are graphically output to the user using the CMS. The CMS is the *Water Expert* user portal and provides the extensibility necessary to update and expand the knowledgebase. Fact and rule generation can occur on demand for the user, and engages the expert system inference engine.

Water Expert preliminarily addresses contaminants contained within the United States Environmental Protection Agency National Primary Drinking Water Regulations. The concentrations or amount of contaminants that provide warnings and recommendations by *Water Expert* are those believed to have an effect on both human (ATSDR, 2013; USEPA, 2013b) and environmental health (USEPA, 2014a). Technology recommendations come from findings of the USEPA. These findings address treatment effectiveness on specified species of contaminants, as previously depicted in

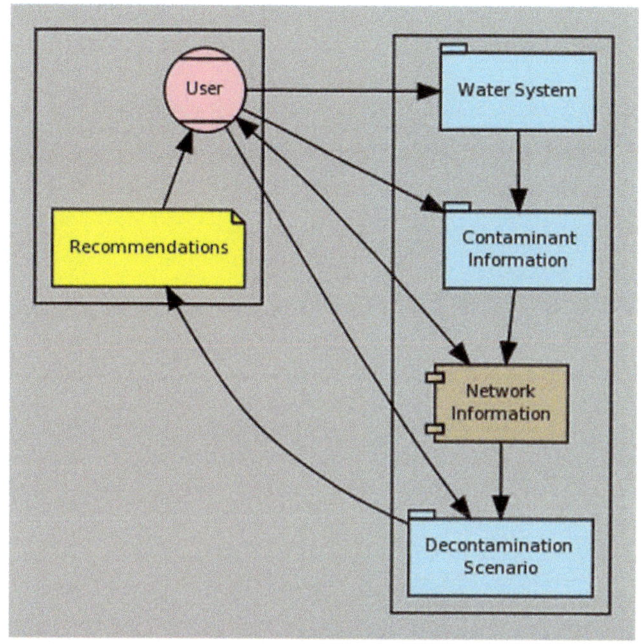

Figure 6. Water Expert decontamination system flowchart.

Table 1. Optionally, outputs from the NDM export to a CSV file for upload into *Water Expert*. These outputs upload into the MySQL RDBMS, and generate additional facts to compare against the knowledgebase. For example, this information interprets the possibility of contaminated piping material interacting with the contaminant(s) of interest.

The University of Missouri literature review and the University of Louisville table top exercise found that decisions made during contamination events are multi-faceted and multi-disciplinary. Concerns alluded to in Table 2 include inputs that consider contaminate characteristics, recovery timetables, infrastructure impact, public relations, government regulations, financial burden, logistics, mutual aid agreements, public health, environmental, social impact, and economic impact. Interaction with utilities throughout the development of *Water* Expert emphasized that water providers struggle to coalesce each of these concerns into informed actions. In many cases, water utilities will simply flush contaminated water from the system without considering secondary affects.

Though the current core incarnation of *Water Expert* does not fully address all of these concerns, the following sections detail that the system is capable of providing recommendations on contaminant characteristics, government regulations, public health, environmental health, infrastructure impact, and treatment alternatives. A condensed and easily readable report provides the recommendations. The researchers envision that such a report will guide the decision maker in addressing a subset of the issues detailed in Table 2.

If the user develops a KYPipe model, he or she can begin to gain a greater understanding of a particular contamination

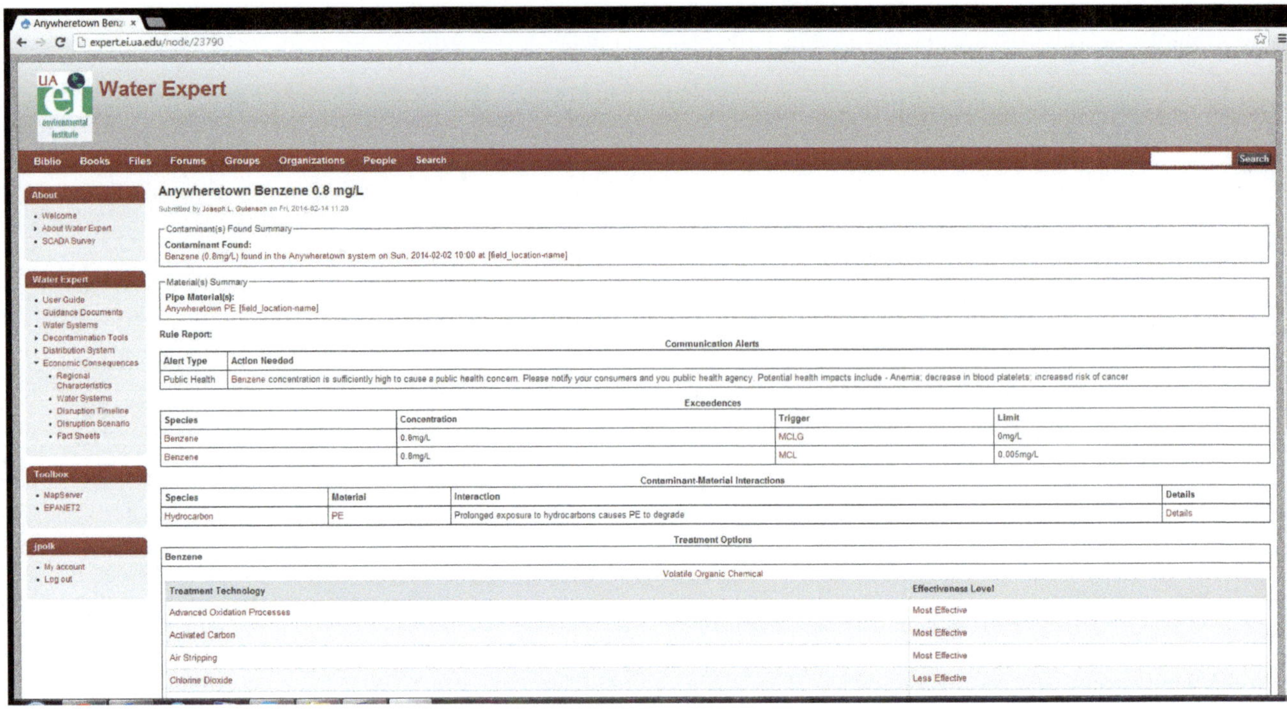

Figure 7. Example recommendation page generated by *Water Expert*.

event. With the NDM users can understand the spatial nature of a contamination event and may also estimate the number of pipes and volume of water which has been affected. This further drives the inference engine within *Water Expert* and augments the final report that *Water Expert* generates.

4.1 User input and system output

A four-step user-input process initiates through the CMS on a local web browser (Fig. 6). This process begins with guiding the user through developing a profile of her or his distribution system including name, administrative contact, address, population served, etc. These "Water System" data are not used by the analysis portion of the current system, but are an indexing mechanism to gather information pertinent to the contamination event. Information on the specific contaminant collects at the "Contaminant Information" portal where the user provides a detailed review of the contaminants found during the contamination event, including the location and concentration of the contaminant. *Water Expert* allows the user to either enter the network information manually or utilize the NDM. The system uses these data to determine whether network infrastructure will be degraded by the contaminant(s) of concern.

Along the data gathering pathway, facts are generated on demand and conglomerated at the "Decontamination Scenario" portal. Once this information is submitted, MySQL ports the generated facts to a server-side instance of CLIPS. CLIPS executes pattern matching between the "local facts" and "global facts". Once the system concludes its analysis, a "Recommendations" page instantiates and renders to the user on the CMS. Here the system warns the user of both the public and environmental health concerns, regulatory concentration exceedance, and contaminant-network material interaction. *Water Expert* additionally supplies a weighted list of treatment technologies to decontaminate the system. For a particular contaminant, the weighted list of treatment technologies communicates to the end user the uncertainty associated with the use of a particular treatment technology. Figure 7 is an example of these recommendations.

When the user has access to a KYPipe hydraulic model of their distribution system, the NDM augments *Water Expert* recommendations. The NDM output exports as a comma separated value (CSV) file which is then read by *Water Expert*. This NDM export file automatically feeds *Water Expert* with piping material for contaminant-pipe interaction. Additionally, the NDM export file provides summarization of the anticipated volume of contaminated water and number of pipes affected or the extent of the contamination in the "Recommendations" page. The system does not currently address workforce shortfalls, aside from providing a weighted listing of possible remediation measures.

4.2 Fact and rule creation

Facts created by both the user (local facts) and site administrators (global facts) through the CMS provides *Water Expert* with a guided protocol that facilitates the extensibility required to update and maintain the evolving knowledgebase. Using PHP code and the Drupal Content Creation Kit (CCK) (Drupal, 2013), CCK content types exist for each type of fact. For example, a *Contaminant* content type generates facts on the contaminants of interest. Data entered are then stored in the RDBMS and analyzed and synthesized by PHP into the fact format necessary for the CLIPS inference engine. Rules are also created in a similar fashion. The CMS also removes facts and rules. The CMS is this mechanism that provides flexibility to update and use knowledge in an expedited manner.

4.3 Training, education and guidance

Prescribed pathways within the CMS represent pre-defined endpoints for typical use cases. These pathways streamline the knowledge retrieval process and simplify user interaction. Fact sheets provide information on USEPA-regulated contaminants and consist of pre-packaged content, which can be retrieved through point and click interaction with the CMS. These fact sheets provide rapid access and address specific aspects of a contaminant(s) of concern. Additionally, interactive guidance documents allow the end-user to peruse decontamination information in a self-determined sequence. Information in guidance documents present hierarchical access with increasing detail. An installation of GraphViz (2014) engages interactive guidance, presenting the user with flowchart visualization. Figure 8 is an example of the interactive GraphViz charts used to navigate the guidance documents.

5 Case study

In order to illustrate the functionality of the system in its current capacity, envision a city named Anywheretown, located adjacent to a major interstate. In this town, the risk of possible chemical spills caused by vehicular wrecks is high. In several US states, a large portion of accidents involving hazardous chemical spills involve those with petroleum oils and fuels (Becker et al., 2000; Golla et al., 2012). Therefore a spill, intentional or unintentional, involving gasoline is a plausible scenario.

Assuming that such an event occurs and that, insufficient cleanup activities are performed at the spill site allow the gasoline to remain within the soil matrix and interact with nearby water distribution system piping. Ong et al. (2010) and the American Water Works Association (AWWA) and Economic & Engineering Services, Inc. (2002) demonstrate that this incident will cause thermoplastic piping (polyethylene (PE), polyvinyl chloride (PVC), etc.) to degrade and allow the substance to permeate the piping. Assuming that the Anywheretown distribution system is entirely comprised of polyethylene and the piping is relatively new, this degradation process would usually occur on the order of several weeks.

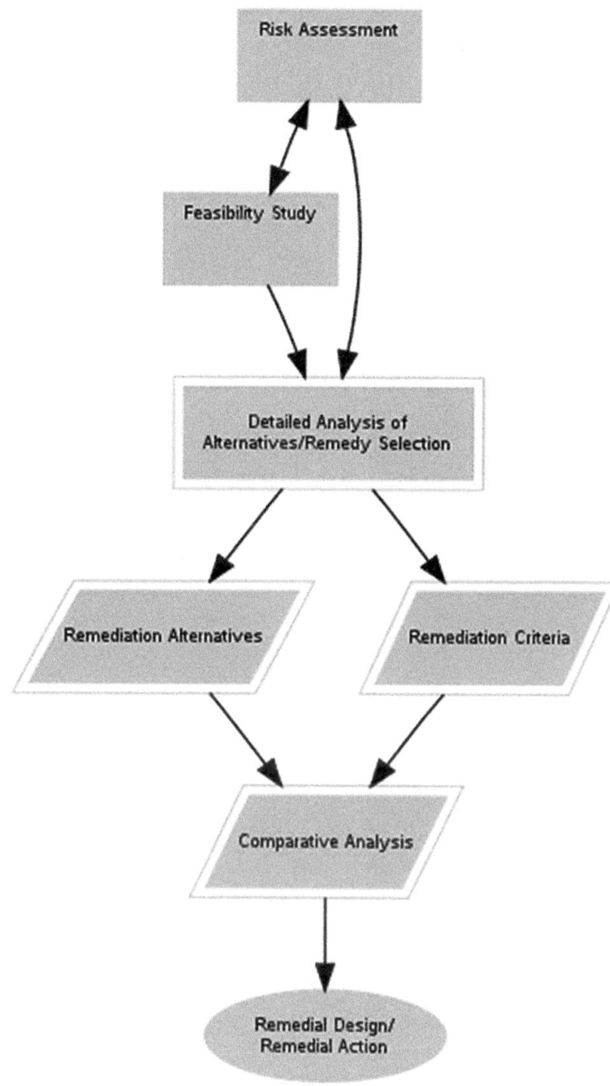

Figure 8. Interactive flowcharts used to navigate the guidance documents. This particular example is based on content derived from USEPA (2004c).

However, for the substance to enter the distribution system, a sufficient drop in pressure at the point of contamination must have occurred. Such a drop in pressure may be the result of a system ill-prepared to supply fire flow, a part of the system being taken offline which isolates sections of the network from service, system pumps going down, and/or intrusion events (Fleming et al., 2005; Besner et al., 2010). Thus, one of these events drove the network to zero or negative pressure at the point of injection, allowing the contaminant to permeate the polyethylene piping and enter the distribution system.

The Anywheretown Water Department then begins receiving taste and odor complaints from customers in the general vicinity of the spill location. Testing reveals that Benzene, a major constituent of gasoline is present at a 0.8 mg L^{-1} con-

Table 3. Anywheretown example data inputs into *Water Expert*.

Water System Location	Anywheretown Interstate	
	Contaminant Information	Network Information
Contaminant	Benzene	
Concentration	0.8	
Concentration Units	$\mathrm{mg\,L^{-1}}$	
Date and Time	2/14/2014 10:00 CST	
Material		Polyethylene

Table 4. Benzene contamination alert.

Alert Type	Action Needed
Public Health	Benzene concentration is sufficiently high to cause a public health concern. Please notify your consumers and you public health agency. Potential health impacts include – Anemia; decrease in blood platelets; increased risk of cancer

centration. Unsure of what to do, the Anywheretown decision makers consult the *Water Expert* system about the situation, entering into *Water Expert* the data found in Table 3. The content type categorizes these inputs. Water System and Decontamination Scenario information are excluded from this table as they are only used as an indexing mechanism. Network Information was added manually, in lieu of no available network information.

The inputs generate the summarized recommendation in Tables 3, 4, 5, and 6. These tables demonstrate *Water Expert*'s ability to summarize the data found to be critical to the decontamination process by the University of Missouri literature review. The decision points found as part of the tabletop exercises conducted by the University of Louisville.

6 Further research

Present capacity to offer decontamination assistance will be complimented with the development of tools which will perform comprehensive examination of the economic effects of disruptions in water service (Alva-Lizarraga and Johnson, 2012; Gutenson et al., 2013). Additionally, tools to help utilities develop monitoring and control systems for their distribution network are under development. Supplementary projects intend to address capacity development issues for infrastructure systems, such as Asset Management and ERP development. Additionally, embedded hydraulic modelling and integrated GIS will be used to expand *Water Expert*'s capabilities. These upgrades will work to address a broader range of issues detailed in Table 2.

Each individual summary prepared in the literature review will extend the knowledgebase of *Water Expert* in the form of facts and rules. These additions will augment the existing fact and rule base and provide drill down capabilities within *Water Expert* for users who desire more descriptive information. An example of this functionality is the system's current ability to recognize the interaction between hydrocarbons, such as diesel and gasoline fuels, with thermoplastic piping.

Table 5. Benzene exceedence alert.

Species	Concentration	Trigger	Limit
Benzene	$0.8\,\mathrm{mg\,L^{-1}}$	MCLG	$0\,\mathrm{mg\,L^{-1}}$
Benzene	$0.8\,\mathrm{mg\,L^{-1}}$	MCL	$0.005\,\mathrm{mg\,L^{-1}}$

Table 6. Benzene contaminant-material interactions.

Species	Material	Interaction
Hydrocarbon	Polyethylene (PE)	Prolonged exposure to hydrocarbons causes PE to degrade

The system is moving from a web-deployed environment to a localized, enterprise-wide cloud based system, similar to that depicted in Fig. 9. Concerns voiced by utility and trade association representatives who hesitate to place sensitive data in a remote database, drive this migration. Thus, future iterations of *Water Expert* will take advantage of the concept's adaptive nature and will transition to fat-client applications installed on user's local machines and interact with local databases. This new format will be designed to interact with a utility's local databases and supply them with recommendations at an institutional level. Inputs and outputs supplied to the system through utility representative's mobile and desktop hardware facilitate immediate feedback and response among all members of the organization.

Abroad, similar on-going efforts in the European Union include *ISIS*, *TAWRA_RTM*, *CATO*, *Safewater Project* and *Secureau*. *Secureau* and the *Safewater Project* focus on developing methods, tools, and mathematical models by which contaminant mitigation can best be achieved (Science Daily, 2013; Safewater Project, 2014). *CATO* is a project designed to provide computerized decision support through integration with water networks existing systems (CATO, 2012). *TAWARA_RTM* focuses on developing real time radionuclide sensor networks (TAWARA_RTM, 2015) ISIS is developing a sensor-based risk analysis system for intelligent remediation following contamination events (ISIS, 2015). Each of these provide valuable collaboration opportunities to enhance the core decontamination module of *Water Expert*.

Further, the nature of the system allows for its use in other areas including natural hazards, such as flooding events, to provide decision makers with recommendations and summaries to complement their actions during these calamities.

Table 7. Benzene Treatment Options based on USEPA (2004c).

Treatment Technology	Effectiveness Level
Advanced Oxidation Process	Most Effective
Activated Carbon	Most Effective
Air Stripping	Most Effective
Chlorine Dioxide	Less Effective
Chlorination	Less Effective
Ozonation	Less Effective
Ultraviolet (UV) Disinfection	Less Effective
Microfiltration, Ultrafiltration	Not Effective
Reverse Osmosis (RO) and Nanofiltration (NF)	Not Effective
Coagulation/Filtration	Not Effective
Ion Exchange	Not Effective
Activated Alumina (AA)	Insufficient Data
Chloramination	Insufficient Data
Direct Filtration	Insufficient Data

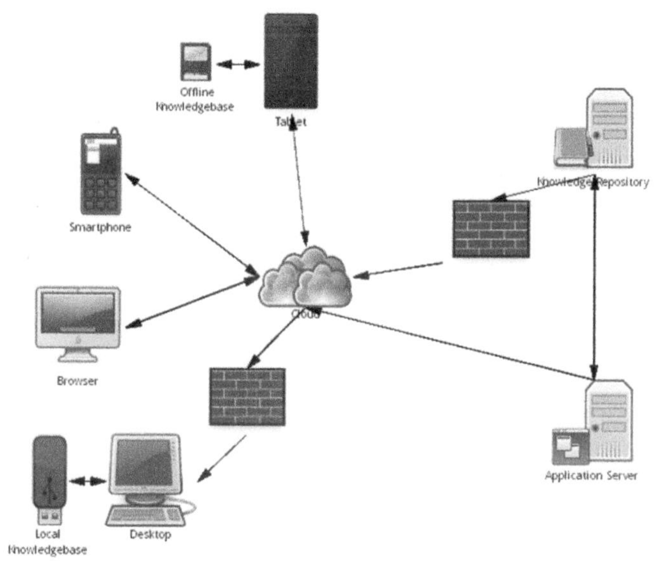

Figure 9. Future cloud deployed environment of *Water Expert*. It is envisioned that this environment will exist as a local, enterprise level system for individual utilities.

7 Conclusions

Decisions made during contamination events are multi-faceted and multi-disciplinary. Utilities detail that they face a broad spectrum of challenges when in the midst of a contamination event. During these episodes, water utilities must consider contaminate characteristics, recovery timetables, infrastructure impact, public relations, government regulations, financial burden, logistics, mutual aid agreements, public health, environmental, social impact, and economic impact. In time sensitive environments, water providers struggle to coalesce each of these concerns into informed actions. In many cases, water utilities will absolve to utilize the most available means of mitigating contamination, without considering all aspects of the decision made.

The functionality of *Water Expert* serves to provide a comprehensive set of tools for which the user can explicitly determine the most viable pathway for decontamination following a contamination event. *Water Expert* provides recommendations by generating a local fact base derived from the user and compares this information to the knowledgebase of global facts and rules using the CLIPS pattern matching, inference engine. This system facilitates interaction with the user through an online instance of the Drupal 6 CMS, with an embedded CLIPS expert system shell. *Water Expert* generates public and environmental health warnings, determines if concentrations exceed regulatory limits, and assesses effectiveness of a wide range of commonly utilized treatment technologies. Additionally, the system provides pre-defined endpoints representing the typical use cases. These pre-defined end points currently exist in the form of fact sheets provided for all USEPA regulated contaminants and interactive guidance documents that guide the user through the decontamination process.

The development of this tool has involved most, if not all, key stakeholder groups in every step of the creative process. From the literature review which tapped into the academic and research aspects of the decontamination process, to the series of tabletop and demonstration workshops that provided hands-on opportunities for practitioners in the water industry. The schema allows the researchers to identify the critical needs of the industry and align them with the overarching development goals for *Water Expert*. This form of development has created a living application that can grow and adapt with the dynamics of the industry.

Key to the functionality of *Water Expert* is its adaptability. The conceptual framework of the system has allowed researchers to integrate *Water Expert* with third-party applications, like the NDM, which opens the door for more robust coupling of models from an assortment of academic disciplines. Intuitively, with this expansion will come an increase in the data available for the inference engine to process, likely promoting a more attuned set of recommendations. To this end, the functionality of *Water Expert* is also completely independent of any software that it currently utilizes. Thus, while a functional system, the schematic of the system can adapt to any necessary changes that arise.

Like many previous hybrid DSS/expert system applications, the system demonstrates the ability to meld heuristic decision making with procedural models. *Water Expert* is novel in that it utilizes the outputs of the NDM to inform the user on decontaminating their distribution system. No prior DSS applications have provided the ability to inform WDNs during contamination events. This aspect of the system is modular as the spectrum of end users is encompassed by the platform. Thus, the system is not limited to WDNs heavily invested in cyber infrastructure. The *Water Expert* framework comprises an initial step at providing decontamination assistance to all WDNs. Though not novel in concept, the framework is the first of its kind to attempt at providing decontamination assistance for WDNs.

Acknowledgements. The US Department of Homeland Security (DHS), Science & Technology Directorate, provided funding for this project through a technology development and deployment program managed by The National Institute for Hometown Security (NIHS).

References

Agency for Toxic Substances & Disease Registry (ATSDR): http://www.atsdr.cdc.gov/mrls/index.asp (last access: 30 March 2014), 2013.

Ahn, B. S., Cho, S. S., and Kim, C. Y.: The integrated methodology of rough set theory and artificial neural network for business failure prediction, Expert Syst. Appl., 18, 65–74, doi:10.1016/S0957-4174(99)00053-6, 2000.

Alva-Lizarraga, S. and Johnson, T. G.: A dynamic demand driven, supply constrained input output approach to modeling economic impacts of water disruption events, 2012 Conference Proceedings, Mid-Continent Regional Science Association, Ottawa, Canada, 6–8 June 2012, 159–177, 2012.

American Water Works Association (AWWA) and Economic and Engineering Services, Inc.: Permeation and leaching, US Environmental Protection Agency, Washington, DC, 22 pp., 2002.

Bahrammirzaee, A.: A comparative survey of artificial intelligence applications in finance: Artificial neural networks, expert system and hybrid intelligent systems, Neural Comput. Appl., 19, 1165–1195, doi:10.1007/s00521-010-0362-z, 2010.

Becker, S. M., Pitt, R., and Clark, S.: Environmental health, public safety, and social impacts associated with transportation accidents involving hazardous substances, Project Summary, University Transportation Center for Alabama, Tuscaloosa, AL, 192 pp., 2000.

Besner, M. C., Ebacher, G., Jung, B. S., Karney, B., Lavoie, J., Payment, P., and Prévost, M.: Negative pressures in full-scale distribution system: field investigation, modelling, estimation of intrusion volumes and risk for public health, Drink. Water Eng. Sci., 3, 101–106, doi:10.5194/dwes-3-101-2010, 2010.

Drupal: https://drupal.org/drupal-6.0 (last access: 31 March 2014), 2008.

Drupal: https://drupal.org/project/cck (last access: 30 March 2014), 2013.

CATO: http://www.cato-project.eu/ (last access: 10 June 2015), 2012.

Chang, N., Pongsanone, N. P., and Ernest, A.: A rule-based decision support system for sensor deployment in small drinking water networks, J. Clean. Prod., 29–30, 28–37, doi:10.1016/j.jclepro.2012.02.010, 2012a.

Chang, N., Prapinpongsanone, N., and Ernest, A.: Optimal sensor deployment in a large-scale complex drinking water network: Comparisons between a rule-based decision support system and optimization models, Comput. Chem. Eng., 43, 191–199, doi:10.1016/j.compchemeng.2012.03.012, 2012b.

CLIPS: A Tool for Building Expert Systems: http://clipsrules.sourceforge.net/ (last access: 31 March 2014), 2013.

Delen, D. and Pratt, D. B.: An integrated and intelligent DSS for manufacturing systems, Expert Syst. Appl., 30, 325–336, doi:10.1016/j.eswa.2005.07.017, 2006.

Engelmore, R. S. and Feigenbaum, E.: Expert systems and artificial intelligence, in: Knowledge Based Systems in Japan, edited by: Engelmore, R. S., Loyola College in Maryland, Baltimore, Maryland, 1993.

Feinsten, J. L. and Hadden, S. G.: Introduction to expert systems, J. Policy Anal. Manag., 8, 182–187, doi:10.1002/pam.4050080202, 1989.

Fleming, K. K., Gullick, R. W., Dugandzic, J. P., and LeChevallier, M. W.: Susceptibility of potable water distribution systems to negative pressure transients, New Jersey Department of Environmental Protection, Trenton, NJ, 55 pp., 2005.

Fonseca, D. J., Navaresse, D. O., and Moynihan, G. P.: Simulation metamodeling through artificial neural networks, Eng. Appl. Artif. Intel., 16, 177–183, doi:10.1016/S0952-1976(03)00043-5, 2003.

Golla, V., Taylor, R., Suhl, J., Eagleson, J., Chavan, P., Grigsby, R., and Bottom, J.: Louisville/ Jefferson county hazardous material commodity flow analysis, Kentucky Division of Emergency Management, Frankfort, KY, 142 pp., 2012.

Graphviz – Graph Visualization Software: http://graphviz.org/, last access: 30 March 2014.

Gutenson, J. L., Ernest, A., Oubeidillah, A. A., and Zhang, X.: Development of a python enabled software application for assessing economic impacts of disruptions in water service, Alabama Water Resources Association Conference, Orange Beach, AL, #46, September 2013.

ISIS: http://isis-project.eu/news-items, last access: 10 June 2015.

Kentucky Infrastructure Authority (KIA): http://kia.ky.gov/wris/data.htm (last access: 30 March 2014), 2013.

Kisi, O., Akbari, N., Sanatipour, M., Hashemi, A., Teimourzadeh, K., and Shiri, J.: Modeling of dissolved oxygen in river water using artificial intelligence techniques, Journal of Environmental Informatics, 22, 92–101, doi:10.3808/jei.201300248, 2013.

KYPipe: http://kypipe.com/decon, last access: 31 March 2014.

KYPipe and University of Kentucky: NDM: Network decontamination model user's manual. KYPipe, LLC & The University of Kentucky, Lexington, KY, 84 pp., 2012.

Margulies, N. and Maal, N.: Mapping Inner Space: Learning and teaching visual mapping, 2nd Edn., Corwin, Newbury Park, CA, 160 pp., 2002.

McCorduck, P. and Feigenbaum, E. A.: The Fifth Generation: Artificial intelligence & Japan's Computer Challenge to the World, Addison Wesley Publishing Company, Reading, MA, 288 pp., 1983.

Moynihan, G. P.: Web-based expert system for foundry pollution prevention, SPIE Proceedings 5262, Environmentally Conscious Manufacturing III, Providence, RI, 27 February 2004, 2004.

Moynihan, G. P., Saxena, P., and Fonseca, D. J.: Development of a decision support system for procurement operations, International Journal of Logistics Systems and Management, 2, 1–18, 2006.

Murray, R., Janke, R., and Uber, J.: The threat ensemble vulnerability assessment (TEVA) program for drinking water distribution system security, Proceedings of the 2004 World Water and Environmental Resources Congress, 2004 World Water Congress, Salt Lake City, Utah, 27 June to 1 July 2004, 1–8, 2004.

Murray, R., Haxton, T., Janke, R., Hart, R., Berry, J., and Phillips, C.: Sensor network design for drinking water contamination warning systems: A compendium of research results and case studies using the TEVA-SPOT software, US Environmental Protection Agency, Cincinnati, OH, 80 pp., 2010.

MySQL: http://www.mysql.com/, last access: 31 March 2014.

Nemati, H. R., Steiger, D. M., Iyer, L. S., and Herschel, R. T.: Knowledge warehouse: An architectural integration of knowledge management, decision support, artificial intelligence and data warehousing, Decis. Support Syst., 33, 143–161, doi:10.1016/S0167-9236(01)00141-5, 2002.

Nourani, V., Baghanam, A. H., and Gebremichael, M.: Investigating the ability of artificial neural network (ANN) models to estimate missing rain-gauge data, Journal of Environmental Informatics, 19, 38–50, doi:10.3808/jei.201200207, 2012.

Ong, S. K., Gaunt, J., and Mao, F.: Impact of hydrocarbons on PE/PVC pipes and pipe gaskets, AWWARF, Denver, CO, 232 pp., 2010.

Panda, S. S., Garg, V., and Chaubey, I.: Artificial neural networks application in lake water quality estimation using satellite imagery, Journal of Environmental Informatics, 4, 65–74, doi:10.3808/jei.200400038, 2004.

Purkait, B., Kadam, S. S., and Das, S. K.: Application of artificial neural network model to study arsenic contamination in groundwater of malda district, eastern india, Journal of Environmental Informatics, 12, 140–149, doi:10.3808/jei.200800132, 2008.

Rene, E. R., Joo, K. S., and Park, H. S.: Experimental results and neural prediction of sequencing batch reactor performance under different operational conditions, Journal of Environmental Informatics, 11, 51–61, doi:10.3808/jei.200800111, 2008.

Safewater Project: https://www.safewater-project.eu/ (last access: 10 June 2015), 2014.

Sandeep, K. and Rakesh, K.: CLIPS based decision support system for water distribution networks, Drink. Water Eng. Sci., 4, 37–50, doi:10.5194/dwes-4-37-2011, 2011.

Sandia National Laboratories: https://software.sandia.gov/trac/spot (last access: 31 March 2013), 2010.

Science Daily: www.sciencedaily.com/releases/2013/07/130704094851.htm (last access: 9 June 2015), 2013.

Shen, W., Guo, X., Wu, C., and Wu, D.: Forecasting stock indices using radial basis function neural networks optimized by artificial fish swarm algorithm, Knowl.-Based Syst., 24, 378–385, doi:10.1016/j.knosys.2010.11.001, 2011.

Stephens, P. and Hermus, C.: Making art connections with graphic organizers, The Art Education Magazine for Teacher, 106, p. 55, 2007.

Suribabu, C. R. and Neelakantan, T. R.: Particle swarm optimization compared to other heuristic search techniques for pipe sizing, Journal of Environmental Informatics, 8, 1–9, doi:10.3808/jei.200600072, 2006.

TAWARA_RTM: http://www.caen.it/csite/CaenProfList.jsp?parent=_173&Preview=_Y, last access: 10 June 2015.

US Department of Homeland Security: Homeland Security Exercise and Evaluation Program (HSEEP), US Department of Homeland Security, Washington, DC, 2013.

US Department of Homeland Security: Homeland Security Exercise and Evaluation Program (HSEEP), US Department of Homeland Security, Washington, DC, 2013.

US Environmental Protection Agency (USEPA): EPA/State joint guidance on sanitary surveys, US Environmental Protection Agency, Washington, DC, 1995.

US Environmental Protection Agency (USEPA): Implementation guidance for the interim enhanced surface water treatment rule, EPA 816-R-01-011, US Environmental Protection Agency, Washington, DC, 2001.

US Environmental Protection Agency (USEPA): Instructions to assist community water systems in compliance with public health security and bioterrorism preparedness and response act of 2002, EPA 810-B-02-001, US Environmental Protection Agency, Washington, DC, 2002a.

US Environmental Protection Agency (USEPA): http://water.epa.gov/infrastructure/watersecurity/lawsregs/bioterrorismact.cfm (last access: 30 March 2014), 2002b.

US Environmental Protection Agency (USEPA): Vulnerability assessment factsheet, EPA 816-F-02-025, US Environmental Protection Agency, Washington, DC, 2002c.

US Environmental Protection Agency (USEPA): Overview and application, EPA 817-D-03-007, US Environmental Protection Agency, Washington, DC, 2003a.

US Environmental Protection Agency (USEPA): Large water system emergency response plan outline: Guidance to assist community water systems in complying with the public health security and bioterrorism preparedness and response act of 2002, EPA 810-F-03-007, US Environmental Protection Agency, Washington, DC, 2003b.

US Environmental Protection Agency (USEPA): Module 1: Water utilities planning guide, EPA 817-D-03-001, US Environmental Protection Agency, Washington, DC, 2003c.

US Environmental Protection Agency (USEPA): Module 2: Contamination threat management guide, EPA 817-D-03-002, US Environmental Protection Agency, Washington, DC, 2003d.

US Environmental Protection Agency (USEPA): Module 3: Site characterization and site guide, EPA 817-D-03-003, US Environmental Protection Agency, Washington, DC, 2003e.

US Environmental Protection Agency (USEPA): Module 4: Analytical guide, EPA 817-D-03-004, US Environmental Protection Agency, Washington, DC, 2003f.

US Environmental Protection Agency (USEPA): Emergency response plan guidance for small and medium community water systems, EPA 816-R-04-002, US Environmental Protection Agency, Washington, DC, 2004a.

US Environmental Protection Agency (USEPA): Module 5: Public health response guide, EPA 817-D-03-005, US Environmental Protection Agency, Washington, DC, 2004b.

US Environmental Protection Agency (USEPA): Module 6: Remediation and recovery guide, EPA 817-D-03-006), US Environmental Protection Agency, Washington, DC, 2004c.

US Environmental Protection Agency (USEPA): Response guidelines, EPA 817-D-04-001, US Environmental Protection Agency, Washington, DC, 2004d.

US Environmental Protection Agency (USEPA): Decontamination and recovery planning: Water and wastewater utility case study, EPA 817-F-08-004, US Environmental Protection Agency, Washington, DC, 2008.

US Environmental Protection Agency (USEPA): Containment and disposal of large amounts of contaminated water: A support guide for water utilities, EPA 817-B-12-002, US Environmental Protection Agency, Washington, DC, 2012.

US Environmental Protection Agency (USEPA): http://water.epa.gov/scitech/datait/databases/wcit/index.cfm (last access: 31 March 2014), 2013a.

US Environmental Protection Agency (USEPA): http://water.epa.gov/drink/contaminants/ (last access: 31 March 2014), 2013b.

US Environmental Protection Agency (USEPA): http://water.epa.gov/scitech/swguidance/standards/criteria/current/index.cfm, last access: 31 March 2014a.

US Environmental Protection Agency (USEPA): http://cfpub.epa.gov/si/si_public_record_report.cfm?address=_nhsrc/&dirEntryId=_218488, last access: 31 March 2014b.

US Environmental Protection Agency (USEPA): http://cfpub.epa.gov/si/si_public_record_report.cfm?dirEntryId=_253555&fed_org_id=_1253&address=_nhsrc/si/&view=_desc&sortBy=_pubDateYear&showCriteria=_1&count=_25&searchall=_%27water%20security%27, last access: 31 March 2014c.

Water Research Foundation: Guidance for decontamination of water system infrastructure, No. 2981, Water Research Foundation, Denver, CO, 2009.

Water Sector Coordinating Council: Critical infrastructure and key resources sector-specific plan as input to the national infrastructure protection plan, EPA 817-R-07-001, US Environmental Protection Agency, Washington, DC, 2007.

Water Sector Decontamination Working Group: Recommendations and proposed strategic plan: Water sector decontamination priorities, American Water Works Association, Denver, CO, 2008.

22

Optimization of coagulation-flocculation parameters using a photometric dispersion analyser

S. R. Ramphal and M. S. Sibiya

Rand Water, Process Technology Department, Glen Vista, South Africa

Correspondence to: S. R. Ramphal (sramphal@randwater.co.za)

Abstract. The size and structural characteristics of floc particles are important design and control parameters in water treatment and should be rapidly monitored with a reasonable amount of accuracy. In this study, a photometric dispersion analyser (PDA) coupled to standard jar test experiments was used to optimize coagulation-flocculation parameters while monitoring floc size and structure as well as the rate of floc formation during coagulation using alum. The optimal coagulation conditions were as follows: sample pH 8; alum dosage, 3 mg L^{-1} as Al^{3+}; G value, 172 s^{-1}; rapid mixing time, 20 s. These conditions resulted in unstable treated water having a calcium carbonate precipitation potential (CCPP) of −15 mg L^{-1} as $CaCO_3$ and required a slaked lime dosage of 17 mg L^{-1} as $CaCO_3$ to equilibrate CCPP to acceptable levels. PDA data revealed that aggregation rate and steady-state variance are primary parameters as both have substantial influence on coagulation-flocculation efficiency. However, the average steady state ratio, although an important parameter, had a lessened impact on coagulation-flocculation efficiency. The results of this study showed that the PDA instrument is an important tool in coagulation kinetic studies and can be employed as an additional tool in the optimization of coagulation conditions.

1 Introduction

The treatability of a water source, including the determination of optimal conditions for specific parameters (most effective coagulant, dosage rates, pH, velocity gradient and mixing time), is conventionally performed at bench-scale using standard jar test experiments. The performance of these experiments is generally determined through measuring turbidity and/or colour removal. Although this method of performance measure can provide sufficient information to permit pilot- or full-scale implementation, it does possess inherent limitations as it provides no information on floc particle size and structure.

The size distribution of floc particles is an important design and control parameter during coagulation-flocculation (Spicer et al., 1996). In combination with floc particle structure and density, the size distribution of floc particles is of great significance in solid-liquid separation processes such as sedimentation (Selomulya et al., 2001). Floc size and structure are influenced by several operational parameters and directly impacts floc density which controls solid removal efficiency during sedimentation (Spicer and Pratsinis, 1996).

During coagulation-flocculation, both floc particle size and structure undergo transitional phases. Initially, upon coagulant dosage, there is a period of rapid floc growth in which the size and structure of the floc aggregates are dynamic in nature. This is followed by a steady- state region or equilibrium between particle growth and fragmentation for a specific period, during which, the floc particle size does not change. The shape of the particle size distribution curve is critical as it influences coagulation-flocculation kinetics, the rate of floc growth, floc characteristics and solid removal processes (Selomulya et al., 2001; Spicer et al., 1996). It is therefore very important to rapidly quantify both size and structure of aggregates to generate a particle size distribution curve with a reasonable amount of accuracy (Selomulya et al., 2001).

One of the most widely used techniques employed to determine aggregate structure and size distribution are light scattering techniques. These techniques are advantageous as

Table 1. Raw water characteristics.

Parameter	pH	Turbidity (NTU)	UV_{300} (cm^{-1})	Temperature (°C)
Average	7.93	55.57	0.4223	20

they offer an estimation of a wide range of particle sizes in a rapid and non-destructive manner (Selomulya et al., 2001). A photometric dispersion analyser (PDA) has become an important instrument that has been widely used in monitoring coagulation-flocculation kinetics (Mixon et al., 2013; Ball et al., 2011; Staaks et al., 2011; Zoubolis and Tzoupanos, 2009, 2010; Xiao et al., 2009; Huang and Liu, 1996).

A PDA is a fibre-optical monitor which measures the fluctuation in the intensity of light transmitted through a flowing suspension (Huang and Liu; 1996). More specifically, it measures the root mean square of the fluctuating signal (V_{rms}) and the average transmitted light intensity (d_c). The ratio of V_{rms} to d_c is a valuable tool in coagulation-flocculation studies as it permits quantification of aggregation or disaggregation of the suspension to be monitored (Huang and Liu; 1996). This ratio can be derived as follows (Xiao et al., 2009):

$$\text{Ratio} = \left(\frac{L}{A}\right)^{1/2} \left(\sum N_i C_i^2\right)^{1/2} \qquad (1)$$

where L is the optical path length, A is the effective cross sectional area of the light beam, N_i is the number and concentration class of size class i and C_i is particle scattering cross-section of size class_i respectively. The ratio values obtained from the PDA is affected only by $\left(\sum N_i C_i^2\right)^{1/2}$, and is almost entirely unaffected by contamination of the tube walls in the flow cell and by drift in electrical components (Xiao et al., 2009). This study is an extension of the one presented at the 12th edition of the International Conference on "Computing and Control for the Water Industry – CCWI 2013" (Ramphal and Sibiya, 2014). Similar to the CCWI 2013 study, this study used a PDA instrument to optimize coagulation conditions and monitor floc size and structure as well as the rate of floc formation using alum as a coagulant. Standard jar test experiments were performed concurrently with PDA experiments to determine the degree of particle destabilization. In addition, this study applied operational guidelines used at a large water utility in Africa for dosage optimization and assessed its impact on floc size and structure as well as the rate of aggregation. Further, the impact of alum dosage on the stability of treated water was also evaluated.

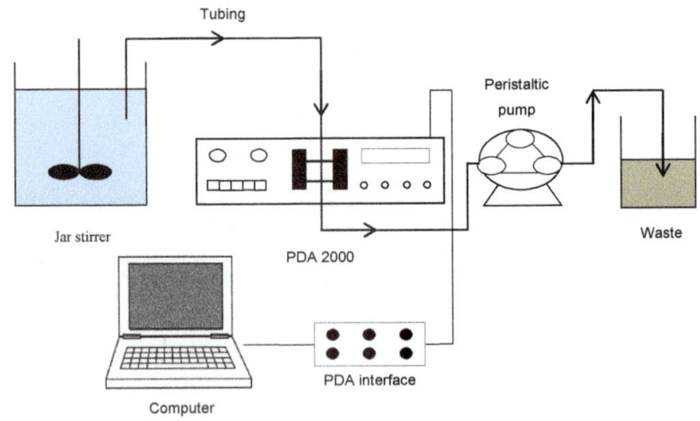

Figure 1. Experimental set up.

2 Methodology

2.1 Jar tests

Jar test experiments were performed on raw water obtained from the Vaal Dam (South Africa) (Table 1). Jar tests were carried out using a variable speed ZR 4–6 Jar tester (Zhongrun Water Industry Technology Development Co. Ltd, China) fitted with six flat paddle impellors. For each jar test the following procedure was performed: one litre of raw water was added to each Perspex jar; the pH of samples were adjusted to the relevant pH using either 0.1 M hydrochloric acid or 0.1 M sodium hydroxide; appropriate dosages of alum were dosed into each jar using plastic syringes; various velocity gradients were applied to the sample using electronically controlled programmes; upon completion of electronically controlled programme samples were left to settle for 15.5 min.

2.2 Instrumentation

2.2.1 Photometric dispersion analyser

Coagulation-flocculation kinetics was determined using the PDA 2000 (Rank Brothers, England). Clear plastic tubing of 3 mm in diameter was used to transport samples from the jar to the fibre optic probes at a flow rate of $22\,\text{mL}\,\text{min}^{-1}$. Sample flow rate was controlled by a peristaltic pump which was placed after the PDA instrument to prevent floc breakage prior to analysis (Fig. 1). To ensure comparison of PDA experiments, the direct current (DC) output of source water was maintained at a constant voltage by adjusting the DC gains on the instrument. For all experiments, the limit and filter modes were used as this reduced the effects of large particles and air bubbles and increased the averaging time to obtain smoother output values, respectively. All data was recorded using an ADC-20 data-logger coupled to the Picolog Software.

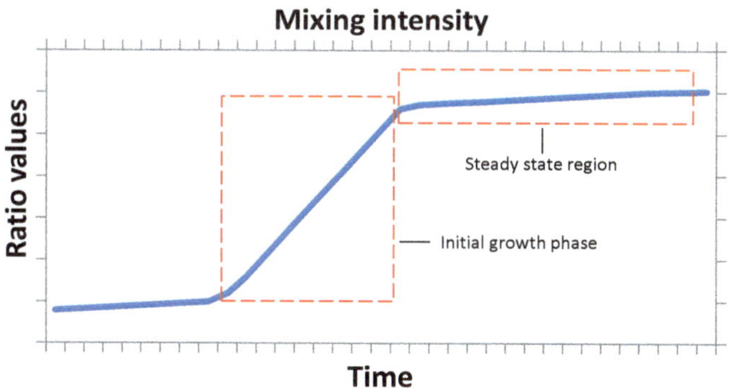

Figure 2. Time dependent ratio curve.

2.2.2 Other instrumentation

A HACH 2100Q turbidimeter and a Cary 50-Conc Varian UV-Vis Spectrophotometer was used to measure turbidity and UV_{300} values respectively. UV_{300} was used a measure of non-visible colour and is hereby referred to as colour. The temperature and pH were measured using a Metrohm 692 pH/ion meter. All titrations were performed using a Schott Titronic 96 auto-titrator while the conductivity was measured using a Labotec 197i conductivity meter.

2.3 Analysis of PDA data

The time dependent ratio values collected during each coagulation-flocculation experiment generated a *ratio* curve of similar pattern (Fig. 2). Figure 2 shows that there are two distinct regions with each indicating the evolution of aggregate size with time at various shear rates (Selomulya et al., 2001). In the initial growth region, fluid shear promotes particle collision which results in aggregate growth. As these aggregates grow, breakage becomes a more significant factor as the size range of the aggregate is more susceptible to fragmentation by turbulent eddies. This results in the attainment of steady-state region. During this phase, due to the attainment of a dynamic equilibrium between aggregate growth and aggregate fragmentation, aggregates are of equal size, since ratio values remain constant over time (Hopkins and Ducoste, 2003; Spicer and Pratsinis, 1996; Oles, 1992).

In this study, three calculations were used to analyse the data obtained during the coagulation-flocculation experiments. These calculations included the aggregation rate during rapid mixing, time-weighted average steady-state ratio and time-weighted steady-state variance as described by Mixon et al. (2013); Staaks et al. (2011); Xiao et al. (2009) and Hopkins and Ducoste (2003). The aggregation rate (AR) during coagulation was calculated by:

$$\mathrm{AR} = \frac{\mathrm{Ratio}_i - \mathrm{Ratio}_0}{\mathrm{time}_i} \tag{2}$$

where Ratio_0 is the initial ratio value at t_0, time_i is the when the PDA output reaches its maximum value of Ratio_i. In accordance to work performed by Staaks et al. (2011), initial calculations determined the aggregation rate of the first minute, first to fifth minute and the entire first five minutes. It was found that the aggregation rate for the first minute yielded the best results; hence, this parameter was used for the rest of the study. The time-weighted average steady-state ratio was determined by:

$$\overline{\mathrm{ratio}} = \frac{\sum_{i=1}^{N} (\mathrm{ratio}_i \times \mathrm{time}_i)}{\sum_{i=1}^{N} \mathrm{time}_i} \tag{3}$$

Equation (3) represents the state of aggregation during the steady state period of coagulation-flocculation. The final calculation used in this study was the time-weighted steady-state variance, which was also computed using the data from the steady state period during coagulation-flocculation as follows:

$$\mathrm{variance} = \frac{\sum_{i=1}^{N} \left[(\mathrm{ratio}_i - \mathrm{average} \ldots \mathrm{ratio})^2 \times \mathrm{time}_i \right]}{\sum_{i=1}^{N} \mathrm{time}_i} \tag{4}$$

Equation (4) was used as a measure of floc size and structural differences. According to Hopkins and Ducoste (2003), smaller variance is indicative of a tighter floc size distribution and a more homogenous, dense and less porous floc structure.

2.4 Assessment of calcium carbonate precipitation potential

The stability of treated water was determined by measuring the calcium carbonate precipitation potential (CCPP) of specific samples. The CCPP was determined using the Stasoft4© software (version 4) which is a tool for designing, modelling and controlling water treatment processes that involve calcium carbonate chemistry. To measure the CCPP, specific water quality parameters, inclusive of temperature, conductivity, dissolved calcium, pH and total alkalinity were determined.

3 Results and discussion

3.1 Optimization of pH

Jar test experiments were performed to assess turbidity and colour removal as well as coagulation-flocculation kinetics at various pH values. This was achieved by adding a constant alum dosage of $10\,\mathrm{mg\,L^{-1}}$ as Al^{3+} to samples adjusted to pH 6–9. According to Pernitsky and Edzwald (2006), favourable pH conditions for alum coagulation generally occur between

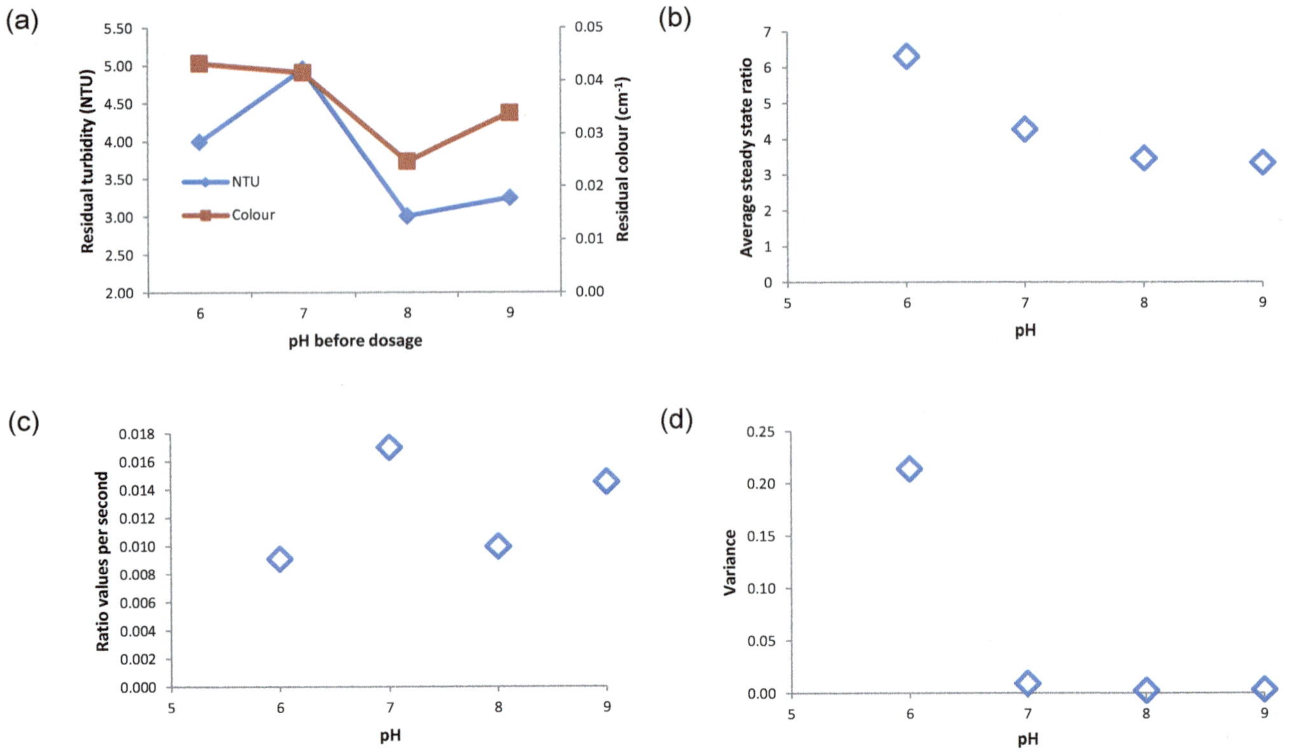

Figure 3. Optimization of pH – **(a)** residual turbidity and colour; **(b)** average steady-state ratio; **(c)** aggregation rate; **(d)** steady-state variance.

a pH of 6–7. Figure 3a indicates that the lowest residual turbidity and colour was observed at an initial sample pH of 8; corresponding to a coagulation pH of 6.29.

The optimal pH also coincided with the minimal solubility of alum, indicating that sweep flocculation is the primary coagulation mechanism through the formation of aluminium hydroxide precipitates (Pernitsky and Edzwald, 2006; Gregor et al., 1997). One would expect that the largest floc size would correspond to highest coagulation efficiency; in other words, the largest floc size and floc growth rate should result in the lowest residual turbidity and colour. However, this was not observed as the lowest residual turbidity and colour (Fig. 3a) corresponded to the lowest average steady state ratio and aggregation rate (Fig. 3b and c). Further, from Fig. 4b, one can see that at the optimal coagulation pH of 6.29, sweep flocculation should be the primary coagulation mechanism; hence, one can expect the largest floc sizes to form under these conditions. However, from Fig. 3b, a trend of increasing ratio values was observed with decreasing pH, which according to Fig. 4a should shift the aluminium hydrolysis species to ionic forms. This strongly suggests the presence of interfering contaminants within the source water.

Figure 3d shows that the lowest steady-state variance was obtained at a pH of 8. According to Mixon et al. (2013) and Hopkins and Ducoste (2003), low steady-state variance values indicates the existence of a narrow range of floc sizes and stronger flocs at a particular location. This can be expected as the coagulation pH corresponds to the minimum solubility, indicating the presence of solid aluminium hydroxide precipitates. In contrast, an initial sample pH of 6 displayed the highest steady-state variance, indicating the occurrence of a wide range of floc sizes and placing further significance on the existence of interfering contaminants within the source water. Although there was a lack of correspondence between residual and PDA data, the residual data provided sufficient evidence to indicate that an initial sample pH of 8 was optimal. This sample pH was used for the rest of the study.

3.2 Optimization of dosage

The optimization of alum dosage was performed by adjusting sample pH to 8 while varying the alum dosage from 2–12 mg L^{-1} as Al^{3+}. Figure 5a indicates that an alum dosage of 6 mg L^{-1} as Al^{3+} resulted in the lowest residual turbidity and colour. This dosage reduced the pH from 8 to 6.69 which favoured sweep flocculation through the formation of aluminium hydroxide precipitates (Srinivasan et al., 1999). Further, the isoelectric point of alum occurs at pH 8 indicating that pH values less than 8 yield positively charged precipitates, which are able to neutralize the negatively charged particles in the water sample (Gregory and Duan, 2001).

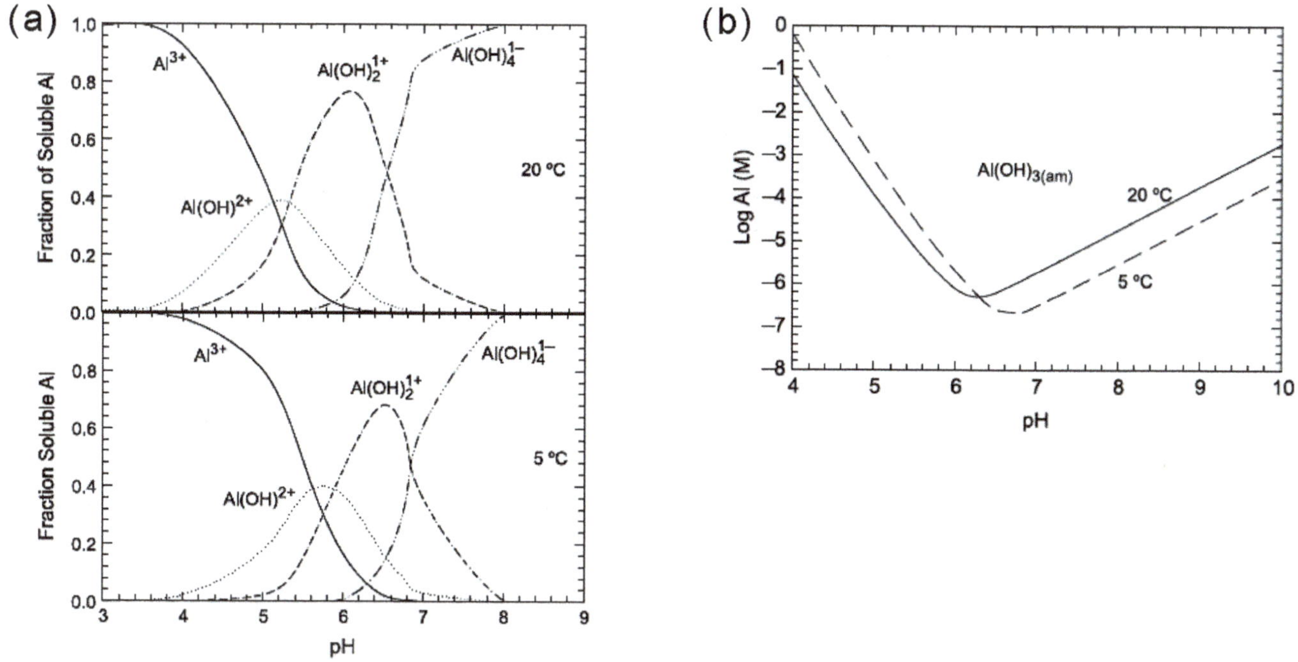

Figure 4. Distribution of aluminium hydrolysis products; **(a)** dissolved species; **(b)** amorphous precipitate (adapted from Pernitsky and Edzwald, 2006).

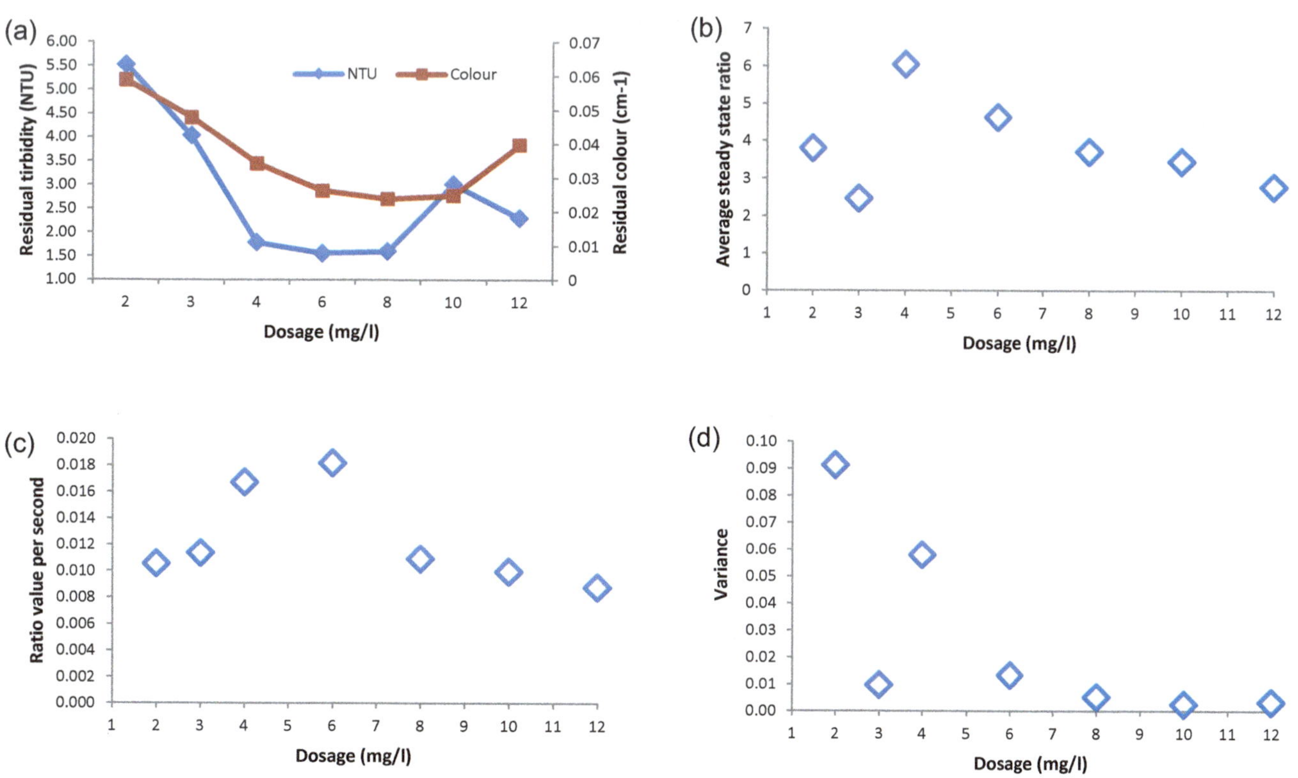

Figure 5. Optimization of dosage – **(a)** residual turbidity and colour; **(b)** average steady-state ratio; **(c)** aggregation rate; **(d)** steady-state variance.

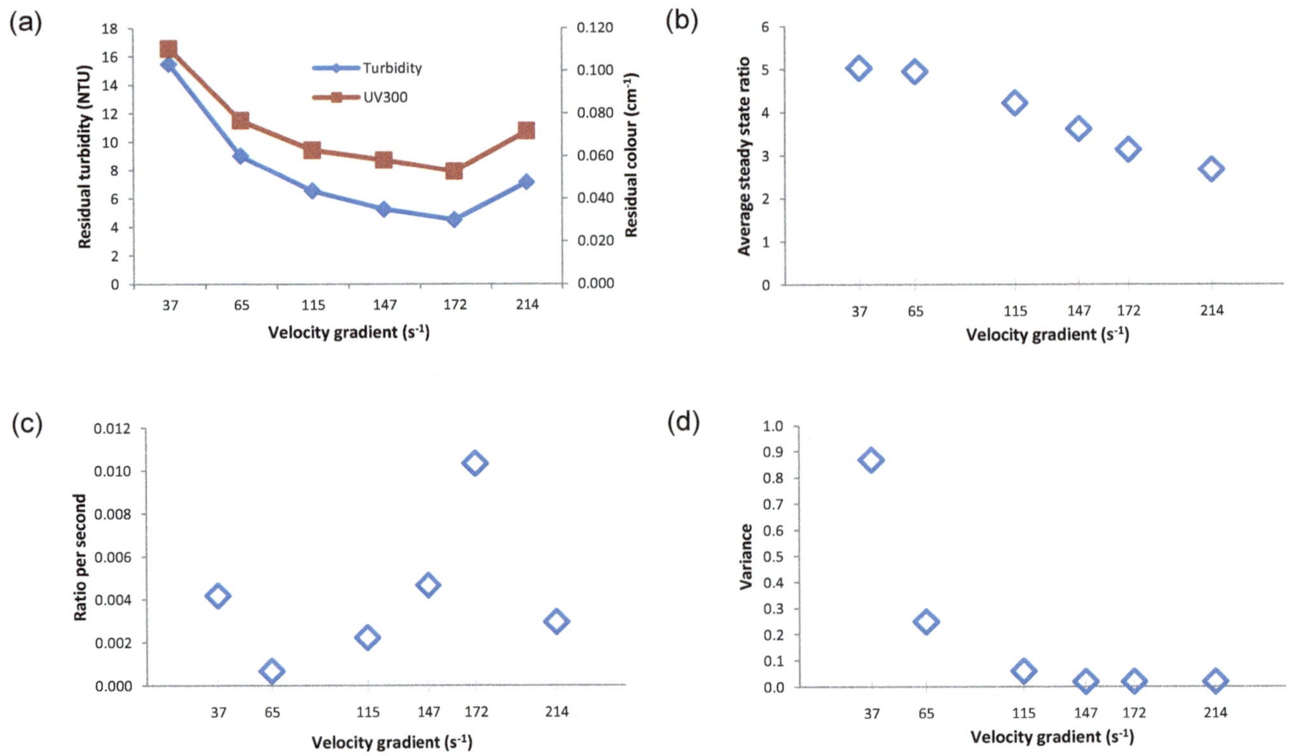

Figure 6. Optimization of velocity gradient – **(a)** residual turbidity and colour; **(b)** Average steady-state ratio; **(c)** aggregation rate; **(d)** steady-state variance.

Figure 5b shows that the largest average steady-state ratios were obtained for alum dosages of 4 and 6 $mg\,L^{-1}$ as Al^{3+}. Although an alum dosage of 6 $mg\,L^{-1}$ as Al^{3+} did not represent the largest average steady-state ratio value, it displayed the highest efficiency in terms of turbidity and colour removal. This indicates that floc size, although important in coagulation-flocculation processes, is not the most important parameter. Figure 5c and d show that an alum dosage of 6 $mg\,L^{-1}$ as Al^{3+} displayed a higher aggregation rate and lower steady-state variance to that of 4 $mg\,L^{-1}$ as Al^{3+} dosage. This indicates that an alum dosage of 6 $mg\,L^{-1}$ as Al^{3+} resulted in a higher floc growth rate and that the flocs size displayed greater degree of strength and floc size homogeneity. Based on these results, it is possible that these two parameters contributed to the enhanced performance of the 6 $mg\,L^{-1}$ as Al^{3+} alum dosage and may possibly be primary parameters for consideration in coagulation-flocculation kinetic studies.

However, at a large water utility in Africa, to reduce chemical costs, an operational guideline using the lowest coagulant dosage which yielded a post-sedimentation turbidity of less than 5 NTU is applied. Under these circumstances, an alum dosage of 3 $mg\,L^{-1}$ of Al^{3+} was optimal. This dosage displayed inferior PDA data to that of the 6 $mg\,L^{-1}$ as Al^{3+} dosage as it presented the lowest average-steady ratio and aggregation rate values signifying the presence of smaller floc particles formed at moderate rate. However, it did exhibit a lower steady-state variance value, indicating that floc particle sizes were more homogenous, denser and less porous in nature. A dosage of 3 $mg\,L^{-1}$ of Al^{3+} was used for subsequent experiments.

3.3 Optimization of velocity gradient

In this experiment, the rapid mixing intensity was varied while applying the optimal alum dosage of 3 $mg\,L^{-1}$ as Al^{3+} and initial sample pH of 8. Figure 6a shows that the lowest residual turbidity and colour was obtained during application of a G value of 172 s^{-1}. At this velocity gradient, an optimal level of energy was dissipated which warranted efficient coagulation through the instantaneous and even distribution of alum (Degremont, 2007). It may seem appropriate to use high G values during rapid mixing to maximise efficient use of a coagulant; however, each application has an upper limit (Bratby, 2006). Once G values exceed this limit, there is a delay in floc formation and size.

This trend can be observed in Fig. 6b and c in which the average steady state ratio and aggregation rate peaked at a specific G value; with further increases in G values resulting in lower values. This occurred as higher energy dissipation increases the shear rate and the number of particle collisions; hence, increasing the rate of floc break-up (Selomulya

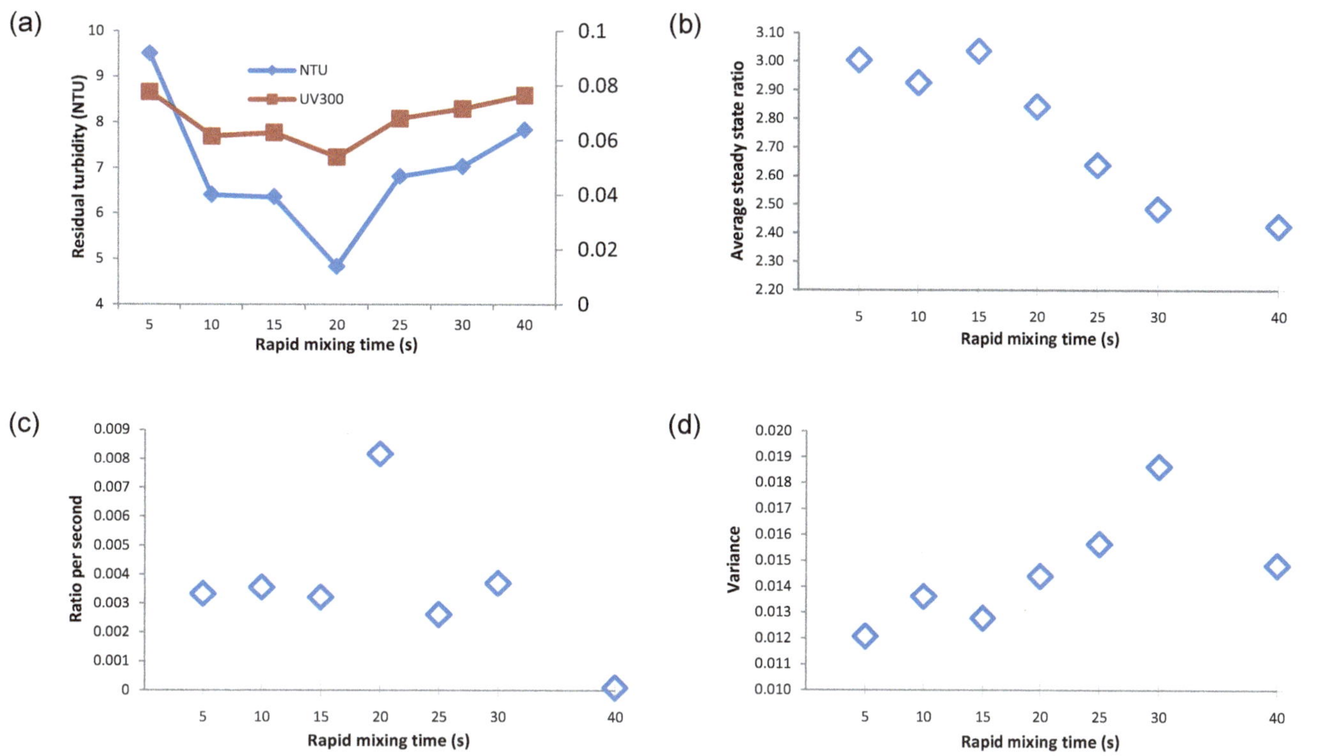

Figure 7. Optimization of rapid mixing time – **(a)** residual turbidity and colour; **(b)** average steady-state ratio; **(c)** aggregation rate; **(d)** steady-state variance.

et al., 2001). This resulted in the formation of smaller; denser and less porous aggregates which are structurally tolerant to higher shear forces, leading to a lower steady state mean floc size attained in a shorter period of time (Hopkins and Ducoste, 2003; Spicer and Pratsinis, 2001, Oles, 1992). This trend has been observed by various researchers (Hopkins and Ducoste, 2003; Selomulya et al., 2001; Spicer and Pratsinis, 1996).

Figure 6d shows that the steady-state variance was substantially higher at lower velocity gradients ($\leq 115\,s^{-1}$). Mixon et al. (2013) and Hopkins and Ducoste (2003) explained that high steady-state variance at lower G values was due to larger aggregates being sheared due to settling into the impellor region. Further, in this study, it was observed that as the G value increased, the steady-state variance decreased. This indicates that at higher shear rates, the formation of a more homogenous floc suspension was favoured (Mixon et al., 2013). As observed in prior experiments, the largest average steady-state ratio obtained at a velocity gradient of $37\,s^{-1}$ did not result in the highest efficiency in terms of turbidity and colour removal. Rather, a velocity gradient of $172\,s^{-1}$ displayed the highest efficiency as it exhibited the highest aggregation rate and lowest steady-state variance when compared to other velocity gradients. A velocity gradient of $172\,s^{-1}$ was used for the rest of the study.

3.4 Optimization of rapid mixing time

For this phase of the study, rapid mixing time was varied while applying the optimal alum dosage of $3\,mg\,L^{-1}$ as Al^{3+}, sample pH of 8 and a G value $172\,s^{-1}$. Figure 7a shows that the lowest residual turbidity and colour was obtained when a rapid mixing time of 20 s was applied. This signifies that a velocity gradient of $172\,s^{-1}$ was applied for an optimal period to ensure efficient coagulation through the instantaneous and even distribution of alum.

Similar to rapid mixing intensity, it may be desirable to apply an optimal velocity gradient for an extended period of time to maximise efficient use of coagulants. However, according to Bratby (2006), extended periods of rapid mixing may give rise to deleterious effects in the coagulation-flocculation process. As such, each application will have an optimal rapid mixing time which is dependent on the rapid mixing intensity and coagulant concentration. The negative effect of prolonged rapid mixing time can be seen in Fig. 7a as higher residual turbidity and colour values were obtained during longer rapid mixing times.

Figure 7b shows that the average steady state ratio values gradually decrease with increasing rapid mixing times. This was anticipated as prolonged rapid mixing times will expose aggregates to shear rates for longer periods of time. This will delay the floc growth process and allow for a longer period of particle collision; increasing the rate of floc breakup.

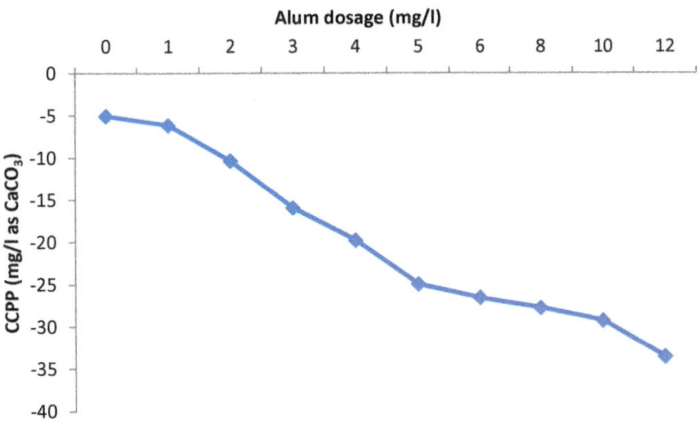

Figure 8. Impact of alum dosage on CCPP of treated water.

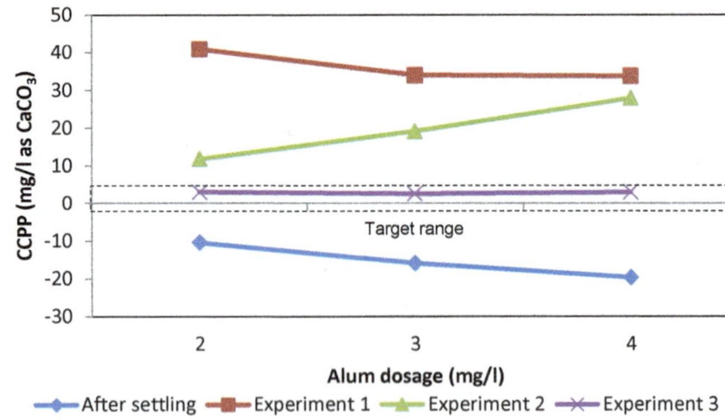

Figure 9. Equilibration of CCPP.

Selomulya et al. (2001) explained that higher circulation times reduces the time aggregates have to form larger sizes due to high shear stress, hence, a decrease in average steady-state ratio was observed.

In line with observations from previous experiments, the highest average steady-state ratio value, obtained from a rapid mixing time of 15 s, did not result in the highest coagulation-flocculation performance in terms of turbidity and colour removal. Rather, a rapid mixing time of 20 s displayed the highest removal efficiency. PDA data represented in Fig. 7c and d indicate that a rapid mixing time of 20 s produced a higher aggregation rate and lower steady-state variance than that of 15 s. This places further significance on the importance of the aggregation rate and steady-state variance in coagulation-flocculation studies. A rapid mixing time of 20 s was used for subsequent experiments.

3.5 Impact of alum dosage on CCPP

The impact of alum dosage on CCPP was investigated at an alum dosage range of 2–4 mg L^{-1} as Al^{3+}, sample pH of 8; G value of 172 s^{-1} and rapid mixing time of 20 s. Figure 8 shows that the CCPP of treated water decreased with increasing alum dosage. To ensure stability of treated water, the production specification for CCPP at a large water utility is −2 to +3 mg L^{-1} as $CaCO_3$. However, Fig. 9 indicates that at an alum dosage range of 2–4 mg L^{-1} as Al^{3+} and a sample pH of 8, the coagulation, flocculation and sedimentation process reduces the CCPP to −10.4 to −19.8 as mg L^{-1} as $CaCO_3$, respectively. These values demonstrate the corrosive nature of treated water and a high probability of damage to infrastructure occurring. As a result, it is imperative that water treated with alum is stabilized using appropriate processes.

Lime is used widely during potable water treatment as a coagulant, to adjust coagulation pH or for stabilization of treated water. As a result, this study used slaked lime to equilibrate the CCPP of treated water. Figure 9 illustrates that three experiments were performed to optimise the slaked lime dosage required to increase the CCPP of treated water to +3.0 mg L^{-1} as $CaCO_3$. Experiment 3 demonstrates that for alum dosages of 2, 3 and 4 mg L^{-1} as Al^{3+}, the CCPP was adjusted to 3, 2.5 and 2.9 mg L^{-1} as $CaCO_3$, respectively. This corresponded to slaked lime dosages of 13, 17 and 23 mg L^{-1} for alum dosages of 2, 3 and 4 mg L^{-1}, respectively.

3.6 Impact of dosage sequence

This experiment was performed to assess the effect of dosing slaked lime on water quality during coagulation, as well as the sequence of chemical dosage on the turbidity of settled water. It should be reiterated that slaked lime was used primarily for pH/CCPP adjustment in this study. Two experimental scenarios were performed:

1. Alum dosed first, followed by slaked lime
2. Slaked lime dosed first, followed by alum

Figure 10a and b show that lower residual turbidity and colour levels were obtained when slaked lime was dosed before alum. Additionally, one should note that under both experimental scenarios, both turbidity and colour removals were enhanced at higher slaked lime dosages. As such, it appears that slaked lime functions as a coagulant chemical in addition to a pH adjustment chemical.

To gain further understanding on the coagulation-flocculation kinetics, samples from both experimental scenarios were passed through the PDA monitor. When alum was dosed first, the ratio output values were very low and did not produce a ratio profile as shown in Fig. 2. As a result, the root mean square (RMS) gain on the instrument was adjusted accordingly to amplify the signal. These instrumental settings were kept constant for both experimental scenarios to allow for data comparison. However, in the case of experimental scenario two (lime addition first); the amplified instrumental settings resulted in a ratio overload during the steady-state growth region. A ratio overload occurs when the

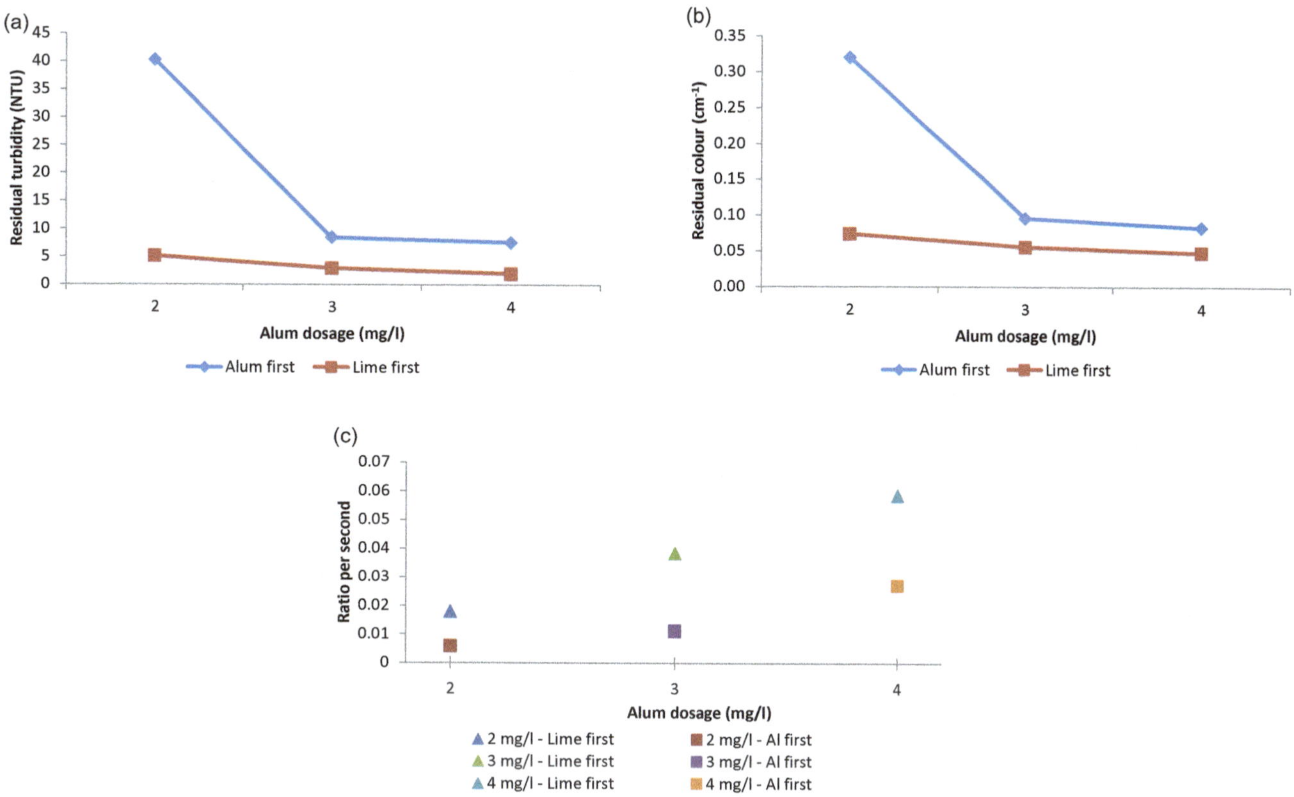

Figure 10. Impact of dosage sequence; **(a)** residual turbidity; **(b)** residual colour; **(c)** aggregation rate.

RMS output is greater than the DC output and can be corrected by reducing the RMS gain. However, this was not possible as any decrease in the RMS gain would have resulted in a loss of the ratio profile of experimental scenario one (alum addition first). As a consequence, this prevented the comparison of PDA data obtained during the steady-state, namely the average steady-state ratio and the steady-state variance. However, it did permit the comparison of the aggregation rates, as during this time the RMS output was less than DC output; hence, the ratio overload signal was not activated.

Figure 10c indicates that the addition of lime first resulted in a higher aggregation rate at the dosage levels investigated. These superior aggregation rates may have been contributing factors to the improved coagulation-flocculation efficiency of adding slaked lime first as shown by the lower residual turbidity and colour levels in Fig. 10a and b.

4 Conclusions

This study used a PDA to optimize coagulation conditions and monitor floc size and structure as well as floc growth rate during coagulation-flocculation using alum. The results of this study indicate that optimal coagulation-flocculation conditions were as follows: sample pH of 8, alum dosage of $3\,\mathrm{mg\,L^{-1}}$ as Al^{3+}, rapid mixing intensity of $172\,\mathrm{s^{-1}}$ and rapid mixing time of 20 s. Additional PDA parameters, namely, the aggregation rate, average steady-state ratio and steady-state variance were generated using PDA ratio data to gain further understanding of the coagulation-flocculation process.

The derived PDA data revealed that in all optimization experiments, the average steady-state ratio, although an important parameter, was a secondary parameter as it has a lessened impact on coagulation-flocculation efficiency. It was found that the aggregation rate and steady-state variance were more suitable parameters to assess coagulation-flocculation performance as they had direct significance on coagulation-flocculation efficiency. These parameters may be seen as primary kinetic parameters and should be given greater attention during coagulation-flocculation studies. The results of this study showed that the PDA instrument is an important tool in coagulation kinetic studies and can be employed as an additional tool in the optimization of coagulation conditions.

Acknowledgements. The authors would like to express gratitude towards both the Experimental Facility at the Process Technology Department as well as the Analytical Services Department at Rand Water for their role in this project.

References

Ball, T., Carriere, A., and Barbeau, B.: Comparison of two online flocculation monitoring techniques for predicting turbidity removal by granular media filtration, Environ. Technol., 32, 1095–1105, 2011.

Bratby, J.: Coagulation and flocculation in water and wastewater treatment, 2nd Edn., IWA Publishing, London, United Kingdom, 2006.

Degremont: Water treatment handbook, 7th Edn., Lavoisier SAS, France, 2007.

Gregor, J. E., Nokes, C. J., and Fenton, E.: Optimizing natural organic matter removal from low turbidity waters by controlled pH adjustment of aluminium coagulation, Water Res., 31, 2949–2958, 1997.

Gregory, J. and Duan, J.: Coagulation by hydrolyzing metal salts, Pure Appl. Chem., 73, 2017–2026, 2001.

Hopkins, D. C. and Ducoste, J. J.: Characterizing flocculation under heterogeneous turbulence, J. Colloid Interf. Sci., 264, 184–194, 2003.

Huang, C. and Liu, C.: Automatic control for chemical dosing in laboratory scale coagulation process by using an optical monitor, Water Res., 30, 1924–1929, 1996.

Mixon, M., Staak, C., Fabris, R., Vimonses, V., Chow, C. W. K., Panglisch, S., van Leeuwen, J., and Drikas, M.: The impact of optimized coagulation on fouling for coagulation/ultrafiltration process, Desalin. Water Treat., 51, 2718–2725, 2013.

Oles, V.: Shear induced aggregation and break up of polystyrene latex particles, J. Colloid Interf. Sci., 154, 351–358, 1992.

Pernitsky, D. J. and Edzwald, J. K.: Selection of alum and polyaluminium coagulants: principles and applications, J. Water Supply Res. T., 55, 121–141, 2006.

Ramphal, S. and Sibiya, S. M.: Optimization of time requirement for rapid mixing during coagulation using a photometric dispersion analyzer, Proc., 12th Int. Conf. on "Computing and Control for the Water Industry – CCWI2013", Perugia, Elsevier, Proc. Engin., 70, 1401–1410, 2014.

Selomulya, C., Amal, R., Bushell, G., and Waite, T. D.: Evidence of shear rate dependence on restructuring and breakup of latex aggregates, J. Colloid Interf. Sci., 236, 67–77, 2001.

Spicer, P. T. and Pratsinis, S. E.: Shear induced flocculation: The Evolution of Floc Structure and Size Distribution during Shear-Induced Flocculation, Water Res, 30, 1049–1056, 1996.

Spicer, P. T., Pratsinis, S. E., and Trennepohl, M. D.: Coagulation and fragmentation: The variation of shear rate and the time lag for attainment of steady state, Ind. Eng. Chem. Res., 35, 3074–3080, 1996.

Srinivasan, P. T., Viraraghavan, T., and Subramanian, K. S.: Aluminium in drinking water: An overview, Water SA, 25, 47–55, 1999.

Staaks, C., Fabris, R., Lowe, T., Chow, C. W. K., van Leeuwen, J. A., and Drikas, M.: Coagulation assessment and optimization with a photometric dispersion analyzer and organic characterization for natural organic matter removal performance, Chem. Eng. J., 168, 629–634, 2011.

Xiao, F., Huang, J. C. H., Zhang B. J., and Cui, C. W.: Effects of low temperature on coagulation kinetics and floc surface morphology using alum, Desalination, 237, 201–213, 2009.

Zoubolis, A. I. and Tzoupanos, N. D.: Polyaluminium silicate chloride – A systematic study for the preparation and application of an efficient coagulant for water and wastewater treatment, J. Hazard. Mater., 162, 1379–1389, 2009.

Zoubolis, A. I. and Tzoupanos, N. D.: Alternative cost-effective preparation methods of polyaluminium chloride (PAC) coagulation agent: Characterization and comparative application for water/wastewater, Desalination, 250, 339–344, 2010.

23

Estimating fast and slow reacting components in surface water and groundwater using a two-reactant model

Priyanka Jamwal[1], **M. N. Naveen**[1,2], **and Yusuf Javeed**[2]

[1]Centre for Environment and Development, Ashoka Trust for Research in Ecology and the Environment (ATREE), Jakkur 560064, India

[2]Department of Civil engineering, The National Institute of Engineering, Mysuru, Karnataka 570008, India

Correspondence to: Priyanka Jamwal (priyanka.jamwal@atree.org)

Abstract. Maintaining residual chlorine levels in a water distribution network is a challenging task, especially in the context of developing countries where water is usually supplied intermittently. To model chlorine decay in water distribution networks, it is very important to understand chlorine kinetics in bulk water. Recent studies have suggested that chlorine decay rate depends on initial chlorine levels and the type of organic and inorganic matter present in water, indicating that a first-order decay model is unable to accurately predict chlorine decay in bulk water. In this study, we employed the two-reactant (2R) model to estimate the fast and slow reacting components in surface water and groundwater. We carried out a bench-scale test for surface water and groundwater at initial chlorine levels of 1, 2, and 5 $mg\,L^{-1}$. We used decay data sets to estimate optimal parameter values for both surface water and groundwater. After calibration, the 2R model was validated with two decay data sets with varying initial chlorine concentrations (ICCs). This study arrived at three important findings. (a) We found that the ratio of slow to fast reacting components in groundwater was 30 times greater than that of the surface water. This observation supports the existing literature which indicates the presence of high levels of slow reacting fractions (manganese and aromatic hydrocarbons) in groundwater. (b) Both for surface water and groundwater, we obtained good model prediction, explaining 97 % of the variance in data for all cases. The mean square error obtained for the decay data sets was close to the instrument error, indicating the feasibility of the 2R model for chlorine prediction in both types of water. (c) In the case of deep groundwater, for high ICC levels ($>2\,mg\,L^{-1}$), the first-order model can accurately predict chlorine decay in bulk water.

1 Introduction

The presence of $0.2\,mg\,L^{-1}$ of residual chlorine in drinking water is known to reduce public health risks significantly (Arnold and Colford, 2007; Pattanayak et al., 2005). One of the key tasks of water managers worldwide, especially in developing nations, is to maintain residual chlorine levels of drinking water in distribution systems. This requires a higher concentration at entry point in order to ensure that a minimal residual chlorine concentration of $0.2\,mg\,L^{-1}$ is retained at any point of time before reaching the end consumers.

To maintain $0.2\,mg\,L^{-1}$ of residual chlorine in water, sodium hypochlorite or liquid chlorine is added to the treated water. Before water reaches the consumers, part of this chlorine is lost due to reactions with the organic and inorganic matter present in water after treatment, as well as to reactions with biofilms and corrosion products present within the distribution network (Al-Jasser, 2007; Hallam et al., 2002; Helbling and VanBriesen, 2007). Any excessive addition of chlorine to water leads to harmful by-products which pose risks to public health. Therefore it becomes very important for water managers to optimize the dose of chlorine added to water, ensuring that the right level of residual chlorine is retained in distribution networks (Hrudey, 2009; Richardson, 2003; Singer, 1999).

A first-order decay process is generally employed to simulate chlorine decay within the distribution network (Hua et al., 1999; Rossman et al., 1994; Vasconcelos et al., 1997). The model assumes that chlorine decay rate is a function of chlorine levels within the bulk water. However, recent studies have shown that in addition to chlorine levels, other factors like the type of organic/inorganic matter, temperature, and pipe material also affect chlorine decay rates in the distribution system (Al-Jasser, 2007; Mutoti et al., 2007; Hallam et al., 2002).

The total chlorine decay within the water distribution network is caused by chlorine reaction (a) in bulk water and (b) with the biofilm attached to the pipe surface, also termed wall decay (Al-Jasser, 2007; Hallam et al., 2002; Powell et al., 2000). For prediction of chlorine decay over time, models require accurate estimation of chlorine reaction in bulk water and with the biofilm on the pipe wall surface. Pilot loop setups/simulators are employed to estimate the contribution of wall reactions to chlorine decay (Frias et al., 2001; Lehtola et al., 2006; Rossman et al., 1994, 2006). This is achieved by subtracting the bulk reaction rate from the total chlorine decay rate in the pipe loop. Various authors have argued about the need to accurately model bulk decay before making an attempt to estimate the contribution of wall decay (Fisher et al., 2011).

Chlorine decay kinetics depends on the type and amount of dissolved organic matter (DOM) and inorganic matter present in bulk water. DOM is derived primarily from decaying organisms such as plants or algae, and is often classified into humic and non-humic substances such as proteins, carbohydrates, and lipids. Generally speaking, DOM from marine and aquatic sources is more enriched in aliphatic structures, while DOM from terrestrial/higher plant sources is rich in aromatic compounds (Chen et al., 2010). The inorganic compounds in surface water are derived from the dissolution of minerals present in the bedrock as water flows, whereas groundwater dissolves minerals as it percolates through the vadose zone as well as during its stay in the saturated zone. Therefore, in comparison to surface water, groundwater contains high levels of inorganic substances such as nitrates, manganese, arsenic, and iron.

Several studies have shown that in homogeneous systems, chlorine exhibits faster reaction rates with ammonia, sulfates, nitrates, nitrites, arsenic, and iron as compared to manganese (Mn (II)) (Deborde and Von Gunten, 2008). Therefore, given different characteristics of the components (organic and inorganic) present in surface water and groundwater, the precursors to chlorine reactions can be divided into fast and slow reacting fractions (Gallard and von Gunten, 2002). Table 1 presents the list of possible fast and slow reacting components in different types of water.

The two-reactant model (2R model) is a simplified second-order decay model which uses notional fast and slow reacting agents involved in second-order reactions with chlorine over long travel periods within the distribution network (Fisher et al., 2011). The second-order reaction rates and the resulting 2R model are given by the following equations:

Table 1. Fast and slow reacting components present in different types of water.

Compounds	Organic	Inorganic
Fast reacting	Aliphatic hydrocarbons	Nitrates, sulfates, ammonia, and nitrites
Slow reacting	Aromatic hydrocarbons	Manganese Mn (II)

$$\frac{dC_f}{dt} = -K_f \cdot C_f \cdot C_{cl}, \tag{1}$$

$$\frac{dC_s}{dt} = -K_s \cdot C_s \cdot C_{cl}, \tag{2}$$

$$\frac{dC_{cl}}{dt} = \frac{dC_f}{dt} + \frac{dC_s}{dt}, \tag{3}$$

where C_{cl} is the concentration of residual chlorine levels ($mg\,L^{-1}$), C_f and C_s are the concentrations of fast and slow reducing agents ($mg\,L^{-1}$), and K_f and K_s are the fast and slow reaction rate coefficients ($L\,mg^{-1}\,h^{-1}$).

The initial concentration of fast and slow reactants (C_{os} and C_{of}) and their respective decay coefficients (K_s and K_f) can be estimated using the AQUASIM software (Fisher et al., 2011).

So far, first- and second-order chlorine decay models have been calibrated and tested for different types of test water under varying conditions such as temperature, type of treatment, and re-chlorination (Fisher et al., 2012; Mutoti et al., 2007; Rossman, 2006). All prior studies have estimated the chlorine decay parameters for water from different sources (variable water quality). The results obtained are specific and suitable for predicting chlorine decay in water supply networks of the source water for which the studies were conducted. This study aims to estimate the chlorine decay parameters using the 2R model relevant to local conditions found in southern India, where hard rock aquifers are abundant, and compare them with surface water parameters. We employed the 2R model to estimate optimal parameters for prediction of residual chlorine in both surface water and groundwater. The model was calibrated separately for test water with initial chlorine levels ranging from 1 to 5 $mg\,L^{-1}$. To establish the suitability of the 2R model, we also validated it against two chlorine decay data sets.

2 Methodology

We first calibrated the 2R model against data sets for two types of water, surface water and groundwater, to obtain a single invariant set of four parameters that characterize the water. The model was then validated by comparing the model estimates with the decay test data. The characteristics of the experimental data set are presented in Table 2.

Table 2. Characteristics of the data set.

Water	Treatment	ICC range ($mg\,L^{-1}$)	Number of experimental runs	Temperature range (°C)
Surface water	Conventional treatment	1–5	4	25–30
Groundwater	No treatment	1–5	4	25–30

2.1 Chlorine decay data sets

Chlorine decay tests and data for calibration and validation of the 2R model were obtained by conducting bench-scale residual chlorine decay tests at the ATREE water and soil laboratory.

Groundwater and surface water samples were obtained from the respective sources and water quality characteristics were determined as presented in Table 3. The test water was stored in amber glass bottles, while sodium hypochlorite was added to attain the desired initial chlorine concentrations (ICCs). The bench-scale tests for both types of water were run at ICC-1, 2, 4.5, and $5\,mg\,L^{-1}$. The amber glass bottles were kept in an incubator set at ambient temperature (25 to 30 °C). During the day, hourly samples were drawn from the bottles and free chlorine levels were measured for 72 h. Free chlorine in the test water was measured using a Merck Spectroquant® Picco colorimeter. The free chlorine measurement range of the instrument is 0.05 to $6.00\,mg\,L^{-1}$ (APHA, 2005). The decay test results obtained were within the measurement range of the instrument.

2.2 Model parameter estimation

The AQUASIM software was used to estimate optimal parameter values for the 2R model to obtain the best fit for the three decay data sets (ICC-1, 2, and $5\,mg\,L^{-1}$; bench-scale laboratory experiments). Optimal parameters are the values that allow the model to predict chlorine decay in each type of water as accurately as possible. As explained by Fisher et al. (2012), the AQUASIM software calculates the sum of squared differences between each experimental data point and the corresponding model prediction, assuming an initial set of parameter values. This sum is derived by the variance of the data to form a chi square (χ^2). Using the simplex technique it then systematically varies the parameter values to search for a set that produces the best fit (minimizes χ^2). The coefficient of determination (R^2) was also obtained which indicates the total variance in the data set explained by the 2R model regardless of the distance between the decay curves. To measure the 2R model accuracy for chlorine prediction, we also estimated the root mean square error (RMSE) for the data sets.

We calibrated the 2R model using chlorine decay data sets for initial chlorine levels of 1, 2, and $5\,mg\,L^{-1}$. The optimal parameters for groundwater and surface water were determined separately. After calibration, the 2R model was validated using two data sets, i.e. of ICC-$2\,mg\,L^{-1}$ (a subset of the calibration data set) and ICC-$4.5\,mg\,L^{-1}$ (independent chlorine decay data set).

3 Results

3.1 Chlorine decay and water quality

The results of bench-scale chlorine decay tests are presented in Fig. 1. The graph presents the fraction of chlorine remaining in test water at different ICCs. The data present two interesting findings: (i) the decay rate decreases with an increase in initial chlorine levels and (ii) the chlorine decay rate in surface water is greater than that of groundwater. The above results are in conjugation with the other studies that suggest that the first-order decay model is unable to predict chlorine decay in bulk water, and the chlorine decay rate depends on the ICC and level of organic/inorganic matter present in the test water (Fisher et al., 2011; Hua et al., 1999; Rossman, 2006).

As shown in Table 2, the ambient temperature during this study fluctuated between 25 and 30 °C. The temperature alone could not explain the variation in the first-order decay constant at different ICCs. Therefore we employed the 2R model to predict the chlorine decay in bulk water. As per the literature review, none of the earlier studies have used the 2R model to estimate fast and slow reacting components of groundwater from deep hard rock aquifers. Fisher et al. (2011) estimated decay parameters for shallow and deep groundwater from the Wanneroo aquifer (Fisher et al., 2015; Warton et al., 2006), which could not be applied as-is to predict chlorine decay in deep groundwater from hard rock aquifers. The decay parameter, estimated by the 2R model, depends on the quality of the source water. Therefore in this study, an attempt has been made to estimate decay parameters specific to the local water sources and conditions.

In the next section, we employ the 2R model to estimate fast and slow reacting components in test waters. In addition, we will also test the feasibility of the 2R model in predicting chlorine decay in test waters.

3.2 2R model calibration and validation

The 2R model for groundwater and surface water was calibrated separately using the data sets for ICC-1, 2, and $5\,mg\,L^{-1}$. Figure 2 presents the calibrated data sets for groundwater and surface water.

Table 3. Water quality parameters of test water.

Parameter	Surface water ($n = 3$)	Standard deviation	Groundwater ($n = 3$)	Standard deviation
pH	7.19	0.47	6.75	0.01
Conductivity (muS cm^{-1})	434	30	1150	17
Nitrates (mg L^{-1})	3	1	177	20
Hardness (mg L^{-1})	180	56	352	8
Alkalinity (mg L^{-1})	196	4	165	3

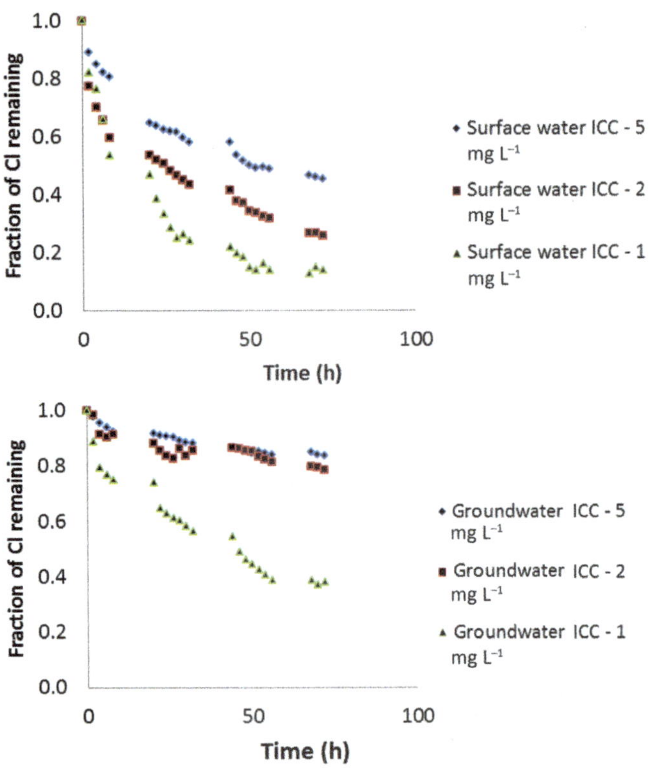

Figure 1. Fraction of chlorine remaining in surface water and groundwater at different initial chlorine levels.

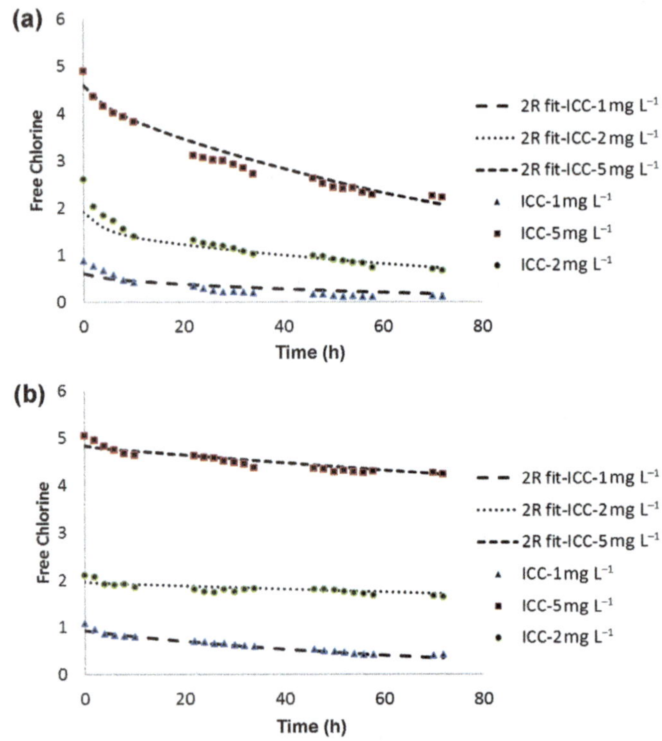

Figure 2. Chlorine decay in surface water **(a)** and groundwater **(b)**: markers – measured chlorine values; curves – values simulated by 2R models.

The 2R model predicted optimal values for four parameters simultaneously by minimizing the chi-squared (χ^2) value. Minimization of χ^2 was readily achieved using the optimization technique available within the AQUASIM parameter estimation procedure. The optimal values with the associated χ^2 value are presented in Table 4.

Figure 2 presents the decay curves obtained from the simulation using optimized parameter values for the calibration data sets. We observed good agreement between chlorine residual data and the model estimates for all ICCs, both for surface water and groundwater.

The χ^2 values obtained for surface water and groundwater were 0.18 and 0.48, respectively. The R^2 values were greater than 0.98 for surface water and 0.99 for groundwater, indicating that only 2 and 1 % variance in the calibration data sets remained unexplained by the model. We also checked the validity of our results by comparing optimal parameter values obtained from our data sets with those of the other studies. Table 4 presents the optimal parameters obtained for other test waters. The χ^2 values obtained for our data sets were comparable to those of the experimental data sets from other studies (Fisher et al., 2012). For surface water, the 2R model underestimated chlorine levels for $0 < t < 60$ h for decay sets ICC-5 and ICC-1 mg L^{-1}, whereas for ICC-2 mg L^{-1} the chlorine levels were underestimated for $0 < t < 20$ h. In the case of groundwater, chlorine decay was underestimated for ICC-5 and ICC-2 mg L^{-1} for $0 < t < 60$ h, and for ICC-1 mg L^{-1} chlorine was underestimated for $0 < t < 5$ h.

Table 4. Comparison of optimal parameter values obtained for different test waters.

Source	Treatment type	ICC range ($mg\,L^{-1}$)	C_{of} ($mg\,L^{-1}$)	C_{os} ($mg\,L^{-1}$)	K_f ($L\,mg^{-1}\,h^{-1}$)	K_s ($L\,mg^{-1}\,h^{-1}$)	C_{os}/C_{of}	χ^2	R^2
Surface water[1]	Conventional treatment	1–5	0.51	2.88	2.55	0.0069	6	0.18	0.978
Groundwater[1]	No treatment	1–5	0.003	0.67	8.07	0.013	223	0.48	0.998
Surface water[2]	Conventional treatment	1–4	0.808	3.88	0.261	0.0102	5	0.18	0.992
Surface water[2]	No treatment	1–4	0.761	2.69	0.199	0.0066	4	0.97	0.987
Surface water[2]	Not known	1–4	0.917	1.98	6.18	0.085	2	0.844	0.987

[1] Bench-scale test (our study). [2] Bench-scale test (Fisher et al., 2012).

In groundwater, the ratio of slow to fast reacting components was 30 times greater than that of the surface water. This suggests that, as groundwater travels through the vadose zone, most of the fast reacting organic matter is consumed by microorganisms, leaving behind the non-biodegradable organic/inorganic matter (slow reacting component) (He et al., 2006; McCarty et al., 1981). The slow reacting components are difficult to oxidize; therefore, over time, high levels of residual chlorine were observed in groundwater. It also suggests that in the absence of fast reacting components, first-order decay models could accurately predict chlorine decay in groundwater at ICCs greater than $2\,mg\,L^{-1}$.

3.3 Validation of the 2R model

The 2R model was fitted for experimental data sets – **(a)** surface water (ICC-2.6 and $4.2\,mg\,L^{-1}$) and **(b)** groundwater (ICC-2 and $4.5\,mg\,L^{-1}$), as shown in Fig. 3.

First we calibrated the 2R model with decay data sets, commencing with the highest and lowest ICCs, as the extrapolation outside the calibration range is less reliable. Then we validated the model by using two chlorine decay data sets, i.e. one from the calibration data set (ICC-$2\,mg\,L^{-1}$) and one from the other data set that has not been used for the model calibration (ICC-$4\,mg\,L^{-1}$).

The R^2 value obtained for both the data sets of surface water was greater than 0.98, indicating that only 2 % of the variance remains unexplained by the 2R model. This suggests the suitability of the 2R model for chlorine prediction in surface water. In the case of groundwater, R^2 values of 0.89 and 0.94 were obtained for decay tests at ICC-2 and ICC-$4.5\,mg\,L^{-1}$, respectively. The reason for the lower R^2 values for the validated data for groundwater is that R^2 is a measure of fit involving the error relative to the variance in the data. The groundwater data have lower variance than surface water data and at lower ICC groundwater data have lower variance than the higher ICC data. Therefore, even with the same level of error in all data, the groundwater chlorine decay data sets showed low R^2 values as compared to the surface water chlorine decay data sets.

To check the prediction accuracy of the 2R model, we estimated the mean square error for both surface water and groundwater data sets. The values obtained were ±0.07 and ±$0.05\,mg\,L^{-1}$ for surface water and groundwater, respectively, which is close to the instrument measurement error (±$0.05\,mg\,L^{-1}$). This indicates that the parameter estimates obtained could accurately predict chlorine decay in surface water and groundwater.

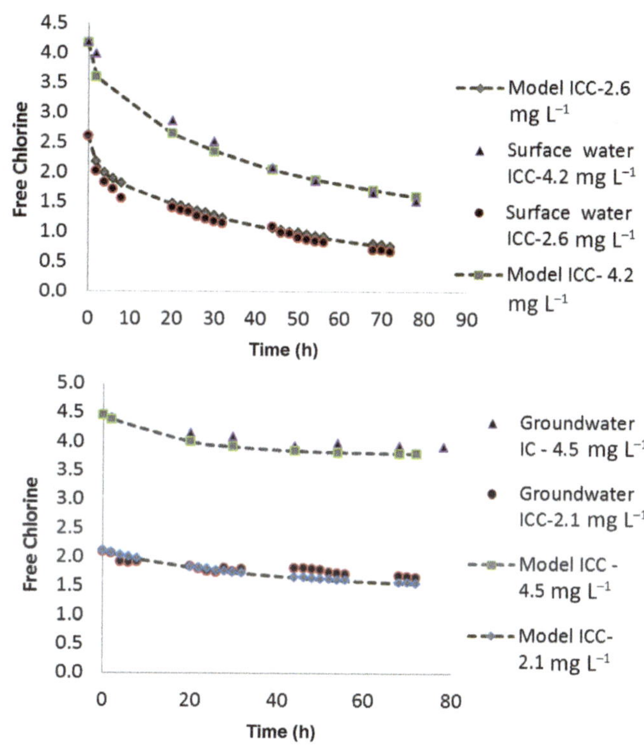

Figure 3. Validation of the 2R model for surface water and groundwater at different ICCs.

4 Conclusions

Through this study, we estimated the fast and slow reacting components and tested the feasibility of the 2R model in predicting residual chlorine in bulk water from different sources, i.e. for surface water (rivers) and groundwater (deep aquifers). The 2R model was calibrated and validated using the AQUASIM software.

For both types of water, the observed chlorine residual values in bulk water closely matched the 2R model predicted values (ICC-1 to $5\,\mathrm{mg\,L^{-1}}$). The 2R model calibration and validation range was well within the chlorine residual range encountered in water distribution networks (0–$6\,\mathrm{mg\,L^{-1}}$). Therefore, we can conclude with a reasonable degree of confidence that these invariant sets of model parameters obtained for surface water and groundwater if used in conjunction with the EPANET-MSX (Multi-Species extension) model will significantly assist public utilities in dealing with water quality issues in water distribution networks.

This study is also important in the context of urban areas in developing countries where 50 % of the water demand (Deccan plateau region – southern India) is met by groundwater pumping (Grönwall et al., 2010). The groundwater is pumped to overhead tanks, chlorinated, and supplied through a piped water connection. As stated above, the results of this study in conjunction with a sufficiently accurate prediction of wall decay will enable water management agencies to determine the ICC levels that allow residual targets at system extremities to be met.

Acknowledgements. Financial support for this research came from the Department of Science and Technology, Government of India grant number SR/FTP/ETA-91/2010 titled "Water Distribution Network Study on Fate and Transport of Contaminant and its Decontamination by Pilot-Scale Tests". The authors are grateful to the referees for valuable suggestions and comments which have led to the profound discussion of the results. We are also grateful to Praveen Raje Urs – Lab Analyst from the ATREE Water and Soil Lab – for helping with analysis of water samples.

References

Al-Jasser, A. O.: Chlorine decay in drinking-water transmission and distribution systems: Pipe service age effect, Water Res., 41, 387–396, 2007.

APHA: Standard Methods for examination of water and wastewater, 21st Edn., American Public Health Association (APHA), USA, 2005.

Arnold, B. F. and Colford, J. M.: Treating water with chlorine at point-of-use to improve water quality and reduce child diarrhea in developing countries: a systematic review and meta-analysis, Am. J. Trop. Med. Hyg., 76, 354–364, 2007.

Chen, M., Price, R. M., Yamashita, Y., and Jaffé, R.: Comparative study of dissolved organic matter from groundwater and surface water in the Florida coastal Everglades using multi-dimensional spectrofluorometry combined with multivariate statistics, Appl. Geochem., 25, 872–880, 2010.

Deborde, M., and Von Gunten, U. R. S.: Reactions of chlorine with inorganic and organic compounds during water treatment–kinetics and mechanisms: a critical review, Water Res., 42, 13–51, 2008.

Fisher, I., Kastl, G., and Sathasivan, A.: Evaluation of suitable chlorine bulk-decay models for water distribution systems, Water Res., 45, 4896–4908, 2011.

Fisher, I., Kastl, G., and Sathasivan, A.: A suitable model of combined effects of temperature and initial condition on chlorine bulk decay in water distribution systems, Water Res., 46, 3293–3303, 2012.

Fisher, I., Kastl, G., Sathasivan, A., Cook, D., and Seneverathne, L.: General Model of Chlorine Decay in Blends of Surface Waters, Desalinated Water, and Groundwaters, J. Environ. Eng., doi:10.1061/(ASCE)EE.1943-7870.0000980, 2015.

Frias, J., Ribas, F., and Lucena, F.: Effects of different nutrients on bacterial growth in a pilot distribution system, Antonie Van Leeuwenhoek, 80, 129–138, 2001.

Gallard, H. and von Gunten, U.: Chlorination of natural organic matter: kinetics of chlorination and of THM formation, Water Res., 36, 65–74, 2002.

Grönwall, J. T., Mulenga, M., and McGranahan, G.: Groundwater, self-supply and poor urban dwellers: A review with case studies of Bangalore and Lusaka, Human Settlements Working Paper, International Institute for Environment, London, UK, ISBN 978-1-84369-770-1, 87 pp., 2010.

Hallam, N. B., West, J. R., Forster, C. F., Powell, J. C., and Spencer, I.: The decay of chlorine associated with the pipe wall in water distribution systems, Water Res., 36, 3479–3488, 2002.

He, P., Xue, J., Shao, L., Li, G., and Lee, D.-J.: Dissolved organic matter (DOM) in recycled leachate of bioreactor landfill, Water Res., 40, 1465–1473, 2006.

Helbling, D. E. and VanBriesen, J. M.: Free chlorine demand and cell survival of microbial suspensions, Water Res., 41, 4424–4434, 2007.

Hrudey, S. E.: Chlorination disinfection by-products, public health risk tradeoffs and me, Water Res., 43, 2057–2092, 2009.

Hua, F., West, J. R., Barker, R. A., and Forster, C. F.: Modelling of chlorine decay in municipal water supplies, Water Res., 33, 2735–2746, 1999.

Lehtola, M. J., Laxander, M., Miettinen, I. T., Hirvonen, A., Vartiainen, T., and Martikainen, P. J.: The effects of changing water flow velocity on the formation of biofilms and water quality in pilot distribution system consisting of copper or polyethylene pipes, Water Res., 40, 2151–2160, 2006.

McCarty, P. L., Reinhard, M., and Rittmann, B. E.: Trace organics in groundwater, Environ. Sci. Technol., 15, 40–51, 1981.

Mutoti, G., Dietz, J. D., Arevalo, J., and Taylor, J. S.: Combined chlorine dissipation: Pipe material, water quality, and hydraulic effects, J. Am. Water Works Ass., 99, 96–106, 2007.

Pattanayak, S. K., Yang, J. C., Whittington, D., and Kumar, K. C. B.: Coping with unreliable public water supplies: averting expenditures by households in Kathmandu, Nepal, Water Resour. Res., 41, W02012, doi:10.1029/2003WR002443, 2005.

Powell, J. C., Hallam, N. B., West, J. R., Forster, C. F., and Simms, J.: Factors which control bulk chlorine decay rates, Water Res., 34, 117–126, 2000.

Richardson, S. D.: Disinfection by-products and other emerging contaminants in drinking water, TrAC-Trend. Anal. Chem., 22, 666–684, 2003.

Rossman, L. A., Clark, R. M., and Grayman, W. M.: Modeling chlorine residuals in drinking-water distribution systems, J. Environ. Eng., 120, 803–820, 1994.

Rossman, L. A.: The effect of advanced treatment on chlorine decay in metallic pipes, Water Res., 40, 2493–2502, 2006.
Singer, P. C.: Humic substances as precursors for potentially harmful disinfection by-products, Water Sci. Technol., 40, 25–30, 1999.
Vasconcelos, J. J., Rossman, L. A., Grayman, W. M., Boulos, P. F., and Clark, R. M.: Kinetics of chlorine decay, J. Am. Water Works Ass., 89, 54–65, 1997.
Warton, B., Heitz, A., Joll, C., and Kagi, R.: A new method for calculation of the chlorine demand of natural and treated waters, Water Res., 40, 2877–2884, 2006.

24

Natural manganese deposits as catalyst for decomposing hydrogen peroxide

A. H. Knol[1], K. Lekkerkerker-Teunissen[1], and J. C. van Dijk[2]

[1]Dunea N.V., Strategy department, P.O. Box 756, 2700 AT Zoetermeer, the Netherlands
[2]Delft University of Technology, Stevinweg 1, 2628 CN Delft, the Netherlands

Correspondence to: A. H. Knol (t.knol@dunea.nl)

Abstract. Drinking water companies (are intending to) implement advanced oxidation processes (AOP) in their treatment schemes to increase the barrier against organic micropollutants (OMPs). It is necessary to decompose the excessive hydrogen peroxide after applying AOP to avoid negative effects in the following, often biological, treatment steps. A drinking water company in the western part of the Netherlands investigated decomposition of about 5.75 mg L^{-1} hydrogen peroxide in pre-treated Meuse river water with different catalysts on pilot scale.

In down flow operation, the necessary reactor empty bed contact time (EBCT) with the commonly used granulated activated carbon (GAC) and waste ground water filter gravel (MCFgw) were found equal with 149 s, corresponding with a conversion rate constant r of 0.021 s^{-1}. The EBCT of the fine coating of ground water filter gravel (MC) was significantly shorter with a little more than 10 s ($r = 0.30\,s^{-1}$).

In up flow operation, with a flow rate of 20 m h^{-1}, the EBCT of coating MC increased till about 100 s ($r = 0.031\,s^{-1}$), from which can be concluded, that the performance of this waste material is better compared with GAC, in both up and down flow operation.

The necessary EBCT at average filtration rate of full scale dual layer filter material (MCFsw) amounted to 209 s ($r = 0.015\,s^{-1}$). Regarding the average residence time in the full scale filters of 700 s, applying AOP in front of the filters could be an interesting alternative which makes a separate decomposition installation superfluous, on the condition that the primary functions of the filters are not affected.

1 Introduction

All over the world surface water is to some extend contaminated with organic micropollutants (OMPs) (Houtman et al., 2010). It is expected that the amount and concentrations of OMPs will increase, due to population growth, aging and global warming (Wuijts et al., 2013). A drinking water company in the western part of the Netherlands identified the threat of OMPs in their source the Afgedamde Maas, a side branch of the Meuse river. The managed aquifer recharge (MAR) by dune filtration and the dosing of powdered activated carbon (PAC) are currently the main barriers against OMPs. After careful consideration advanced oxidation process (AOP) was selected as the most optimal technique to extend the treatment scheme (Abrahamse et al., 2007) and research was carried out with ozone and UV based AOP, in combination with hydrogen peroxide. When AOP is installed before MAR it is expected that this two processes will provide a synergistic, hybrid system (Lekkerkerker-Teunissen et al., 2012).

A drawback of applying AOP is the remaining hydrogen peroxide in the treated water. After dosing 6 mg L^{-1} hydrogen peroxide in the AOP influent water, about 5.75 mg L^{-1} hydrogen peroxide is remaining in the AOP effluent water. This AOP effluent water flows out in infiltration ponds which recharge the dunes. It is established that even a concentration of 2 mg L^{-1} affects about 80 % of the organisms in the infiltration ponds (Reeze et al., 2010). Because the infiltration ponds are situated in a protected nature area (Natura 2000), it is of utmost importance to lower the hydrogen peroxide concentration till a level that will not affect the ecology. As a company standard, the maximum allowed concentration of hydrogen peroxide in the infiltrated water is established at $0.25\,mg\,L^{-1}$.

Hydrogen peroxide in water has a tendency to decompose in water and oxygen, because the reaction products are more stable than the hydrogen peroxide itself (Petrucci, 2007):

$$2H_2O_2 \rightarrow 2H_2O + O_2$$

$$\Delta H\ [kJ\ mol^{-1}]\ 2\times -134.1 \rightarrow 2\times -237.2 + 0.0$$

Overall the chemical drive of this so-called decomposition reaction is $206.2\,kJmol^{-1}$. However, the decomposition is very slow, the reason why hydrogen peroxide solutions are commercially available. Decomposition is strongly affected by light and catalysts as catalase, (spores of) metal oxides and activated carbon. In full-scale application, homogeneous catalysis (enzymatic with for example catalase or iron) is not practical. The most promising technique is therefore heterogeneous catalysis. Common catalysts include manganese oxide, silver, platinum and activated carbon. The surface of these catalysts provides a favorable environment to catalyze the decomposition, though the mechanism is not well understood. According to Masel et al. (2001) the reaction rate is increased as this alternative route has a lower activation energy than without the catalyst. Decomposition starts with adsorption of hydrogen peroxide on the catalyst. The rate of adsorption, and with that the decomposition rate, is higher at higher water temperature and with a larger catalyst surface.

Filtration over granulated activated carbon (GAC) is a proven technology to decompose hydrogen peroxide (Kruithof et al., 2007). The decomposition reaction with pure manganese dioxide in granular form is slow and time demanding (Bazri, 2008). However, Miller et al. (1999) reported decomposition of hydrogen peroxide in the presence of sandy material from an aquifer and riverbed. The sand was naturally coated with metal oxides, from which manganese oxide was most active. It was known that dissolved manganese in surface and groundwater deposits as natural manganese oxide compounds on grains in sand filters of drinking water treatment plants. Merkle et al. (1996) measured a high specific surface area of naturally coated filter material, which increases with the amount of manganese coating. Decomposing hydrogen peroxide with existing or waste filter material will contribute to a sustainable society. As a proof of principle, 1 % hydrogen peroxide solution was added to grains of a full-scale double layer sand filter of a drinking water treatment plant. Microscopic examination yielded promising results: (oxygen) bubbles were formed at the surface of the grains.

This research focused on the decomposition of about 5 and $10\,mg\,L^{-1}$ hydrogen peroxide till $0.25\,mg\,L^{-1}$ in the pilot plant installation. Three different catalysts were investigated: commonly applied GAC and two types of manganese coated filter material. The main parameter determined for the three decomposing materials was the reactor empty bed contact time (EBCT), based on the conversion rate of the decomposition.

2 Materials and methods

2.1 Column reactors

Catalytic decomposition was investigated in vertical column reactors, with outside diameters of 0.20, 0.40 and 0.60 m, called R20, R40 and R60, respectively. Sample points over the height of the columns made it possible to analyse the hydrogen peroxide after different contact times. The two reactors R60 were fed with pre-treated Meuse river water, abstracted before the full-scale double layer sand filters. The reactors R20 and R40 were fed with the effluent of these full-scale double layer sand filters. The flow through the reactors could be varied between 1.0 and $3.0\,m^3\,h^{-1}$, which corresponds with flow rates between 4 and $100\,m\,h^{-1}$. The standard operation was down flow, but up flow operation also was applied with flow rates up till $40\,m\,h^{-1}$. Before the water flowed in the reactors, 10 % hydrogen peroxide was dosed into the water with a membrane pump with a maximum capacity of $3.0\,L\,h^{-1}$. The applied concentrations were about 5 and $10\,mg\,L^{-1}$ in the influent water.

2.2 Decomposing material

Three different catalysts were investigated: fresh GAC and two types of used manganese coated filter material. The filter materials were collected from full-scale filters at drinking water treatment plants.

The GAC type was extruded activated carbon with a diameter of 0.8 mm, especially suitable for catalytic processes. The manganese coated filter material were obtained from two different drinking water treatment facilities.

The first manganese coated filter material was obtained from a ground water treatment plant (MCFgw). This filter material is regularly replaced by fresh sand as a result of the rapid growth of the grain size by the coating, resulting in a lower purification performance. A sample is taken at the end of the filter life time of 7 years. Especially ground water may contain a high content of manganese, which may be makes the replaced grains suitable for decomposing hydrogen peroxide. The second manganese coated filter material was obtained from a surface water treatment plant (MCFsw). The MCFsw was collected from a double layer sand filter, with a filter life time of 31 years. In order to clean the grains after filling the reactor, MCFgw was backwashed with water before the first tests. During backwashing, a part of the coating was separated from the grains and formed a top layer. This top layer of fine coating was removed from the reactor and also used as catalyst (MC).

Table 1. Specifications of different tested filter gravels.

Parameter	Unit	Course gravel MCFgw	Coating MC	Anthracite/sand MCFsw
10 % Grain	mm	1.01	0.25	2.06/0.97
Uniformity 60/10 %	–	2.26	2.02	1.18/1.13
Manganese content	$g\,kg^{-1}$	14.0	100.0	0.15/0.09
Specific surface area*	$m^2\,g^{-1}$	4.0	64	1.9/2.1

* BET surface area by N_2 adsorption.

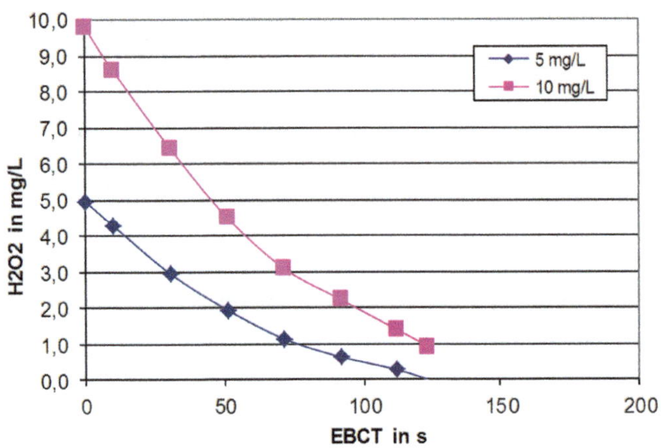

Figure 1. Decomposition of hydrogen peroxide with ROW 0.8 cat in R20 (bed height 1.4 m, flow rate $40\,m\,h^{-1}$, water temperature 1.8 °).

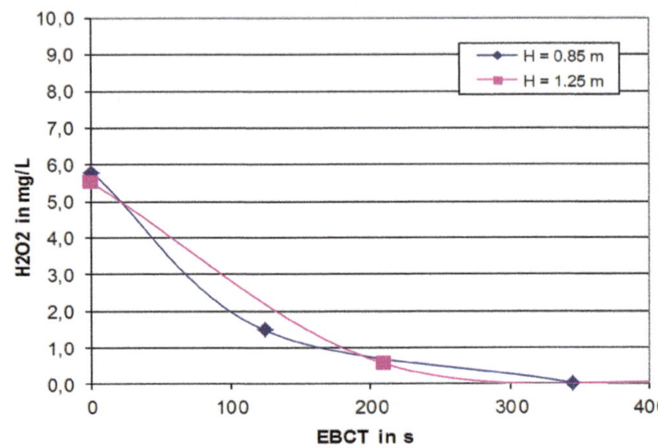

Figure 2. Decomposition over MCFsw in R60s with a bed height of 0.85 m (0.6 m gravel 0.8–1.2 mm and 0.25 m anthracite 1.4–2.4 mm) and a bed height of 1.25 m (1.0 m gravel 0.8–1.2 mm and 0.25 m anthracite 1.4–2.4 mm), a flow velocity of $4.4\,m\,h^{-1}$, and a water temperature of 12.4 °C.

2.3 Analyses

The hydrogen peroxide concentration in water was analysed on site with a spectrophotometer. The measurement is based on the reaction of hydrogen peroxide with titanium(IV)oxysulphate solution, following DIN 38409 H15. Samples were collected and measured in a volumetric flask, after adding 5.0 mL of the titanium(IV)oxysulphate solution by a pipette. When higher concentrations were expected than $6\,mg\,L^{-1}$, the samples were diluted with milli-q water. The samples were measured at a wavelength of 420 nm and corrected for background absorbance, which were determined by analysing the samples without the addition of titanium(IV)oxysulphate solution. At a hydrogen peroxide concentration of $5.8\,mg\,L^{-1}$ the standard deviation σ was $0.02\,mg\,L^{-1}$.

3 Results and discussion

3.1 Activated carbon

With GAC in down flow operation, $5\,mg\,L^{-1}$ hydrogen peroxide was decomposed completely in just more than 120 s empty bed contact time (EBCT), see Fig. 1. In the same EBCT, $10\,mg\,L^{-1}$ hydrogen peroxide was decomposed for about 90 %. This experiment confirmed that activated carbon is well-functioning catalyst for decomposing hydrogen peroxide, even at the low water temperature of 1.8 °C.

3.2 Manganese coated filter material

3.2.1 Specifications

The manganese content of MCFgw is about 100 times higher than from MCFsw due to the higher manganese concentration in groundwater (Table 1). The specific surface area is twice as high. MC is characterized by a high manganese content and specific surface area and small grain size.

With XRD analysis, besides calcium and iron compounds, the manganese containing compounds ramsdellite and birnessite were detected in MC (Hendrix, 2014).

3.2.2 Decomposition rate

The decomposition rate of the different manganese containing materials are investigated in down flow operation. The decomposition of $5\,mg\,L^{-1}$ hydrogen peroxide with anthracite and gravel MCFsw is plotted in Fig. 2.

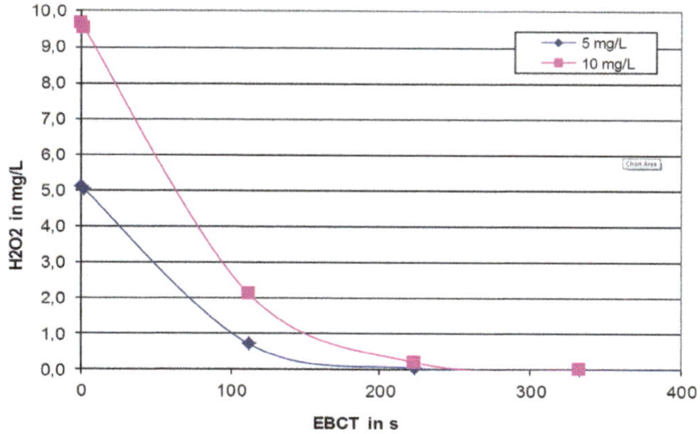

Figure 3. Decomposition over MCFgw in R40 (bed height 2.3 m, flow rate $25\,m\,h^{-1}$, water temperature 3.3 °C).

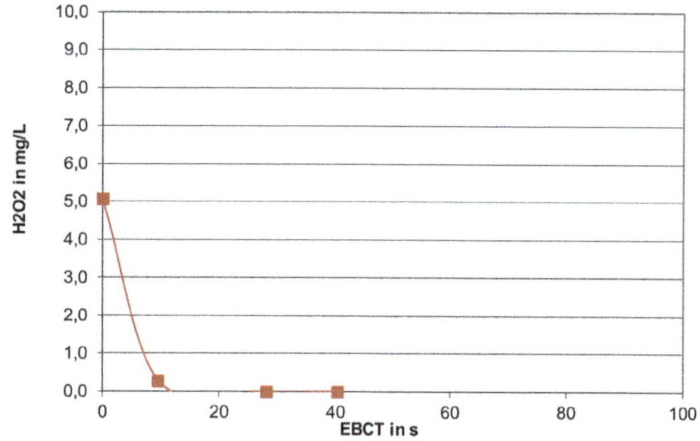

Figure 4. Decomposition over MC in R20 (bed height 0.36 m, flow rate $40\,m\,h^{-1}$, water temperature 11.7 °C).

In about 250 till 350 s the hydrogen peroxide in both reactors was decomposed completely. The EBCT of the 0.25 m anthracite layer was about 200 s, which means that the anthracite had a major contribution to the decomposition. The total EBCT of 700 s of the 0.85 m bed layer was the same as the average EBCT in the full scale filters, which means that at least 400 s EBCT remains for the primary functions of the rapid sand filters (removal of iron and manganese, nitrification and mineralization). This observation confirms that applying AOP in front of the rapid sand filters can be an interesting option in practice and could make a separate decomposing installation superfluous, provided that the primary filter functions are not disturbed.

Decomposing with the coarse gravel MCFgw was investigated with a dose of 5 and $10\,mg\,L^{-1}$ (Fig. 3).

A dose of $5\,mg\,L^{-1}$ hydrogen peroxide was completely decomposed in an EBCT of 180 s. A dose of $10\,mg\,L^{-1}$ was decomposed in 260 s. Despite the manganese content of MCFgw of $14\,g\,kg^{-1}$, about 100 times higher compared to MCFsw, and twice as large specific surface area, $4.0\,m^2\,g^{-1}$ compared to $1.9/2.1\,m^2\,g^{-1}$, the difference in the decomposing time between MCFgw and MCFsw was no more than 30 %. Possibly a part of the manganese oxides in MCFgw were enclosed and did not contribute to the decomposition.

The results of decomposition with the fine coating MC was investigated with a dose of $5\,mg\,L^{-1}$ hydrogen peroxide and plotted in Fig. 4.

Already within 10 s 95 % of the hydrogen peroxide was decomposed. Clearly the coating with a relatively high manganese content of $100\,g\,kg^{-1}$ had a positive effect on the rate constant of decomposing hydrogen peroxide, conform literature (Bazri, 2008). In addition, the small particles and high specific surface area (Table 1) form a large catalyst surface.

Decomposition of hydrogen peroxide is a first order reaction in a heterogeneous environment. Because the reaction does not take place under homogenous circumstances, the decomposition rate is not expressed in a reaction rate con-

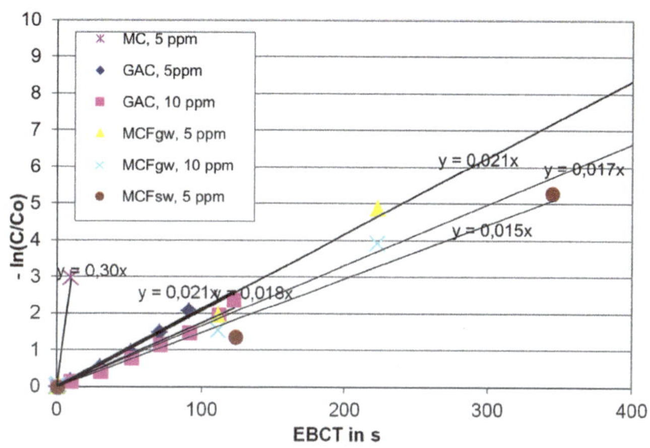

Figure 5. Conversion rate constants of the different tested catalysts.

stant, but in the conversion rate constant r (Coulson et al., 1996). The conversion rate constant is not only depending of temperature and pressure (as in homogeneous milieu), but also on contact surface of the catalyst and hydraulic conditions in the reactor. In case of a first order reaction, a linear relation exist between the negative natural logarithm of the quotient of the effluent hydrogen peroxide C and influent concentration C_0 and the reaction time ($-\ln C / C_0$ vs. t). The relations of the different investigated catalysts are plotted in Fig. 5.

MC had the highest conversion rate constant (note that the value of r is only based on one sample point, because the water in the second sample point did not contain hydrogen peroxide anymore) and much higher compared to the other catalysts. The high specific surface area of $64\,m^2\,g^{-1}$, the high manganese content in the form of (partly) ramsdellite and small grain size, all three factors contributed to the high conversion rate. Páez et al. (2011) also reported high conversion rate constants with synthesized small nano-needle particals ramsdellite ($\varnothing = 10\,nm$, $L = 180\,nm$). In batch opera-

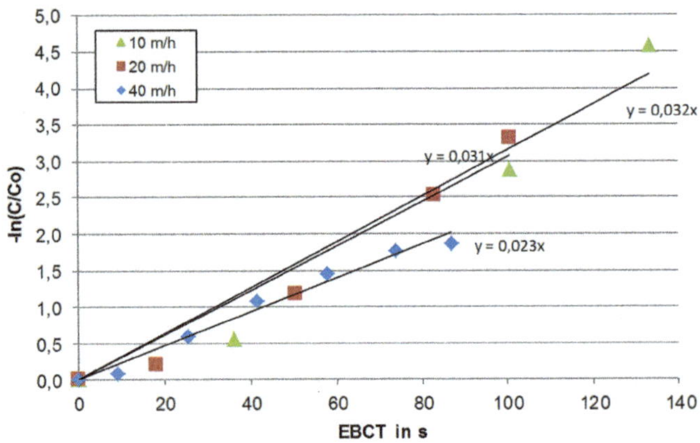

Figure 6. Up flow decomposition with MC in R20 (static bed height 0.36 m, water temperature 5.8 °C).

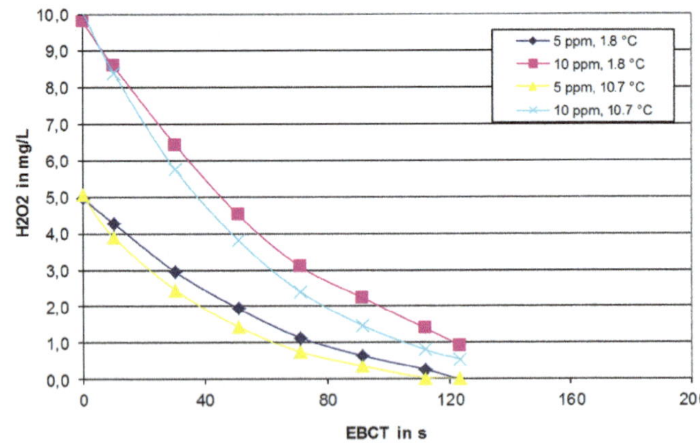

Figure 7. Decomposition with GAC at different water temperatures.

Table 2. EBCT of catalysts to decompose a peroxide concentration of $C_0 = 5.75\ \mathrm{mg\,L^{-1}}$ till $C = 0.25\ \mathrm{mg\,L^{-1}}$.

Catalyst	r ($\mathrm{s^{-1}}$)	EBCT (s)
MCFsw	0.015	209
GAC	0.021	149
MCFgw	0.021	149
MC	0.30	10.4

tion a conversion rate constant was measured of $0.015\ \mathrm{s^{-1}}$ after addition of only $0.374\ \mathrm{g\,L^{-1}}$ particles.

The conversion rate constant of $0.015\ \mathrm{s^{-1}}$ of MCFsw is in line with the reported conversion rate constant of $0.007\ \mathrm{s^{-1}}$ during column experiments with sandy aquifer material with different metal deposits, from which manganese appeared to have the biggest contribution to the decomposition (Miller et al., 1995). Not only the manganese content of $0.06\ \mathrm{g\,kg^{-1}}$ of the sandy aquifer material was comparable with MCFsw ($0.15/0.09\ \mathrm{g\,kg^{-1}}$), but also the specific surface area of $1.75\ \mathrm{g\,m^{-2}}$ (MCFsw $1.9/2.1\ \mathrm{g\,m^{-2}}$).

In addition, the conversion rate of the same catalyst depended on the initial hydrogen peroxide concentration. This observation is reported earlier under heterogeneous conditions by Coulson et al. (1996).

In Table 2 the necessary EBCT is calculated for the investigated catalysts with an initial concentration $C_0 = 5.75\ \mathrm{mg\,L^{-1}}$ (corresponding to a dose in the AOP influent of $6\ \mathrm{mg\,L^{-1}}$) and an effluent concentration $C = 0.25\ \mathrm{mg\,L^{-1}}$ (the allowed concentration in the infiltration ponds), which means that $\ln C_0/C = 3.135$ and $t = 3.135/r$. Here, it was assumed that the calculated conversion rate constants at about $5.0\ \mathrm{mg\,L^{-1}}$ hydrogen peroxide will not differ much from the conversion rate constants at $5.75\ \mathrm{mg\,L^{-1}}$.

The necessary EBCT in the full scale dual layer filters (MCFsw) amounts about 200 s. The EBCTs of GAC and MCFgw equals about 150 s. The EBCT of MC is much lower with approximately 10 s.

3.2.3 Mode of operation

For full scale operation down flow operation over fine material has the disadvantage of fast clogging of filter material with suspended particles and potential increase of the resistance due to oxygen bubble formation, depending of the oxygen saturation degree of the water. These disadvantages could be solved by up flow operation.

Up flow operation of the coating MC was investigated using three different flow rates. A flow velocity of $40\ \mathrm{m\,h^{-1}}$ appeared to be the maximum velocity, without carrying the coatings with the effluent of the reactor. The coatings became in fluidised state at a flow velocity of $4\ \mathrm{m\,h^{-1}}$. The decomposition of $5.8\ \mathrm{mg\,L^{-1}}$ hydrogen peroxide is plotted in Fig. 6.

From Fig. 6 it was concluded, that the decomposition was sufficient, which means $\ln C_0/C \geq 3.135$, with up flow velocities of 10 and $20\ \mathrm{m\,h^{-1}}$. The necessary EBCT was approximately 100 s. At a velocity of $40\ \mathrm{m\,h^{-1}}$ the decomposition goal could not be achieved. Presumably, the adsorption of hydrogen peroxide on the catalyst was inadequate due to the increased expansion of up to 170 %, which resulted in less catalyst surface in the same reactor height.

Pelleting MC till a diameter of about 2 mm would make higher flow velocities possible, and with that a smaller reactor footprint. However, changing the diameter or the surface area of the catalyst or the hydraulic conditions in the reactor, will have impact on the value of the conversion rate constant and consequentially on the design criteria. In addition, the decomposition of hydrogen peroxide depends on the water temperature and will be faster with increasing water temperature. This is visible in Fig. 7, wherein the decomposition is plotted of experiments with GAC at 1.8 and 10.7 °C. The concentration of hydrogen peroxide was always lower after the same contact time at a higher water temperature.

4 Conclusions

In down flow operation, the necessary EBCT to decompose 5.75 mg L^{-1} hydrogen peroxide in pre-treated Meuse river water with the commonly used GAC and waste groundwater filter gravel (MCFgw) were the same with about 150 s, corresponding to a conversion rate constant r of 0.021 s^{-1}. The EBCT of the coating of groundwater filter gravel (MC) was much shorter with a little more than 10 s ($r = 0.30\ s^{-1}$).

MC was suitable for up flow operation till a flow velocity of about 20 m h^{-1}. The necessary EBCT was about 100 s ($r = 0.031\ s^{-1}$). At an up flow rate of 40 m h^{-1} the decomposition goal could not be achieved, probably due to the increased porosity in the fluidized bed. Pelleting MC till a diameter of about 2 mm would make higher up flow velocities possible.

The necessary EBCT at average filtration velocity of dual layer filter material (MCFsw) amounted to about 200 s ($r = 0.015\ s^{-1}$). Regarding the average residence time in the full scale filters of 700 s, applying AOP in front of the filters could be an interesting alternative making a separate decomposition installation superfluous, on the condition that the primary functions of the filters are not affected.

Acknowledgements. The authors would like to thank the staff of treatment location Bergambacht for their technical support.

References

Abrahamse, A. J., IJpelaar, G. F., and Knol, A. H.: Project Uitbreiding Zuivering DZH, optionele technieken en locatie keuze, KWR, Dunea, 2007.

Bazri, M. M.: Assimilable Organic Carbon Formation during Ultraviolet-Hydrogen Peroxide Advanced Oxidation Treatment of Surface Water, University of British Columbia, Vancouver, 2010.

Coulson, J. M. and Richardson, J. F.: Chemical Engineering – Part 1, Oxford, Butterworth-Heinemann Ltd, 1996.

Hendrix, R. W. A: XRD analysis of sand grains, Materials Science and Engineering TUDelft, January 2014.

Houtman, C. J.: Emerging contaminants in surface waters and their relevance for the production of drinking water in Europe, J. Integrat. Environ. Sci., 7, 271–295, 2010.

Kruithof, J. C., Kamp, P. C., Bram, J., and Martijn, B. J.: UV/H_2O_2 Treatment: A Practical Solution for Organic Contaminant Control and Primary Disinfection, Ozone, Sci. Engin., 29, 273–280, 2007.

Lekkerkerker-Teunissen, K., Knol, A. H., Altena L. P., Houtman, C. J., Verberk, J. Q. J. C., and van Dijk, J. C.: Serial ozone/peroxide/low pressure UV treatment for synergistic and effective organic micropollutant conversion, Separat. Purificat. Technol., 100, 22–29, doi:10.1016/j.seppur.2012.08.030, 2012.

Masel, R. I.: Chemical Kinetics and Catalysis, Wiley-Interscience, New York, 2001.

Merkle, P. B., Knocke, W., Gallagher, D., Junta-Rosso, J., and Solberg, T.: Characterizing filter media mineral coatings, J. AWWA, 62–72, December 1996.

Miller, C. M. and Valentine, R. L.: Hydrogen peroxide decomposition and quinoline degradation in the presence of aquifer material, Water Res., 29, 2353–2359, 1995.

Miller, C. M. and Valentine, R. L.: Mechanistic studies of surface catalyzed H_2O_2, Decomposition and contaminant degradation degradation in the presence of sand, Water Res., 33, 2805–2816, 1999.

Páez, C. A., Liquet D. Y., Calberg, C., Lambert, S. D., Willems, I., Germeau, A., Pirard, J. P., and Heinrichs, B.: Study of photocatalytic decomposition of hydrogen peroxide over ramsdellite-MnO_2 by O_2-pressure monitoring, Catal. Commun., 15, 132–136, 2011.

Petrucci, R. H., Harwood, W. S., Herring, G. E., and Madura, J.: General Chemistry: Principles and Modern Applications, 9th Edn., Harlow, Prentice Hall, 2007.

Reeze, A. J. G., Matthijs, J. C. P., and Pancras, T.: Waterstof peroxide (H_2O_2) in infiltratieplassen, Eco(toxico)logische effecten en afbraaksnelheid, Arcadis, 14 December 2010.

Wuijts, S., van der Grinten, E., Meijers, E., Bak-Eijsberg, C. I., and Zwolsman, J. J. G.: Impact of climate change on surface water as a resource for drinking water: From problem areas to measures, Dutch Ministry of Infrastructure and Environment, 2013.

Permissions

All chapters in this book were first published in DWES, by Copernicus Publications; hereby published with permission under the Creative Commons Attribution License or equivalent. Every chapter published in this book has been scrutinized by our experts. Their significance has been extensively debated. The topics covered herein carry significant findings which will fuel the growth of the discipline. They may even be implemented as practical applications or may be referred to as a beginning point for another development.

The contributors of this book come from diverse backgrounds, making this book a truly international effort. This book will bring forth new frontiers with its revolutionizing research information and detailed analysis of the nascent developments around the world.

We would like to thank all the contributing authors for lending their expertise to make the book truly unique. They have played a crucial role in the development of this book. Without their invaluable contributions this book wouldn't have been possible. They have made vital efforts to compile up to date information on the varied aspects of this subject to make this book a valuable addition to the collection of many professionals and students.

This book was conceptualized with the vision of imparting up-to-date information and advanced data in this field. To ensure the same, a matchless editorial board was set up. Every individual on the board went through rigorous rounds of assessment to prove their worth. After which they invested a large part of their time researching and compiling the most relevant data for our readers.

The editorial board has been involved in producing this book since its inception. They have spent rigorous hours researching and exploring the diverse topics which have resulted in the successful publishing of this book. They have passed on their knowledge of decades through this book. To expedite this challenging task, the publisher supported the team at every step. A small team of assistant editors was also appointed to further simplify the editing procedure and attain best results for the readers.

Apart from the editorial board, the designing team has also invested a significant amount of their time in understanding the subject and creating the most relevant covers. They scrutinized every image to scout for the most suitable representation of the subject and create an appropriate cover for the book.

The publishing team has been an ardent support to the editorial, designing and production team. Their endless efforts to recruit the best for this project, has resulted in the accomplishment of this book. They are a veteran in the field of academics and their pool of knowledge is as vast as their experience in printing. Their expertise and guidance has proved useful at every step. Their uncompromising quality standards have made this book an exceptional effort. Their encouragement from time to time has been an inspiration for everyone.

The publisher and the editorial board hope that this book will prove to be a valuable piece of knowledge for researchers, students, practitioners and scholars across the globe.

List of Contributors

M. Walker and J. Newman
University of Nevada, Department of Natural Resources and Environmental Science, MS 370/FA 132, 1664 N. Virginia Street, Reno, NV 89557, USA

J. Tomperi and K. Leiviskä
University of Oulu, Control Engineering Laboratory, P.O. Box 4300, 90014 University of Oulu, Finland

M. Pelo
Finnsugar Ltd, Sokeritehtaantie 20, 02460 Kantvik, Finland

F. De Paola and M. Giugni
Department of Hydraulic, Geotechnical and Environmental Engineering, University of Naples "Federico II", Italy

A. Grefte and L. C. Rietveld
Delft University of Technology, Faculty of Civil Engineering, P.O. Box 5048, 2600 GA Delft, The Netherlands

M. Dignum
Waternet, P.O. Box 94370, 1090 GJ Amsterdam, The Netherlands

E. R. Cornelissen
KWR Watercycle Research Institute, P.O. Box 1072, 3430 BB Nieuwegein, The Netherlands

L. Mezule, S. Larsson and T. Juhna
Department of Water Engineering and Technology, Riga Technical University, Azenes 16/20-263, Riga, Latvia

P. K. Mutiyar and A. K. Mittal
Department of Civil Engineering, Indian Institute of Technology Delhi, 110016 New Delhi, India

A. Pekdeger
Institute of Geological Sciences, Free University of Berlin, 12249 Berlin, Germany

H. K. Shon, S. Phuntsho, D. S. Chaudhary and S. Vigneswaran
School of Civil and Environmental Engineering, University of Technology, Sydney (UTS), P.O. Box 129, Broadway, NSW 2007, Australia

J. Cho
Department of Civil and Environmental Engineering, Yonsei University, Yonsei-ro 50, Seodaemun-gu, Seoul 120-749, Korea

M. A. Tahir and H. Rasheed
Pakistan Council of Research in Water Resources, Khayaban-e-Johar, H-8/1, Islamabad, Pakistan

E. J. Pieterse-Quirijns and E. J. M. Blokker
KWR Watercycle Research Institute, P.O. Box 1072, 3430 BB Nieuwegein, the Netherlands

E. van der Blom
Uneto-VNI, P.O. Box 188, 2700 AD Zoetermeer, the Netherlands

J. H. G. Vreeburg
KWR Watercycle Research Institute, P.O. Box 1072, 3430 BB Nieuwegein, the Netherlands
Delft University of Technology, Faculty of Civil Engineering, P.O. Box 5048, 2600 GA Delft, the Netherlands

L. D. S´anchez and L. M. Marin
Universidad del Valle, Facultad de Ingeniería, Instituto Cinara, Cali, Colombia

J. T. Visscher
JTF-Rotterdam BV, Rotterdam, The Netherlands

L. C. Rietveld
Delft University of Technology, P.O. Box 5084, 2600 GA Delft, The Netherlands

K. Sandeep
Research scholar, IIT Delhi, Delhi, India

K. Rakesh
Faculty, Department of Civil Engineering, Indian Institute of Technology, Delhi, India

T. Y. Jeong and S. D. Kim
Department of Environmental Science and Engineering, Gwangju Institute of Science and Technology (GIST), 1 Oryong-dong, Buk-gu, Gwangju 500-712, Republic of Korea

J. Jeon
Department of Environmental Chemistry, EAWAG, Swiss Federal Institute of Aquatic Science and Technology, Ueberlandstrasse 133, 8600, Duebendorf, Switzerland

I. W. M. Pothof
Deltares/Industrial Hydrodynamics, Delft, The Netherlands
Delft University of Technology/Water management, Delft, The Netherlands

E. J. M. Blokker
KWR Watercycle Research Institute, Nieuwegein, The Netherlands

S. Hajibabania and P. Le-Clech
UNESCO Centre for Membrane Science and Technology, School of Chemical Engineering, The University of New South Wales, Sydney 2052, Australia

A. Verliefde
Particle and Interfacial Technology Group, Faculty of Bioscience Engineering, Ghent University, Coupure Links 653, 9000 Gent, Belgium
Delft University of Technology, Faculty of Civil Engineering and Geosciences, Department of Sanitary Engineering, P.O. Box 5048, 2600 GA Delft, The Netherlands

J. E. Drewes
Environmental Science and Engineering Division, Colorado School of Mines, Golden, CO 80401-1887, USA

L. D. Nghiem
School of Civil Mining and Environmental Engineering, The University of Wollongong, Wollongong, NSW 2522, Australia

J. McDonald and S. Khan
Water Research Centre, The University of New South Wales, Sydney 2052, Australia

S. Tarfasa
Department of Economics, Hawassa University, Hawassa, Ethiopia

K. Neilands, M. Bernats and J. Rubulis
Riga Technical University, Department of Water Engineering and Technology, Latvia

K. Behzadian
Centre for Water Systems, College of Engineering, Mathematics and Physical Sciences, University of Exeter, Exeter, UK
Environmental Research Centre, Amirkabir University of Technology, Tehran, Iran

Z. Kapelan
Centre for Water Systems, College of Engineering, Mathematics and Physical Sciences, University of Exeter, Exeter, UK

G. Venkatesh and S. Sægrov
Department of Hydraulic and Environmental Engineering, Norwegian University of Science and Technology, Trondheim, Norway

H. Brattebø
Department of Energy and Process Engineering, Norwegian University of Science and Technology, Trondheim, Norway

Z. Deng and J. P. van der Hoek
Waternet, P.O. Box 94370, 1090 GJ Amsterdam, the Netherlands
Delft University of Technology, Department of Water Management, Stevinweg 1, 2628 CN Delft, the Netherlands

S. Mol
Waternet, P.O. Box 94370, 1090 GJ Amsterdam, the Netherlands

J. Fernández-Pato and P. García-Navarro
LIFTEC (CSIC) - University of Zaragoza, C/María de Luna 5, 50018 Zaragoza, Spain

C. Capponi, M. Ferrante, B. Brunone and S. Meniconi
Dipartimento di Ingegneria Civile ed Ambientale, The University of Perugia, via G. Duranti 93, 06125 Perugia, Italy

M. Pedroni, M. Zaghini and F. Leoni
TeaAcque Srl, Via Taliercio 3, 46100 Mantova, Italy

J. L. Gutenson, A. N. S. Ernest, A. A. Oubeidillah and X. Zhang
The University of Alabama, Tuscaloosa, Alabama

J. R. Fattic
Western Kentucky University, Bowling Green, Kentucky
Akebono Brake Corporation, Glasgow, Kentucky

L. E. Ormsbee
University of Kentucky, Lexington, Kentucky

S. R. Ramphal and M. S. Sibiya
Rand Water, Process Technology Department, Glen Vista, South Africa

Priyanka Jamwal
Centre for Environment and Development, Ashoka Trust for Research in Ecology and the Environment (ATREE), Jakkur 560064, India

M. N. Naveen
Centre for Environment and Development, Ashoka Trust for Research in Ecology and the Environment (ATREE), Jakkur 560064, India

Department of Civil engineering, The National Institute of Engineering, Mysuru, Karnataka 570008, India

Yusuf Javeed
Department of Civil engineering, The National Institute of Engineering, Mysuru, Karnataka 570008, India

A. H. Knol and K. Lekkerkerker-Teunissen
Dunea N.V., Strategy department, P.O. Box 756, 2700 AT Zoetermeer, the Netherlands

J. C. van Dijk
Delft University of Technology, Stevinweg 1, 2628 CN Delft, the Netherlands

Index